Human Diversity

Its Nature, Extent, Causes and Effects on People

Bernard Charles Lamb

BSc, PhD, DSc, CBiol, FRSB, FRSM, NGWBJ
Emeritus Reader in Genetics,
Imperial College London, UK

World Scientific

NEW JERSEY · LONDON · SINGAPORE · BEIJING · SHANGHAI · HONG KONG · TAIPEI · CHENNAI · TOKYO

Published by

World Scientific Publishing Co. Pte. Ltd.

5 Toh Tuck Link, Singapore 596224

USA office: 27 Warren Street, Suite 401-402, Hackensack, NJ 07601

UK office: 57 Shelton Street, Covent Garden, London WC2H 9HE

Library of Congress Cataloging-in-Publication Data

Lamb, Bernard Charles.

    Human diversity : its nature, extent, causes and effects on people / Bernard Charles Lamb, BSc, PhD, DSc, CBiol, FSB, FRSM, NGWBJ, Emeritus Reader in Genetics, Imperial College London, UK.

      pages cm

    Includes bibliographical references and index.

    ISBN 978-9814632355 (hardcover : alk. paper)

    1. Ethnopsychology. 2. Human genetics. 3. Nature and nurture. 4. Race. I. Title.

    GN270.L36 2015

    155.8'2--dc23

                            2015015474

British Library Cataloguing-in-Publication Data

A catalogue record for this book is available from the British Library.

Typeset by Stallion Press

Email: enquiries@stallionpress.com

Printed in Singapore

Acknowledgements

I am most grateful to my former students at Imperial College London, one of the best three universities in Britain. They filled in questionnaires, wrote well-researched essays on aspects of human diversity, and some did their research project on it. We had many stimulating discussions. During 40 years of teaching all aspects of genetics there, including human and medical genetics, I learnt much from my students.

Since becoming a fellow of the Royal Society of Medicine, I have used its superb library and attended numerous relevant talks and symposia.

This book contains many first-hand accounts by people who are different in some special way. They have written about how their differences have affected them and how their differences have affected others around them. I am so grateful to them.

I really appreciate the assistance of all who contributed information, stories or advice, or who checked drafts of chapters. These helpful people include Ray Ward, with his vast general knowledge, Catherine Robinson, Brenda Lamb, Michael Gorman, John Lamb, and Michael and Annabelle Daiches.

This book would be less useful if written from a solely British perspective, but I have been fortunate to travel to many countries, frequently staying in the homes of friends and ex-students. My wife and I often entertained overseas students, getting to know many of them well.

For facts, I have used my own and my students' research, published papers, review articles, books, census data from several countries, figures from the Office of National Statistics, news articles, websites and literature from patient-support groups and the National Health Service, and many other sources, including pieces by medical correspondents, particularly in *The Daily Telegraph*. I am aware that any source can give inaccurate information and that information can rapidly get out of date.

Most photos are my own. Thousands of people, including total strangers, kindly allowed me to photograph them. Plates were generously provided as follows: Dr Simon Zwolinski — 1A, B, C; Lasantha — 9H; John Lamb — 15D; Dr David Dickinson — 15E, Catherine Robinson — 20E; Julia Bostock— Photos 3.1 and 3.2.

Contents

Chapter 1

Introduction: Scope of the Book; Types of Human Difference and Their Causes

1.1. The aims and scope of this book

Every person is different. How one copes with life's turbulence depends on personality, culture, beliefs, abilities, location and genetic potential. These differences are the subject of this book. It is about human diversity, its nature, extent, causes and effects on people. In words and many photographs (Plates 1–20), it describes the fascinating differences between individuals and the effects those differences have on their possessors and on others. Human diversity involves medicine, human biology, sociology, genetics, psychology, history and geography. This blend of facts, opinions, original research, personal stories and humour should give readers a better understanding of humanity in health and sickness.

Human diversity involves inherited characteristics, the environment and gene/environment interactions. It is thus appropriate to have an author with experience of teaching, research and publication in many fields of genetics treating the subject. My research from 1963 onwards has been on the generation, passing on, maintenance and effects of genetic diversity, including mutation, recombination, the genetic code, adaptation to environment, population genetics and evolution.

The topics include diversity aspects of race, class, culture, height, weight, shape, languages, sex, attraction, reproduction, choice, the brain, intelligence, personality, the senses, the heart and blood, diseases, cancer, eating, lungs, liver, kidneys, excretion, genetics, skin, bones, muscles and names, with background biology where necessary. The sections on disease should help sufferers or those caring for them. The examples come from many countries, as shown in the Plates.

Human diversity can strike in unexpected and even lethal ways. Unsuspected allergies to foods or poison ivy can kill. A British couple were on honeymoon, trekking in Nepal. On one climb, they suffered from altitude sickness and died. The rest of the party survived, illustrating diversity in adaptability to high altitudes. Tibetans rarely show altitude sickness, with 4.5 million people living on the Tibetan Plateau, 'the roof of the world', mostly at altitudes from 10,000 to 15,000 feet (3,500 to 4,600 m), with low oxygen concentrations. Lorenzo *et al.* (2014) (*et al.* means 'and others') found that most Tibetans have a mutation preventing the excessive high-altitude production of red blood cells that thickens the blood and can cause death. This adaptation is found in 85% of Tibetans and only 1% of other races. Adaptation to high altitudes is also found in Ghurkhas and Sherpas in Nepal, and in mountain-dwellers in Ethiopia and the Andes.

Something physically trivial may be a disaster for cultural reasons. An Asian mother of two girls in England developed a small white vitiligo (Chapter 13.6) patch on one arm. She was terrified that if her mother-in-law saw it, it would mean divorce, being sent back to India with no support, where she would have to resort to prostitution or begging to live. She committed suicide.

Human diversity may be benign, as in the thousands of different blood group combinations. It may be tolerable, as in having one or four kidneys (Chapter 19), alarming (Plate 17A) or lethal, like being born with one eye in the centre of the forehead. According to Changing Faces (www.changingfaces.org.uk), more than a million people in Britain have disfigurements such as scarring or marks affecting the appearance of their faces or bodies, and half a million have 'unusual faces'. Interactions between people who are different form the basis of many historical events, novels, plays and operas, often with tragic endings. Those differences are of sex, race, religion, factions, class and even height. Fictional examples of differences in family or culture resulting in death include *Romeo and Juliet* and *Madam Butterfly*. In history, the Roman Emperor Titus wanted to marry

Jewish Queen Berenice in 79 AD. The Romans successfully objected on the grounds of not wanting a foreigner, of a different religion.

A real-life mixture of differences was between motor-racing supremo, Bernie Ecclestone, and his Croatian wife Slavica, a former model. He was 28 years older; she spoke Croatian and Italian while he spoke only English. At 6 feet 2 inches (188 cm), Slavica was nearly a foot taller than Bernie, who was 5ft 2½ in (159 cm). They divorced after 23 years. Another example was British pop star Mick Jagger and his lover, American dress designer L'Wren Scott. She became a catwalk model for Chanel in Paris, famous for her 42 inch (107 cm) legs. She was 6 ft 3 in (190.5 cm) tall and aged 49 when she committed suicide, while he was 70 and more than 5 inches (13 cm) shorter.

Anyone who is strikingly different is subject to curiosity, pointing, comments and intrusive questions. People want to question them about their special feature as it is interesting to the questioner. For those questioned, it may be painful or boring to go over again. Sadly and regrettably, those with an obvious special feature, especially disfigurement, tend to be remembered almost exclusively for it. A warm-hearted, witty, clever, young lawyer might just be remembered as 'that lady with the awful squint'.

While human diversity is good in adding variety and interest, and the possibility of beneficial complementarity between different types, it is bad in being a source of strife. There are conflicts between races, nations, classes, castes, ideologies, religions, language groups, modernisers and traditionalists, age groups, the sexes, etc.

Individual variation arises from genetic and environmental differences. Genetic differences come from pre-existing variation and new variation arising from **mutation** (DNA changes) and **recombination**, bringing together new combinations of existing parental genes. Environmental differences include climate, diet, upbringing and education. Diversity helps a population adapt to changing circumstances, including disease onset. Some individuals may be better able than others to survive those changes.

In group selection, some groups survive better than others, and the existence of diversity within it may help a group survive. It is good to have a mixture of innovators and traditionalists, of brainy and muscular types, of extraverts and introverts. A population entirely of politicians, of lawyers or of building workers could not compete with balanced populations.

Large populations can accommodate more specialists than can small ones. Very small groups who were pioneers crossing America had to become largely self-sufficient, doing their own timber-cutting, building, hunting, trapping, self-medication and trading, as portrayed by Laura Ingalls Wilder (1867–1957) in her Little House on the Prairie books based on her childhood.

Some differences are readily observed, as for sex, height, age, skin colour or wearing glasses, but many are *internal* — intelligence, heart disease or an inability to digest milk. We change over short periods, with different moods affecting our thoughts, sociability and reactions.

This book explains *why* we differ from one another: genetic differences, environmental differences, including diseases as well as climate, culture, upbringing, chance, and personal choice. It explores the effects which differences have on their possessors, from pride to self-loathing, and what effects they have on those observing differences in others, from pity to envy. What is it like to have ME (Myalgic Encephalomyelitis), cystic fibrosis, diabetes, a heart attack, severe food allergies, or to be extremely tall? These are all covered, and many more.

A special feature is a series of personal accounts of what it is like to be different in various ways. These first-hand accounts were written for this book by individuals I know personally. They describe their special attribute, at what age it became apparent, what its physical or mental consequences have been, how others have reacted to it, and how it has affected their lives and happiness. Some writers would not want to lose their disability as it is so much part of them. I was astonished by the frequency and variety of special features shown by my students and their families, such as one intake of 100 students having two cases of pits in the head (Plate 19G) or three cases of a knee problem called Osgood-Schlatter disease, or webbed toes, having attempted suicide, having four kidneys or being born with potentially lethal toxoplasmosis.

Although physical and mental disorders can make people unhappy, people with handicaps such as blindness or being confined to a wheelchair can live useful and fulfilling lives. One ex-student has cystic fibrosis, with serious lung infections, digestive problems and a probable early death, but she remains bright and cheerful. Here is one account from a happy former undergraduate. Most of her conditions are considered later.

Personal account. *I have several physical problems and dyslexia*

Anaemia, like brother and mother; take iron tablets and try to have a balanced diet. Thyroid problems, autoimmune; my thyroid is underactive, with high antibodies, high levels of thyroid-stimulating hormone. I took desiccated pig gland tablets but was told to stop. Now I am on synthetic thyroxin. Left-handed, making me clumsy and dropping things. Might still have a mild heart condition which I had when little, with breathing problems — supposed to go with puberty but I am still short of breath. Polycystic ovary syndrome with large ovaries and irregular periods, but do not have hairiness symptoms. Bad sense of balance. I bump into people and fridges, hit doorways and am often bruised. Get car sick. Dyslexic: am frightened by lots of text; letters often seem mixed up, e.g., from/form. This affects my work with late hand-ins, slow comprehension; I may need to read something 20 times to understand it. Penicillin allergy like mother. In spite of all these problems, I am happy and fine.

She later added this, in her dyslexic English.

i finally got my diagnosed with not only dyslexia, but dysparaxia [dyspraxia]. i have synesthesia [synaesthesia] when i associates colores with number and days. i also have Irlene syndrome, needing a colored filter (pink or aqua green) over white pages for reading, as black on white is too harsh, making reading very slow. my parents are not supportive, my mum completely regected the idea that i was dyslexic. i am defenetly sure that wiull fail these exam again. istill cant separate my "normal" behaviour from "dyslexis or dyspraxic" behaiour, it is a part of me ... anyway istill need time to recognise that i am "desabeled" and i can tell you it feels sooooo strange, sometimes unfair!

Many disorders and diseases are described, so sections of this book can be used as a lay person's medical guide. Professional medical advice should always be sought for diagnosis and treatment. Some common afflictions covered include cancer, diabetes and high blood pressure. Some diseases described in Chapter 11 affect more than a million people. A very rare one is congenital (present at birth) tracheamalacia. Anna Chrostowski was born with a floppy, kinked trachea (wind pipe). When she cries, it contracts, stopping her from breathing. Her parents have to give her immediate mouth-to-mouth resuscitation or she will die. She has to sleep in an elevated position next to her mother, attached to a monitor which activates an alarm if Anna fails to breath within twenty seconds.

This book includes statistics of human interest, such as the amount of blood your heart pumps a day, how much urine is passed, how big human eggs are, and how fast sperm can swim.

The differences covered include common ones such as weight, eye colour and intelligence, to rarer conditions such as dwarfism, haemophilia (lack of blood clotting) and hermaphroditism (having features of both sexes). One effect may have several possible causes. Someone with a finger missing could have been born that way, or lost it in an accident or through leprosy or frostbite.

Care is needed with the term 'normal' as we tend to take those around us as normal. A man of normal height for a Canadian Mountie would be abnormal amongst pygmies. A man wearing a long white robe, a *galabiya*, would be normal amongst Egyptian peasants but not in Edinburgh. Adults being able to digest milk is normal in Holland but not in Thailand. A white boy in England went to school when he was five and sat next to a black boy. The white boy was not used to seeing black children but was very friendly with a black West Indian lady who has very visible vitiligo (Chapter 13.6), with obvious pink patches of skin on her hands, neck and face. When he got home, the white boy said that he had seen a very strange black boy, all black, with no pale patches at all! Consider the term 'natural': is it natural to wear clothes or for men to shave? Is monogamy natural?

Chapter 8 explains some genetic terms and the inheritance of characteristics. You can refer to it when necessary or just ignore those terms and carry on reading. If medical terms or chemical names mean nothing to you, just ignore them; some readers will find them relevant.

Abbreviations used include WHO, World Health Organization; OECD, Organisation for Economic Co-operation and Development; in the UK, NHS, National Health Service; NICE, National Institute for Health and Care Excellence; ONS, Office for National Statistics. In the interests of brevity, I often omit rare exceptions and complications. Different sources sometimes give conflicting information. There are suggestions for further reading, of things to do if you wish, and an index so that a topic can easily be traced.

1.2. The major types of human difference

There are obvious visible differences but some become noticeable only when people do particular things, like limping when walking, or being aggressive when driving. Many differences are not outwardly visible, like having diabetes or being a genius. You cannot tell by looking whether someone's blood group is A, Rhesus positive, or O, Rhesus negative, although that is crucial in blood transfusions.

The more you look, the more differences you see, such as differently shaped ears, noses, eyebrows and heads (see the Plates). There are huge differences in teeth, seen only when people open their mouths. We differ in unexpected ways. Much leg and back pain is related to the sciatic nerve and piriformis muscles of the lower back. One would expect relations between nerves and muscles to be constant but that is not so. This muscle and nerve may each be divided or undivided. The nerve may pass under, through or over and under the muscle.

Differences in beliefs can have practical implications. In 2014, Benjamin Black, a British doctor fighting the lethal outbreak of the Ebola virus in West Africa, said that many infected people were not coming for treatment, allowing the disease to spread. In remote parts of Sierra Leone, Guinea and Liberia, many people think that such diseases are due to witchcraft, so they do not seek medical treatment.

For some characteristics, there is **continuous** variation from one extreme to the other, with all intermediates occurring, as for age and for weight. Other characteristics are sharply **discontinuous**, like blood groups, where for the ABO group, you are A or B or AB or O, with no intermediates. For sex there is a pretty clear distinction into two main types, male and female, with minor variations (Chapters 4 and 10).

Although textbooks often use the terms 'quantitative variation' and 'qualitative variation', I shall use 'continuous variation' and 'discontinuous variation' as they are clearer. There are characteristics where the variation is **partly discontinuous** and **partly continuous**. Thus for human height there is largely discontinuous variation between dwarfs and non-dwarfs (Chapter 18, Photo 1), but there is continuous variation in height within each of these two types.

Characteristics with discontinuous variation are often controlled by single genes (loci) with simple patterns of inheritance (Chapter 8), whilst characteristics with continuous variation are often controlled by more than one — sometimes many — genes, usually with environmental effects as well. See Chapter 8 for more about traits with continuous variation and multifactorial characteristics, which include cleft lip, club foot, high blood pressure, diabetes and schizophrenia.

1.3. Examples of differences and their causes; heritability

We have biological differences caused by our genes. We are influenced physically and psychologically by environments at different life stages, by diet, exercise, diseases, upbringing, education, work and social conditioning. Our personal choices — say about shaving, clothing and hair length — make a large difference to our appearance, as can chance, such as losing a leg.

We have differences between races, nations, populations, sub-populations (e.g., students, the elderly or religious groups), and individuals. We even have differences within individuals. From the fertilised egg there is development and differentiation into liver, brain, eye, bone or eggs or sperm cells, with each gaining specialised functions and losing others.

There are chance 'accidents' in cells, such as spontaneous mutations of genes, or gains, losses or rearrangements of chromosomes (Plate 1C). If the altered cell is viable, it can divide and give many daughter cells, so that the individual is a mosaic of two or more cell types. The lady in Plate 17C is a vivid example of a mosaic, with genetically different freckled/non-freckled skin. Human females are always mosaics from X-chromosome inactivation (Chapter 4).

Genetic factors have a large influence on sex, height, intelligence, the colour of eyes, hair and skin, whether we carry inherited disorders such albinism (giving very white skin and hair; Plate 15E) and our susceptibility to infectious diseases. Genes affect aspects of behaviour, such as alcoholism, schizophrenia or manic-depression, although environmental triggers are partly responsible for such inherited tendencies becoming expressed. Schizophrenia occurs in about 0.8% of the population, but if one identical twin has it, the other has a chance of 45% of getting it, while a non-identical twin — a brother or sister of the sufferer — has a chance of about 13%, showing its strong genetic element. The figure of 45%, not 100%, for concordance between identical twins shows that environmental factors or chance must be involved.

Some characteristics such as blood groups (Chapter 17), eye colour or sex are determined almost exclusively by genes and our 46 chromosomes (Plate 1A). A minority have different chromosomal make-ups, such as people with 47 chromosomes, including Down Syndrome individuals (Plate 19E) with three copies of the very small chromosome number 21, or individuals with two Xs and one Y sex chromosome, giving Klinefelter Syndrome, with mainly male features, such as a penis, but also with some female tendencies, such as breast development (Chapter 4). A *syndrome* is a characteristic group of symptoms associated with a medical condition. Medically, a distinction is made between symptoms observed by the patient and signs detected by the physician, but most people ignore that.

Environmental factors causing differences include climate, what and how much we eat, social and religious customs (which may include male or female circumcision), education, and diseases. Because they are not genetic, age and chance count as environmental factors. We have our internal environments, including the amount of sex hormones, where the concentration of testosterone is a major factor whether people of either sex develop baldness or hair thinning.

Heritability is an important concept. It is the percentage of variation for a characteristic in a population which is due to genetic variation, the rest being due to environmental variation. Thus the ABO blood groups have 100% heritability as all variation in them is due to genes, not environment. A characteristic largely determined by environmental variation would have a low percentage heritability. As the amounts of genetic and environmental variation for a characteristic differ between populations, heritabilities are usually shown as a range of typical values, e.g., height, 80–90% (environmental variation accounts for 10–20% of the variation); autism, 60–80%; IQ, 60–80%; bipolar disorder, 60–80%; weight, 50–70%;

schizophrenia, 50–60%; educational achievement, 40–60%; depression, 30–60%; obsessive-compulsive disorder, 30–55%; epilepsy, about 30%.

Different diseases, such as polio, leprosy, chicken pox and colds, can affect our appearance, mobility, happiness, fitness, alertness, sociability and life span. Their effects vary from death to loss of fingers or toes, to unsightly spots, to minor and temporary irritants.

I prefer the term *disease* for a transmissable condition such as chicken pox, and *disorder* for non-infectious conditions such as diabetes and dementia, but in medicine disease is used for both types of condition. I have vitiligo (white skin patches, Plate 17A, B), which is non-infectious, and do not consider myself diseased.

Types of human disease and disorder include inherited ones (haemophilia), infectious ones transmitted directly between humans (syphilis), infectious ones transmissible over short distances (influenza through sneezing or contact), ones transmitted through other organisms (malaria via mosquitos), ones from dietary deficiencies (rickets from too little vitamin D), and ones due to non-biological agents, such as arsenic poisoning.

Our environment and our genes interact for many characteristics. Our weight depends on age, sex, how much and what we eat, how much we exercise, and on our genes. Intelligence depends on genes and environmental factors such as how well nourished we were in the womb and later, how much mental stimulation and training we received, and how much we use our brains.

Some acquired behavioural differences arise through legislation (e.g., driving on the right or the left), social conditioning or religious edicts. Dietary differences can be religious, such as Muslims and Jews not eating pork and Hindus not eating beef, or from local availability. Inuits are traditionally more likely to eat whale, seal and fish rather than lamb or tropical fruits.

The side-effects of medical treatments can cause dramatic changes, such as chemotherapy causing loss of hair from the head and body. Drugs such as penicillin and aspirin are safe for most people but in sensitive individuals they can cause skin rashes (Plate 16D), breathing difficulties or even death from anaphylactic shock.

Time can bring about huge changes in individuals, from ageing, from chance, individual effort or social and political changes. Nelson Mandela,

Robert Mugabe, Jomo Kenyatta and Martin McGuinness were active members in Africa or Northern Ireland of subversive terrorist groups which carried out acts of violence against buildings and people, yet in time they became members or even leaders of their country's governments.

Reference

Lorenzo, F. R. *et al.*, A genetic mechanism for Tibetan high-altitude adaptation. *Nat Genet* (2014) **46**: 951–956.

Chapter 2

Races and Inter-Mixing, Nationalities, Cultures, Castes, Classes and Religions

2.1. Introduction

The world has a huge diversity of races, nationalities, cultures, sects, castes, classes and religions. These are major divisors of mankind, together with sex and language.

In China, the majority are Han but 8.5% belong to one of 55 officially recognised ethnic minorities. The Han are the world's largest group, with 20% of the world's population, 92% of Chinese, 98% of Taiwanese, 74% of Singaporeans and 25% of Malaysians. Minorities in China include Zhuang (18 million), Uyghur (11 million), Manchu (11 million), Hui (10 million, mainly Muslims), Miao (9 million), and numbers of Yi, Tujia, Mongols, Tibetans and Koreans. The Miao (Plate 12D) have subgroups with different languages, dialects and customs.

The Indian Constitution lists 645 tribes but India is known for its Hindu hereditary castes, with subcastes and local variations. It is usually unacceptable to marry outside one's caste, which is fixed by birth. There are rules forbidding interaction between castes involving food, drink, utensils and even shadows.

The hierarchy has Brahmins, the priestly caste, at the top, followed by Kshatriyas, the warriors, then Vaishyas, the merchants, and below them Shudras, the labourers, servants and artisans. Dalits (untouchables) handle dirty jobs such as rubbish, sewage and corpse disposal. According to the

Manu Smriti, the Hindu text on caste laws, Dalits are 'fierce untouchables' and 'dog-cookers' who 'should be outside the village', wear the clothes of the dead, eat from broken crockery and 'wander constantly'. In 2001, international agencies helping earthquake victims in Gujarat found separate refugee camps for four Hindu castes, untouchables and Muslims.

There are about 900 Dalit sub-castes, the lowest being Musahars, the rat-catchers, who are excluded from Hindu temples, banned from drinking water from wells used by higher castes, and whose children have to eat separately at school. There is 6% literacy among men, 2% among women. Musahars make up nearly half the population of Bihar State. They were delighted when in 2014 Jitan Ram Manjhi, aged 68, was elected that state's chief minister. In earlier days he had survived by catching and eating rats.

Classes are social divisions such as upper class, upper middle class, lower middle class and working class, although work is done by all classes. There is movement between classes by education, marriage, commercial success or individual effort. Members of a class tend to share educational, economic and cultural characteristics.

There are expectations associated with classes, relating to income, education, habits, ways of speaking, etc. Count Unico Wilhelm Van Wassenaer (1692–1766) was a Dutch nobleman who insisted that his music be published anonymously as composing music was beneath the dignity of the nobility.

A big difference is between free people and slaves. The taking of slaves goes back before ancient Egypt, Rome and Greece. In Ancient Athens, slaves outnumbered citizens by three to one. In 2013, the Global Slavery Index estimated that about 29 million people still live in some form of slavery, especially in China, India, Pakistan and Arab nations.

An Indian Hindu's account of castes and sects

By Kamlindra in Rajasthan, 2013.

I am of the warrior caste, the Kshatriya. If one of the untouchable caste visits my home, they will sit on the ground and would not dare to sit in a chair next to me even if invited to do so. There are separate utensils to be used and washed up by untouchables. My grandmother insists on my enforcing the caste rules, but I think that the system will die out within 40 years. The untouchables would go only in the late morning or early afternoon into towns or villagers occupied by higher castes because shadows are shorter then. If the shadow of an untouchable touches a higher class Hindu, the latter feels polluted and can legally beat the untouchable to death. The untouchable might be glad as it gives him the chance of being reborn into a higher caste.

There are only 72 days a year for Hindus to marry, auspicious by the lunar calendar. The couple's horoscopes must match on at least 18 out of 36 points. Marriage partners are chosen by the parents, with advice from those who know them. The couple do not see each other before marriage but exchange horoscopes, photos and CVs, and are allowed to phone each other. If one of the couple does not like the other, they do not say that but claim that the horoscopes do not fit.

The bride's family has to host the bridegroom's stag party dinner for perhaps 250 men, which is a big cost. I did not see my bride until the wedding. With only 72 days a year for weddings, people are often invited to several on the same day and have to choose. Your social standing depends on the number of guests at your wedding: 400 guests, poor standing; 1,100, good; many more than that, excellent.

Marrying someone from a different caste is strongly disapproved of. The whole extended family of the bride and groom will be ostracised. The brothers and sisters will find it very hard to marry if the whole family has been disgraced in this way. Honour killings relieve a family from the disgrace. The killer gets a life sentence in jail but is usually released after four years of good behaviour or if the family has influence.

The Tata clan are Parsees, originally from Persia, and own much of the steel and motor industry. One can only be a Parsee if both parents are Parsees. Dead bodies are exposed in the Zoroastrian 'towers of the dead' for vultures to devour. In built-up areas, there are often too few vultures, leading to complaints about bad smells from rotting bodies in the towers. Many vultures died from eating pesticide-laden animal corpses. The vultures died out completely in Mumbai, so special aviaries were built to raise vultures.

Jains make up less than 1% of the population but own 28% of the wealth. They are on the boards of most big Indian companies. Men can attain salvation if they are good. Even very pious women cannot do so, but can be reborn as a man. Jains do not believe in killing. They do not eat meat or anything from underground, such as garlic, onions or potatoes, in case worms are killed at harvest. The monks eat only what can be held in one hand for a meal. Male and female monks walk long distances on pilgrimages, going bare-foot and wearing a cloth over the mouth so as not to take in and kill insects [Plate 8A].

The Sikh religion goes back to the 15th century, with Gobind Singh in the 17th century as the 10th and last guru [teacher and spiritual leader]. *He recruited men of the warrior caste, including Hindus, to fight the Muslims, and gave his military men the name Singh. Although I am not a Sikh, my second name is Singh. All male Sikhs are Singhs but not all Singhs are Sikhs. Baptised male Sikhs wear turbans of any colour, but with us Hindus, turban colour depends on caste. The Sikh holy book, kept at the Golden Temple at Amritsar* [Plate 8D], *is known as the 11th and perpetual living guru, Guru Granth Sahib. It is used to determine a child's name.*

India looks a very dirty country. It is cleaned up only every four years when an election is looming.

The Vishnoi are a blond tribe claiming descent from Alexander the Great. To get handsome children, the handsomest man in the village was chosen to have the privilege of sleeping with any married woman if he wished. If that man's shoe was outside the door, the husband had to stay out. The man was stoned to death after four years in that role.

Although America considers itself classless, there are the so-called *Boston Brahmins,* upper-class citizens often associated with Harvard University and descendants of the early English colonists. Their exclusiveness is portrayed in the 'Boston Toast' by Harvard alumnus John Collins Bossidy:

> *And this is good old Boston,*
> *The home of the bean and the cod,*
> *Where the Lowells talk only to Cabots,*
> *And the Cabots talk only to God.*

2.2. Cultures and nationalities

A culture is a subdivision of civilisation, largely sharing beliefs, language, literature, values and traditions. It may, but need not, involve a shared religion, food preferences or geographical proximity. It may spread widely through conquest or migration. Many countries have Chinatowns.

Cultures differ from races as people of more than one race may belong to the same culture. Kabira speaks Russian, Armenian, Turkish and English, and has lived in Russia, Turkey and England. She has Russian, Armenian and Turkish ancestry, all Caucasoid. She is clearly multi-cultural and multi-national, but is she multi-racial?

Recent decades have seen cultural shocks from mass immigration. Germany has many immigrants. More than five million Muslims immigrated, legally and illegally, to France from former colonies such as Morocco, Algeria and Tunisia, plus Turks and Kurds. Their different culture and religion cause many problems. France has banned the wearing of burkas and niqab veils in public.

French and English cultural traditions are different, and so are Chinese (ancient, now Communist), Japanese (ancient, a hereditary monarchy with an emperor), Malaysian (Islamic) and Singaporean (republic). Belgium has huge problems from the clash of cultures between French-speaking Walloons of southern Belgium, Dutch-speaking Flemings of northern Belgium, and German speakers.

Cultural differences, in which language and religion often feature, have led to the break-up of countries, as in the partition of India (1947) into Hindu India and Muslim Pakistan, Ireland into Protestant Ulster and Catholic Eire, the *de facto* partition of Cyprus (1974) into Turkish (Muslim, Turkish-speaking) and Greek (Greek Orthodox, Greek-speaking) parts, and the break-up of Yugoslavia from 1992 into Slovenia, Serbia, Croatia, Bosnia, Montenegro, Macedonia and Kosovo. Cultural and racial differences underlie attempts at independence by Tamils in Sri Lanka and Kurds in the Middle East.

Sri Lanka exemplifies the conflict-producing problems of different cultures. There are about 74% Sinhala, 13% Sri Lankan Tamils, 6% Indian Tamils (especially tea estate workers), 7% 'Moors' (Muslims), and 1% Burghers (offspring of Europeans and Asians), Malays and others. The two main cultures differ in religion (Buddhist/Hindu), language (Sinhala/Tamil), origin (Sinhalese from North India, Tamils from South India), and there have been conflicts for more than 2,000 years.

A distinctive subculture is the austere Amish, descended from Anabaptists from Germany and Switzerland who fled to America in the 18th and 19th centuries. They reject modern civilisation and have severe religious and social practices. Typical images portraying Amish show bearded men in big black hats, women in bonnets, large families and horse-drawn buggies. The average of seven children per family accounts for the fact that in two or three hundred years they have increased from 6,000 to 280,000. Most speak Pennsylvania Dutch (a form of German) but in Indiana, many speak a dialect of Swiss German.

Once people are baptised into the Amish church they must marry only within that faith. Those departing from its strict requirements are excommunicated and socially excluded, shunned by their own families. There are restrictions on clothing, telephones, power-line electricity and motorised vehicles. There are at least eight subgroups with different rules and religious views.

Unlike Hindus, Sikhs believe in equality and have no castes. They do not believe in discrimination by sex or creed. Most males have the surname Singh (lion) and most females have the surname Kaur (princess). After initiation, male Sikhs should keep **the five Ks**:

- kesh, uncut hair covered in a turban;
- kara, an iron bracelet;
- kirpan, a small sword tucked into a special strap;
- kachehra, a cotton undergarment;
- kanga, a small comb of wood.

At the Sikh holy Golden Temple at Amritsar, I saw men bathing in the holy lake wearing only their turbans and the cotton shorts prescribed by their religion (Plate 8D). There are about 800,000 Sikhs in Britain, courted by politicians at elections.

At an Indian Jain temple there was a long list of entry prohibitions, including no menstruating women and no leather as that comes from killed cows. This meant removing shoes, leather belts, wallets, purses, handbags, etc.

What is acceptable in one culture is often unacceptable in others. Muslims in Britain celebrated their festivals by ritual slaughter of animals, especially sheep, in their gardens. The appalled British quickly had this banned. In 2013, the Kazakh champion football team, Shakhter Karagandy, slaughtered a sheep before their home game against Celtic as is their tradition. Their spokesman said that they planned to do this in their away game in Glasgow: "As far as we know in Scotland, the agriculture is very developed, so it shouldn't be an issue to find a sheep." Celtic made it clear that that was totally unacceptable.

Nationality differs from race. Boxer Nicola Adams, MBE, was a plucky Gold Medal winner at the 2012 Olympics. She was born in Leeds, England, and her nationality is British, yet with her black skin and African features, no one could take her as being of English ancestry.

Here are factors which help a nation minimise internal and external conflicts:

- Clear natural boundaries, such as a sea, mountains or broad rivers, to minimise border disputes.
- One race.
- One culture.
- One language.
- One religion or none.
- Sufficient resources and land for housing, industry, agriculture and leisure activities, but not enough to provoke envy from other nations.
- Friendly neighbouring countries, if any.
- Enough people to allow for a range of specialised services.
- A balance between adventurous innovators and cautious traditionalists. Not all changes and 'reforms' are good, but neither is stagnation.

In history, most areas that had been conquered had changes to their racial composition and culture. Lebanon was part of Syria for much of its history, with conflicts between its Christian and Muslim heritages. It was a part of Canaan and home to the Phoenicians (c. 1550–539 BC). It has been occupied by the Egyptian, Persian, Assyrian, Hellenistic, Roman, Armenian, Eastern Roman, Arab (Umayyad, Abbasid, Fatimid), Seljuk and Mamluk empires. There was a mandate to France after World War 1, independence in 1943, the Lebanese Civil War (1975–1990) and more troubles since.

When national boundaries are set, there are often anomalies. Between North and South Korea, the border runs through a frontier building. When parts of Africa were divided into countries by people drawing lines on maps, all kinds of problems arose by including in the same nation mutually hostile tribes with different traditions, cultures and religions. With the benefit of hindsight, it is easy to blame politicians but often there is no ideal solution, only a range of lesser and greater evils.

'Ethnic cleansing' means mass killings by one nation, race, sect, religion or tribe, of another. It overlaps with genocide. There are nasty examples from most continents, but ones from Africa are all too common. In Rwanda in 1994, there was a huge genocidal slaughter of the Tutsi people and of moderate Hutus by members of the Hutu majority. In about 100 days, 500,000–1,000,000 Rwandans were killed, about 20% of Rwanda's total population and 70% of the Tutsi living there.

2.3. Race

2.3.1. *Introduction*

Some politically correct types argue that race does not exist because it cannot be defined precisely. That is a false argument because beauty, truth, poverty and illness are hard to define precisely, yet they exist. In big cities, one sees a number of clearly distinguishable physical types, differing in skin colour, eye colour, hair and facial features. They differ by race, even though races may not be sharply defined, with overlaps and hybrids. The differences are genetic, persisting in different environments even where environment has some effect, as on skin colour but not on hair type, nose length and how flaring the nostrils are. Sometimes efforts are made to modify racial features using hair straighteners, skin lighteners or darkeners. Different races are shown in Plates 2 to 20. The word 'ethnicity' is more politically correct than race but means almost the same.

The American Anthropological Association (website accessed 18/4/2014) promotes the view that race is a recent human invention, and that race is about culture, not biology. I disagree strongly. Race is a fundamental *biological* property. Negroid people from the Congo will have Negroid offspring whatever environment or culture they are brought up in. Mongoloid Chinese parents will have Chinese-looking children, even if the babies are adopted and raised in the Congo. If race does not exist, one could not have mixed-race children who look of mixed race. Race is as real as the violence it provokes in many countries.

One dictionary defines race as: 'A group of people of common ancestry, distinguished from others by physical characteristics, such as hair type, colour of eyes and skin, stature, etc. Principal races are **Caucasoid**, **Mongoloid**, and **Negroid**'. According to another dictionary, Caucasians are native to Europe, North Africa, and western and central Asia (a geographical description, not mentioning physical features), while Mongoloids are described as a broad-headed, yellow-skinned, straight-haired, small-nosed race (all physical features and no geographical ones!). Negroids are described as having full lips, tightly curling hair (Plates 4C, 5A), etc. (only physical features and with no mention of skin colour or geography).

The first dictionary states that Caucasoids are a light-complexioned group including the peoples indigenous to Europe (see many Plates), N. Africa, S. W. Asia, and the Indian subcontinent (Plate 11 D, E) and their descendants in other parts of the world. It states that Mongoloids are characterised by yellowish complexion, straight black hair (Plate 4G, H, I), slanting eyes, short nose, and scanty facial hair, including most peoples of Asia, Eskimos and North American Indians. It describes Negroids as having brown-black skin, tightly-curled hair, a short nose and full lips, including indigenous peoples of Africa south of the Sahara, their descendants elsewhere and some Melanesian peoples. One needs to put together the descriptions of both dictionaries to get an idea of the geographical and physical features of the three main races, which can be subdivided in various ways. See also a definition of race in Chapter 15.2. English ancestries include hunter-gatherers from the Continent about 10,000 years ago, invading Romans, Picts, Scots, Irish, Angles, Jutes, Saxons, Vikings and Normans, persecution-fleeing Huguenots and Jews, and recent immigrants.

Bodmer (2015) gave a masterly survey of genetic research on human origins, based on blood groups, HLA genes, enzyme variants, Y chromosomes, mitochondrial DNA, and an enormous amount of DNA sequencing of many individuals in many populations. All the results show the evolution of mankind from an African origin into three main racial groups, African/Negroid, Asian/Mongoloid and European/Caucasian. There is even quite good DNA separation between European countries and even within them. For example, DNA analysis shows the effects in the Orkneys of invasions by Norse Vikings from 873 to 1468, and of Vikings and Anglo-Saxons in South Wales.

In further pioneering work on human genetic diversity in relation to history and geography, the same group sequenced DNA samples from more than 2,000 people in rural areas (excluding recent immigrants), analysing DNA differences at over 500,000 positions within the genome (Leslie *et al.*, 2015). This produced the first fine-scale genetic map of any country, finding striking patterns of genetic variation across the UK, with distinct groups of genetically similar individuals clustered together geographically. Comparisons with sequence information from more than 6,000 Europeans revealed vestiges of population movements into the UK over the past 10,000 years, confirming and expanding on known historical migrations. Surprisingly, Celts in different parts of the UK (Scotland, Northern Ireland, Wales and Cornwall) are genetically very different from one another. There are separate genetic groups in Cornwall and Devon, with a division almost exactly along the modern county boundary between them, the River Tamar.

A map of the tribes in Britain in 600 AD is very similar to one produced today from genetic clusters in this study, suggesting that many local communities mainly stayed put over 1,415 years, maintaining regional identities. The majority of eastern, central and southern England is made up of a single fairly homogeneous genetic group with a significant DNA contribution from Anglo-Saxon migrations (10–40% of total ancestry).

In this book the term 'race' is interpreted broadly, going beyond the super-categories of Caucasoid, Mongoloid and Negroid. Offspring (hybrids) between Negroids, Mongoloids and Caucasoids (Plates 2 and 3), e.g., children from African × English (the mother is written first, and × = mated to), Thai × English, Japanese × English, Chinese × Indian (Indians are darker-skinned Caucasoids), Colombian × Chinese and Japanese × Greek unions are multi-racial, but so surely are Swedish × Pakistan and Colombian × English hybrids, even though both parents are Caucasoids. In those examples all known to me in England, the mother is usually the non-Caucasoid. Marriages are usually *patrilocal*, with the woman moving to the husband's area. Language was not a barrier to mixing for World War 2 GI Brides. Millions of American servicemen were in Britain for the invasion of Europe. Many British men were away from home and the women often socialised with Americans, leading to romances and often marriages. The British men resented this, saying that the Americans were 'oversexed, overpaid and over here'; being able to tempt the women with nylons, cigarettes and other luxuries was an unfair advantage. The GIs replied that the British were 'undersexed, underpaid, and under Eisenhower', and took their GI Brides back to America. Approximate numbers of GI Brides were 100,000 from Britain, 175,000 from the Continent (20,000 German Fräuleins), 15,500 from Australia and 1,500 from New Zealand.

Some US servicemen were black, which led to the birth of inter-racial hybrids. For many people in Britain, including my wife, a black soldier during the World War 2 was the first black man they had seen, surprising though that must seem to youngsters today.

Some well-known people are racial hybrids. President Barack Hussein Obama's mother is mostly English and his father and paternal grandfather were from the Luo tribe in Kenya. Golfer Tiger Woods is more Asian than 'black', being a quarter each of Chinese, Thai and Afro-American, and one eighth each of Dutch and Native American. Actress Halle Berry is half Caucasian and half African American. Racing driver Lewis Hamilton is half white British and half Grenadian. Although these people are by ancestry half or less than half black, they are called black by the media and used as examples of successful black people.

2.3.2. *Racism*

Racism exists in many forms, ranging from smug feelings of superiority to active discrimination, to violence and most extremely, genocide. It overlaps with tribalism and religious prejudice, and exists to different degrees in many multiracial countries. A Malaysian student of Chinese ethnicity, Julia Mei Li Sung, complained bitterly about discrimination in Malaya in favour of ethnic Malays for university admission and government jobs. Widely known past examples of racism were White on Black discrimination in voting rights in the USA and South Africa, with segregation in schools, universities, housing and public transport. An appalling modern example is that of Robert Mugabe in Zimbabwe, who has driven out most of the Whites and dispossessed the white farmers who were the mainstay of the economy. It was a thriving country, a net exporter of food, but Mugabe's racism has ruined it, much to the disadvantage of the average black citizen.

Racism is a big, controversial topic. Some zealots detect racism in almost everything and everyone. Some silly local government employees want to ban the term 'Christmas' and replace it with 'Winterval' on the spurious grounds that 'Christmas' offends those of other religions or none. In Christian countries Christmas is often heartily celebrated by non-Christians, and Christians celebrate non-Christian festivals, with schools enjoying Diwali, the Hindu festival of lights. We receive Christmas cards from Muslims, Hindus, Buddhists, Christians, Jews, agnostics and atheists.

In America there have been complaints that the word 'niggardly' is racist, whereas it means 'stingy' and has nothing to do with 'nigger', a term some Blacks call each other without offence. The film about the Dam Busters in World War 2 used the name of Guy Gibson's dog, Nigger. No one considered it offensive. In 1999, ITV broadcast a censored version of the film with that name removed.

Ardent detectors of racism fail to distinguish between **prejudice** — prejudging before having the evidence — and a **considered opinion**, based on extensive experience. People in Britain are afraid of being accused of racism for making true observations. A female Conservative MP was sacked for saying that motorists found it harder to see dark-skinned people at night. I sympathise with people who grew up in cities such as London, Birmingham, Sheffield, Bristol and Bradford, where their home areas are now unrecognisable from immigration, with 'no-go' areas at night for white people. The fight against real racism is hindered by politically correct false accusations of racism.

In 2009, a chef was discussing food labels and mentioned the old golliwog labels (for decades considered inoffensive) on Robertson's marmalade, which showed the traditional black-faced stuffed doll which children in Europe and Australia often loved greatly. A black employee with a grudge reported him. At the Court of Appeal, Lord Justice Floyd said that the word golliwog was obviously racist and offensive if used in the presence of a black person.

The 'isms' of racism, sexism and ageism have led to a compensation culture where there are genuine victims but in many cases no offence has been committed, yet people claim to be victims to get monetary compensation.

If Britons are called 'limeys' by Americans, 'rosbifs' (roast beefs) by the French or 'whingeing poms' by Australians, no one takes offence. Most Americans don't mind being called 'Yanks', but some from the south, the Confederate States in the Civil War, might object.

Racial stereotypes are deemed offensive but can be a result of accumulated experience, although not all individuals fit their stereotype. Many, but not all, Irish have the 'gift of the gab' (talk fluently); many, but not all, Welsh sing well. Some races or nationalities seem happier, friendlier and more relaxed than others. The English are regarded as cold in nature as opposed to hot-blooded Latin/Mediterranean types.

An American law professor, Amy Chua, coined the term 'Tiger Mother' for Chinese mothers like herself who make better parents and have more successful children because they pressurise their offspring. She claimed that eight groups consistently achieved more success in America than others: Chinese, Jews, Indians, Iranians, Lebanese-Americans, Nigerians, Cubans and Mormons. In the British working class, Chua picked out the Chinese and Indians as much more successful than the Britons. White working-class British boys performed worst in exams. She mentioned that British Jews are disproportionately successful, in 2012 making up 0.5% of the population but accounting for a quarter of the UK's wealthiest people.

A study by the OECD, published in 2014, showed that the children of factory workers and cleaners in parts of the Far East were more than a year ahead in educational achievements than the offspring of British doctors and lawyers. Out of 63 developed nations, the UK was ranked 26th for maths, 23rd for reading and 21st for science, with China's Shanghai city top in each area.

The 'triple package', which Chua thinks accounts for the greater success of some groups, consists first of a sense of superiority, a feeling that your group is exceptional. Second, and paradoxically, is a feeling of insecurity, with a need to prove yourself. The third is impulse control, self-discipline.

Racial or national jokes, based on stereotypes, often have the ring of truth. With jokes about an Englishman, a Welshman and a Scotsman, most people of those nationalities find them mildly amusing, not offensive, and recognise some truth in the behaviour portrayed. One of my favourite nationality jokes is told by Australians and by New Zealanders, with the nationalities reversed. The New Zealand version is: *A New Zealander emigrated to Australia, thereby increasing the average IQ of both countries.*

Although we in Britain pride ourselves on freedom of speech, we do not have it for race. For me, good race relations are where we have friends, acquaintances and colleagues of different races and can joke with them about racial differences, exchanging friendly banter, without anyone taking offence or losing their job.

In 2015, Trevor Phillips, the black former head of the (British) Commission for Racial Equality and of the Equality and Human Rights Commission, appeared in a Channel 4 TV programme, 'Things We Won't Say About Race That Are True'. He said that campaigners like himself had sincerely believed that if they could prevent people expressing prejudiced thoughts, they'd stop thinking them, but that he and the others had been utterly wrong. He conceded that some of the racial stereotypes which the Equalities Commission had tried to prevent were accurate. He mentioned as examples that a third of London pickpockets were Romanian, that black people were six times as likely to be jailed for robbery, that the Chinese were tops at people-trafficking, that for drug dealing, Afro-Caribbeans were pathetic amateurs compared to the Colombians, and that white idiots were the national champions of alcohol-fuelled crime.

By refusing to generalise about the actions of different races or communities for fear of being accused of racism and stereotyping, politicians and the police had let people down. These astonishing comments by Phillips highlighted the mess which over-zealous multiculturalism had got Britain into. Those who had dared to express any concern about the rapidly changing face of their country were shouted down as bigots and racists.

We should accept that races, nationalities and sexes have different interests and different degrees of success in different fields. There are different opportunities, traditions, upbringing, motivation, training facilities and abilities. People from snowy countries are more likely to be good skiers than those from deserts. It is just as well that there are no colour, racial or sexual quotas for top football sides in Europe or for heavyweight boxers, where black men are strongly over-represented. In America, 80% of the players in the National Basketball Association teams are black, and why not? Physique, opportunities, tradition and choices are involved.

2.3.3. *Racial quotas*

If one group is over-represented in one field, they must be under-represented in others. The main race-relations bodies in Britain, which hanker after imposing racial quotas, are themselves unrepresentative, with too few white British employees. Talent and choice should rule, not quotas. If one has racial quotas, what should they be based on? The racial proportions in a country, a district, a city, a suburb, a village? All would give different proportions. Quotas and affirmative action have winners and losers, cause controversy, treat people by category rather than as individuals, and make people more conscious of race, age, sex and sexual orientation.

The imposition of racial or sexual quotas (or affirmative action), say for jobs, directorships or university admission, is controversial. A member of a minority, who may be very able, is liable to be looked down on by members of the majority who think that he or she only got the job to make up that minority's quota. A perceptive black American judge, Supreme Court justice Clarence Thomas, described how law firms took little notice of his Yale law degree, believing he was the beneficiary of the affirmative action programmes that he opposes. He says that Americans are too sensitive about race. I agree.

Priya is a brilliant Sri Lankan statistician who gained a good university job in America before quotas for minorities were established, overtly or covertly. She told me that in the 1970s and 1980s she could have obtained a job in any maths department in America as they were all desperate to increase their proportion of females and 'blacks' (which in America usually includes 'browns') to meet quota expectations.

In 2013 in Britain, the Labour Shadow Transport Secretary described as a "national scandal" the fact that only 4.2% of engine drivers were women. She should think about the unsocial hours involved (difficult if you are bringing up babies and have to spend nights away from home) and remember that women are over-represented in other areas, including nursing and as medical students. Had this woman lived in Ancient Rome, would she have called it a national scandal that women were grossly under-represented amongst the legionnaires, gladiators and galley-slave rowers? Would she have objected to 100% of Vestal Virgins being women?

2.3.4. *Political correctness gone wrong*

After studying the murder of a black youth, retired High Court judge Sir William Macpherson in 1999 issued 'The Macpherson Report', which called London's Metropolitan Police 'institutionally racist'. This inhibited the police force's efforts to combat crime. Police know from experience which groups are most likely to be involved in crimes in particular areas, and are more likely to stop and search people fitting that profile. They were inhibited from doing so by Macpherson's strictures.

In August 2014, a startling report by Professor Alexis Jay showed that for over 16 years more than 1,400 girls in the Rotherham area had been sexually abused by gangs of largely Asian paedophiles, especially Muslims of Pakistani origin. The police and council officers had turned a blind eye for fear of being considered *racist*. Such is the dire Macpherson legacy. Politically correct Labour councillors, child-protection officers and police officers refused to investigate complaints from care workers and girl victims about the abuse. One researcher who reported it was not believed and was sent on an 'ethnicity and diversity course' by child-protection bosses.

People who are less successful than they think they should be sometimes sue for discrimination on grounds of race, gender or age. In 2014, black Police Constable Carol Howard made a claim against the Metropolitan Police for lack of promotion, citing her race and gender. Her white male colleagues might feel equally aggrieved at their lack of promotion but could not use race or gender as weapons.

In contrast to all these problems arising from racial differences, there is a wonderful example of beneficial complementarity between races: that between the Gurkhas of Nepal and the British, in military matters. From the founding of the first Gurkha battalion in 1815, through the Indian Mutiny of 1857 and two world wars to the present day, Gurkha troops have performed incredible feats of gallantry under British officers. What Fromm wrote in 1965 (*The Daily Telegraph*, 25th April) still applies now: 'Two peoples could hardly be more dissimilar in almost every respect — race, religion, traditions and customs, yet, brought together in fighting battalions, they have forged an extraordinary bond of mutual devotion, respect and superhuman courage'.

2.4. Racial mixing

The leader of the 1789 Mutiny on the Bounty, Master Mate Fletcher Christian, and eight fellow mutineers, together with 19 Tahitian friends and lovers, settled on the uninhabited Pitcairn Islands. Their racially mixed and isolated population became a hotbed of feuding, massacres, alcoholism and sexual jealousy. The population reached 233 before World War 2 but dropped to 47, all interrelated. In 2004, half the adult male population of 14 were charged with rape and gross indecency against Pitcairn girls aged five to fifteen. One defence offered was that such activities were a cultural trait there. "It's Polynesian to break your girls in at 12," said an Australian woman who was married to a Pitcairn descendant.

People of mixed race can be extremely attractive, especially Asian/European combinations. One pretty Negroid/Caucasoid hybrid was Dido Elizabeth Belle (1761–1804), the illegitimate daughter of Admiral Sir John Lindsay and an African slave, Maria Belle. There is a charming portrait of Dido and her white cousin Elizabeth; it is easily seen after Googling Dido Belle. Another very attractive lady of mixed race is Helena Christensen, a supermodel and winner of various beauty competitions. She has a Peruvian mother and a Danish father.

Some combinations of characteristics from different races seem harmonious but others look unnatural. For example, ginger hair and blue eyes look odd on people with mainly Negroid or Oriental features.

Mixed-race people are increasing faster than any other group in Britain, with their numbers rising from about 680,000 in the 2001 census to 1.2 million in 2011, a 76% increase in 10 years.

Racial segregation was a huge issue in America well into the last century, especially in schools. In 1924, the State of Virginia's General Assembly passed 'The Racial Integrity Act', requiring a baby's racial description to be recorded at birth with only two classifications, 'White' or 'Colored', which included blacks, American Indians and all other non-whites. Race was defined by the 'one-drop rule', where one drop of any African or Native American ancestry, even the smallest proportion of non-white ancestry, made a person legally classified as coloured. The law banned interracial marriage by criminalising marriage between whites and coloureds. The law was overturned by the United States Supreme Court only in 1967.

Research on what it is like to be of mixed race

Chi Wai Tong (7/8th Chinese, 1/8th Jamaican) did his project on what it is like to be racially mixed. His classifications were: White, Black, Oriental (South East Asians with Mongoloid characteristics), Indian (South Asians with darkish skins and Caucasoid features, including Pakistanis and Sri Lankans), Latino, Middle Eastern. The 81 subjects' feelings about being racial hybrids were extremely encouraging. Table 2.1 shows that none felt bad about being a hybrid and most thought it was excellent.

When asked about factors in choosing a spouse, the majority felt that race was not important, although some would consider it; half thought that religion was not important in marriage. Another question was how the hybrids felt about race. The results were: 26%, race is defined culturally; 21%, defined mentally; 18%, defined by outward appearance; 18%, socially important; 9%, irrelevant; 6%, of no social importance; 2%, does not exist. Only 18% felt that race was defined by outward appearance, which conflicts with dictionary definitions and common perceptions.

When asked with whom or what they associated their **physical appearance**, their replies were: mother, 19%; father, 17%; mixed race, 60%; country, 4%. When asked about with whom or what they associated

Table 2.1. Hybrids' feelings about being racially mixed.

Consequences	Effects of being racially mixed	Feeling about being racially mixed	Effect on accepting other cultures
Dismal	0%	0%	0%
Bad	1%	0%	0%
Neutral	17%	15%	11%
Good	22%	14%	18%
Excellent	60%	71%	71%

their **mentality**, their replies were: mother, 26%; father, 14%; mixed race, 26%; country, 34%. When asked about with whom or what they associated their **culture**, their replies were: mother, 26%; father, 18%; mixed race, 24%; country, 32%.

When asked about how they **generally classified themselves**, their replies were: with mother, 6%; with father, 9%; mixed race, 37%; with a country, 48%, so country is clearly important. When asked **which ethnicity** boxes they ticked, the respondents replied: White, 16; Black African, 2; Black Caribbean, 1; Black Other, 2; Indian, 0; Pakistani, 2; Bangladeshi, 1; Chinese, 1; Other Asian, 5; Other, 43; Prefer Not To Say, 0. The majority opted for 'Other' but quite a number selected only one parent's ethnicity.

When asked about how people viewed or treated them when they visited their parents' countries of origin, the results were: as a foreigner, 37%; as mixed but positively, 45%; as mixed but negatively, 0%; as a resident, 18%.

When asked about how comfortable they felt in their parents' countries of origin, they had to circle numbers from 1, very uncomfortable, to 5, feel at home. **Mother's country**: 1, 1% very uncomfortable; 2, 6%; 3, 24%; 4, 21%; 5, 48% feel at home. **Father's country**: 1, 2%; 2, 5%; 3, 26%; 4, 17%; 5, 51%. About 50% do not feel fully at home in a parent's country, although feeling very uncomfortable was uncommon.

Mr Tong compared the hybrids with photos of their parents. Here are his findings.

Oriental/White: Most hybrids resembled mixes. Dominant Oriental features included dark black hair and dark eyes. Skin was often more coloured than white. Many inherited the facial and physical shapes of their white ancestry.

Indian/White: These hybrids mainly resembled their Indian forebears. Possibly paler than their Indian parent, they usually retained Indian features, especially nose shape. If they do not look Indian, they are often mistaken for Mediterranean.

Black/White: All resembled their black ancestry and would immediately be identified as black. They associated themselves with their black ancestry. In no other group did individuals identify so strongly with just one side of their mixture. Perhaps it is due to the dominant black features that are inherited, such as wiry hair, dark skin and facial structure.

These individual stories of mixed-race students show astonishing degrees of multi-culturalism.

Mario lives in Belgium with his Japanese mother and German father. He was born in Belgium, has joint German and Japanese nationality, and was educated in a Japanese primary school and in Belgium before studying for his degree in England. He is completely fluent in German, Japanese and English, and fairly fluent in French. He defines his appearance, mentality and culture as European, and puts his ethnicity down as Other (Eurasian). He takes both parents' cultures as his own. In both parents' countries he is viewed as a foreigner. In Japan they think he is European and in Europe the reverse. In Japan or Germany, he neither feels at home nor uncomfortable. He feels neutral about being of mixed descent. Race would not be important in a choice of spouse.

Leo Xenarkis (Plate 2D) has a Japanese mother, a Greek father, and is fluent in Japanese, Greek and English. His Greek name, mixed-race origin and appearance did not stop him from becoming chairman of the Imperial College Japanese Society.

Meera (Plate 2C) is a bright, attractive Singaporean with a Chinese mother and an Indian father. She looks entirely Chinese and her father asked me ruefully, "What happened to my genes?"

Yasmine (Plate 3A) was born in England to a Swedish mother and a Muslim Pakistani father, both of whom reside in England although born in Sweden and Pakistan respectively. Yasmine is fluent in English and spoken Swedish, with some French, and thinks and dreams in English. Her mentality is English but is culturally a mix of English, Swedish and Pakistani. She specifies her ethnicity as Other. People usually think she is Greek or Italian. In Sweden she feels at home, but neither at home nor very uncomfortable in Pakistan. She is agnostic but thinks that religion is important in choosing a spouse, not wanting anyone too religious. She accepts and enjoys other cultures, enjoys being of mixed descent and thinks it is beneficial.

2.5. Religion

There is a huge diversity between and within religions. There is no room to describe the differences between them here. Religions can provide comfort, companionship, purpose in life, and explanations (right or wrong, and since they differ, they cannot all be right) of the origin of the universe, life, the human race, the possibility of an afterlife or reincarnation, and so on.

Many Christians believe in miracles, such as that of Saint Dionysius, Bishop of Paris. He was martyred soon after 250 AD and is said to have picked his head up after being decapitated and walked ten kilometres while preaching a sermon of repentance (Plate 8E).

In a British Social Attitudes Survey in 2013, 48% said they had no religion; others were Anglican, 20%; Catholic, 9%; other Christian denominations 17%, and other religions 6%, including Islam, Judaism, Hinduism and Sikhism. Religious fervour differs greatly between religions, with attendance here at mosques roughly equalling that at churches.

Differences over religion are widespread even within cultures or marriages, causing many problems. They divide people over fundamental issues and minor ones. Is there a supreme God, or more than one god? If so, where did He/She/It/They come from and get powers to create a universe? Do miracles occur by divine intervention or are they just extremely rare natural phenomena?

Some people believe literally in every word of their religion's holy book(s). Others pick and choose. Others look for rational explanations for supposed divine interventions, such as the Biblical Ten Plagues of Egypt (the Nile turned to blood; frogs; gnats; flies; cattle killed; boils; lightning and hail; locusts; three days of darkness; killing of the first-born Egyptians). King Herod's Massacre of the Innocents (Matthew 2:16), ordering the killing of all males under the age of two in Bethlehem, is rightly condemned, but surely the killing of first born Egyptians (sons?) at God's command was also a slaughter of the innocents, racist and ageist (and sexist)?

Fighting between religions has cost many millions of lives, e.g., Christians against Muslims (as in the Crusades), Muslims against Hindus, Jews against Muslims, Buddhists against Hindus, etc. Equally bitter have been struggles between branches of the same religion, again with huge losses of life. There were long wars between Protestant and Catholic Christians. In the Middle East there has been and is severe fighting between the Sunni and Shia branches of Islam.

Even where constitutions guarantee religious freedom, minority religions are often discriminated against. In Nigeria, Muslim terrorist groups burn churches and kill Christians. In Communist countries such as China and North Korea, religions are actively discouraged. People who convert from one religion to another are often ostracised or killed.

In America and India in particular, there have been sects which claim to be religions and achieve widespread publicity/notoriety. Leaders who are TV-evangelists often accumulate vast sums of money from believers, and some fashionable gurus amass a fleet of Rolls Royce cars and have sexual relations with many disciples. There are organisations which exist to rescue people from exploitation by such sects.

The Jews have often been persecuted. Their historical claim to be God's chosen people has not endeared them to others. When Judea became part of the Selucid Empire, the laws of 167 BC outlawed Jewish sacrifices, feasts and circumcision. Muhammad expelled the Jewish tribes of Medina. Later, Jews were allowed to live in Muslim lands under restrictions, including paying special taxes. In Muslim Spain, massacres of the Jews took place between 1066 and 1465.

During the Crusades and later, Jews were massacred, expelled or forcibly converted to Christianity in many parts of Europe. In 1290 AD, King Edward I expelled all Jews from England. They were permitted to return by Oliver Cromwell in 1657, in exchange for money. In the Papal States, which existed until 1870, Jews were permitted to live only in specific neighbourhoods called ghettos. Until the 1840s, they had to attend sermons urging their conversion to Christianity, and it was illegal to convert from Christianity to Judaism. Jews often incurred jealousy by accumulating wealth and excelling in professions such as law and medicine. The Western entertainment industry is dominated by people of Jewish origin. The persecution of Jews by German Nazis during World War 2 was racial, not religious. About six million European Jews were killed in the Holocaust between 1941 and 1945.

Famous Jews include actresses and actors such as Barbra Streisand, Harrison Ford, Gwyneth Paltrow, Natalie Portman, Sarah Jessica Parker and Paul Newman; musicians Bob Dylan, Leonard Bernstein and Yehudi Menuhin; scientists Albert Einstein, Carl Sagan and Jonas Salk; businessmen Sir James Goldsmith, George Soros and Bernard Madoff; fashion designers Calvin Klein, Ralph Lauren and Levi Strauss; comedians Sacha Baron Cohen, Bette Midler and the Marx Brothers; directors/producers Steven Spielberg, Woody Allen and Roman Polanski; writer Simon Schama and artists Chagall, Modigliani and Pissarro.

Approximate numbers of Jews today are: USA, 6 million; France, 500,000; Canada, 400,000; UK, 300,000; Russia, 200,000; Argentina, 200,000; Germany, 120,000. There are about 6 million Jews in Israel, making 75% of the population. Certain religions have restrictions on what foods adherents may eat and what they may do on particular days, especially the Sabbath. One Jewish student emailed me to say, 'My sabbath does not allow me to do any work, travel or use electricity. I am concerned as to what written work you will set.' Some Jews leave work early on Fridays to get home before sunset and the start of their Sabbath. The Bevis Marks Synagogue (1701) is the oldest in London and allowed a non-Jew to live in their accommodation rent-free. The Christian was able to open and close doors, switch lights on and off and do basic caretaking activities which the Jews did not want to do on the Sabbath.

Islamic dietary laws (Halal, in the Koran) and Jewish dietary laws (Kashrut, Kosher, in the Torah and Talmud) contain similarities and differences. Both involve ritual slaughter of animals and prohibit eating pig products. For orthodox Jews, shellfish such as crabs, lobsters, shrimps, clams and oysters are forbidden. Dairy and meat products must be kept separate. Fish must have fins and scales, and mammals must be ruminants with cloven hooves, which excludes rabbits. Apart from locusts, invertebrates are forbidden, as are reptiles and amphibians. Jewish, but not Islamic, laws permit wine made by kosher methods. Some Christians obey the food laws in Leviticus 11:3–8 and Deuteronomy 14:3–21. Many do not, feeling that they were made in ancient times for those times, places, people and hygiene practices.

Many priesthoods ascribed natural disasters, such as earthquakes, floods and epidemics, to a failure to worship their gods. The gods could be placated, they said, by giving more money, gifts and devotion to their temples (and therefore the priesthood). Today, science has driven out some old superstitions. I believe that earthquakes are caused by tectonic plate movements, not by a Greek god called Poseidon who needs sacrifices to placate him.

Fatalism is a belief that all events are predetermined so that man is powerless to alter his destiny. It affects a number of religions, especially Islam. If a daughter is sick, saying "It is the will of Allah if she dies," should not preclude calling a doctor. The Indian railways minister from 2004 to 2009 was a Hindu, Lalu Prasad. He exhibited a dangerous fatalism when he blamed India's bad transport safety record on a god: "Indian railways are the responsibility of Lord Vishwakarma, so is the safety of passengers. It is his duty, not mine." Fatalism can be an excuse for defeatism, laziness and not trying.

2.6. Diversity in a school, then and now

In 1953 in my all-boys secondary school in Kingston, Surrey, we were uniform racially — all white Caucasians — but now there are about 42% of coloured boys. We were mainly Church of England, with some Catholics and Jews. What marked out the Jewish boys and masters was that they got Jewish holidays in addition to normal ones. Religion did not affect friendships or whom one played with.

We were not curious about other boys' parental income or status. Friendliness, sense of humour, sporting and academic ability, and the choice of subjects to study, were more important. Even when 16 to 18, few of us had girlfriends. There was no sex education or contraceptive advice, although in biology we saw preserved rabbit reproductive organs. There was almost no homosexuality but private masturbation was frequent.

As far as I know, there were no food allergies, anorexia, bulimia or attention deficit hyperactivity disorder. We had no computers, smart phones, sexting, on-line pornography or social media. In the playground we played football, cricket or other games.

2.7. The origins of racial and population differences

The driving forces in human genetics are **mutation, recombination, natural selection, selection by humans** and **random genetic drift**, where pure chance can change gene frequencies. Very small populations are prone to random drift, where one form of a gene can become frequent by chance, not through selective advantage. If a few individuals colonise a new area, they may be unrepresentative of their parental population, with different gene frequencies.

Humans — *Homo sapiens* — are a young species, evolving from a small group in sub-Saharan Africa about 150,000 to 200,000 years ago. Humans diverged from Neanderthals — *Homo neanderthalis* — about 440,000 to 270,000 years ago. Genetic archaeologists, making deductions from modern DNA sequences, calculate that there was a crash down to about 10,000 people, reducing genetic diversity, before man emerged from Africa.

About 100,000 years ago, colonising groups left southern Africa, moving into the Middle East about 100,000 years ago and then into South Asia about 70,000 years ago, reaching Australia about 50,000 years ago. Later emigrations of *Homo sapiens* were to Europe and thence via a land bridge to Britain about 30,000 years ago. About 15,000 to 20,000 years ago, man crossed from Asia to Alaska, and then through northern North America to the south of North America by about 12,000 years ago, and thence to South America. The continents were much more connected then.

One of our ancestor species, *Homo erectus*, began leaving Africa nearly two million years ago. Members reached Britain about half a million years ago but became extinct. All other types of human, including Neanderthals, were wiped out when *Homo sapiens* spread about 100,000 years ago. Those initial colonists of Britain died out or retreated back to the Continent during the Ice Age which began around 24,000 years ago. When the Ice Age ended, humans crossed back to Britain from about 15,000 BC. The English Channel formed around 9,000 BC, and the British Isles were completely separated from Europe by about 8,300 BC.

Groups differed in mutations, recombination, genetic drift and selection pressures, which is how population and racial differences arose. Larger populations can maintain more genetic diversity than small ones. That diversity enables populations to survive challenges, whether from the environment or disease, speeding adaptation, with less favourable types dying out.

There is only a small amount of DNA sequence diversity in humans, about 0.08% on average (about one nucleotide-pair is different in 1,250 pairs) between two people taken at random, of which 5 to 15% is typically between populations on different continents, 6 to 10% is between populations on the same continent and 75 to 85% is within populations. The average DNA difference between humans is about five times larger than 0.08% when one adds differences in copy number, deletions and insertions. The amount of DNA sequence difference between man and chimpanzees is about ten to fifteen times that between humans.

Although DNA differences between races are not large, they are sufficient to specify the racial differences. A British geneticist in 2013 was working on gene sequences in 27 genes for internal characteristics with no known racial connections. He analysed a large number of sequences from a wide range of humans, finding three distinct sets of sequences. When he looked at the racial backgrounds of those involved, they correlated clearly with Caucasoids, Mongoloids and Negroids. That was not what he was studying but the results showed *three well-separated major groups*, or races.

2.8. Race, adaptation to environment, selection, chance and preferring one's own kind

Ultraviolet light causes skin cancer, and dark skins afford much better protection than do pale skins, which are very good at using light to synthesise essential vitamin D. The strong African sunlight would have been sufficient for vitamin D synthesis, but dark skins would have been a disadvantage in less sunny climates. The compromise between UV-protection and vitamin D synthesis has a different balance depending on how sunny a climate is. This largely accounts for the dark skins of Negroids, the yellowish skins of Orientals and the pale skins of Europeans.

Orientals typically have short noses. Longer noses are better than short ones at warming cold air before it reaches the lungs, which is advantageous in cold winters. Some racial characteristics are explicable in terms of adaptation to local environments at some stage but others are not.

Some group differences have arisen by chance. Humans throughout history have suffered many population crashes, being severely reduced by disease, war, starvation, floods, droughts, etc. The survivors may be the **fittest** (e.g., genetically resistant to a disease outbreak) or the **luckiest** (e.g., ones susceptible to a disease but who were not infected). Suppose a small group of colonisers comes from a larger population with blood groups A and O. By chance, the small group might consist of four people of blood group O. Because O is recessive, all their descendants will be of blood group O (unless mating occurred outside that group). Groups can thus become different by chance, not selection.

There is also selection by humans. In Ancient Greece, babies born with defects were often exposed and left to die. People who are different in some way may be selected against when it comes to marriage, like women with vitiligo (pale skin patches, Plate 17) in some African and Middle Eastern countries. Chapter 10 has more on mate selection. There can be selection in favour of rare types, such as blue-eyed blond-haired women being favoured in a brown-eyed, dark-haired society.

Preferring one's own kind restricts interracial marriage, while the attraction of the different and exotic promotes it. Preferring one's own kind is partly instinctive, partly learned. The preference is not confined to race, religion or nationality but extends to regions, even within towns and cities.

Most of the 7.2 million people in Hong Kong regard themselves as very different from the Chinese nation, and resent the transfer of sovereignty from Britain to China in 1997. At Imperial College London, there was not much mixing between Cantonese-speaking students from Hong Kong and Mandarin-speakers from China or Singapore, although all spoke English. Here are two typical comments by Hong Kong students, who clearly preferred their own kind:

- *I have never been to China. I have been told that the Chinese are dirty and dangerous. We would be robbed if we went there. They think we are all rich.*
- *In politics, Hong Kong people always worry that the Chinese Government will take away the freedom which we are having* [were having under the British]. *In daily lives, we hate some of the Chinese people because they are so dirty. It is always them who sit on the floor of the train, who eat on the bus (this is not allowed in HK), who try to get on a train but not letting others get out first, who jump the queue, who make unreasonable complaints.*

2.9. Races and geography

The preference of ethnic groups to be amongst their own kind is strong, with benefits in social cohesion, concentration of favoured types of shop, places of worship, type of education and recreation, etc. Migrants prefer to settle in an area which has many of their kind already there, making them feel more at home and with less linguistic disadvantage.

In Britain, there are strong concentrations of Pakistanis in Bradford and Leicester. In Greater London, strong associations are of South Koreans in New Malden, Poles in Ealing, French in South Kensington, Bengalis in the East End, and Earls Court was known as Kangaroo Valley from all its Australians. High taxation in France has driven many top earners to England, with an estimated 300,000 French in London.

By 2011, there were 7.4 million immigrants in Britain, 12% of the population. According to the OECD, nearly two-thirds of OECD countries had immigrant numbers exceeding 10% of the population. Birth rates are usually much higher amongst immigrants than among natives.

If many migrants move into an area, residents tend to move out, wanting to be amongst their own kind. 'White flight' has happened in several countries, including America, when previously 'white' areas became dominated by non-whites moving in. There are many parts of London where native Whites feel like strangers. The important points about likely troubles from immigration are the **relative numbers** of immigrants and residents, and the **amount of difference** in appearance, habits and cultures. If the numbers of immigrants in an area are very small, many residents are helpful and friendly, but large numbers are perceived as threatening.

In many countries, people are dependent on immigrants for certain jobs. The NHS would collapse but for Filipina and African nurses, Indian, Pakistani and European surgeons, and other foreigners. Many of our builders and plumbers are East Europeans, and much agricultural labour, such as seasonal fruit-picking, is done by groups of East Europeans or Asians prepared to work outdoors in all weathers at repetitive tasks.

References

Bodmer, W., Genetic characterization of human populations: from ABO to a genetic map of the British People. *Genetics* (2015) **199**: 267–279; doi:10.1534/genetics.114.173062

Chua, A. and Rubenfield, J., *The Triple Package: How Three Unlikely Traits Explain the Rise and Fall of Cultural Groups in America.* (2014) Bloomsbury, London.

Leslie, S. *et al.*, The fine-scale genetic structure of the British population. *Nature* (2015): **519**: 309–314.

Recommended reading

McKie, R., *Face of Britain: How Our Genes Reveal the History of Britain.* (2006) Simon & Schuster, London.

Chapter 3

Height, Weight, Shape and Obesity

3.1. Introduction

Twin studies show that adult height variation is about 85% due to genes and 15% to environment. Giants, dwarfs and midgets are described in greater detail in Chapter 18. Weight variation is about 63% due to genes and 37% to environment, including eating and exercise. Water accounts for about 60% of our weight.

Shape includes fat, skinny, apple-shaped, pear-shaped, 'flat as a pancake', busty, 'legs up to her armpits', squat, beer-bellied and well-proportioned. Photo 3.1 shows a woman in fine shape. Some occupations have prescribed shapes, such as fashion models being tall and thin. Ballerinas should have a small head, long neck, narrow hips, long arms, and legs proportionally longer than the body.

With age, fat is deposited towards the centre of the body but fat layers under the skin decrease, causing wrinkling. Lean mass decreases in old age as muscles, liver and the brain lose cells. Bones become less dense, often causing osteoporosis. Height decreases by about one centimetre (0.4 inch) every ten years after 40. The loss is greater after 70, with a typical loss of one to three inches (2.5 to 6.5 cm). For example, a 73-year-old woman had been 175 cm but reduced to 165 cm. A man of 78 had been 178 cm but became 169 cm.

The National Sizing Survey used 3D body scanners to compare the vital statistics of 11,000 British women in 1951 and 2004, aged 16 to 95, as shown in Table 3.1.

Photo 3.1.   A woman in fine shape.

Table 3.1.   Comparisons of average measurements of British women in 1951 and 2004.

	Imperial		Metric	
	1951	**2004**	**1951**	**2004**
Height	5 ft 3 in	5 ft 4.5 in	160 cm	163 cm
Chest	37 in	38.5 in	94 cm	98 cm
Waist	27.5 in	34 in	70 cm	86 cm
Hips	39 in	40.5 in	99 cm	103 cm
Weight	9 st 10 lb (136 lb)	10 st 3 lb (143 lb)	61.7 kg	65 kg

The change over 53 years was for women to be taller, heavier and larger, with a dramatic enlargement of the waist, a 24% increase. In 2000, its own measurements led to Marks & Spencer changing its clothes labelling, so for example size 14 became 12, boosting women's self-esteem but not making them slimmer. In the 1960s, size 10 was 31-24-33 inches (bust, waist, hips) but now is 34-27-37.

A tall person is usually wider from side to side and from front to back, compared with a small one, so weight increases proportionally more than height, as taken into account in Body Mass Index (BMI) calculations. What is overweight for a small-framed man might be ideal for a large-framed man.

Work out your **Body Mass Index** (BMI), your weight in kilograms divided by the square of your height in metres. If necessary, convert the units:
1 pound = 0.454 kg
1 stone = 6.35 kg
1 inch = 2.54 cm
Example: I weigh 12 stone, 76.2 kg. My height is 6 feet, 1.829 metres. Squaring that height gives 3.345 square metres, so my BMI is 22.8, within the normal range.

The ranges of BMI values accepted as healthy are 20 to 24.9 for men and 18.5 to 23.6 for women. Overweight is a BMI between 25 and 30; over 30 is obese. In 2014 in America, 27% were obese, 35% were overweight and 35% were of normal weight, with a few underweight.

A study published in *The Lancet* (Ng *et al.*, 2014) showed that more young British females were overweight than elsewhere in western Europe. In Britain, 29% of females under the age of 20 were overweight or obese, with only Greeks as bad among 22 countries. Among adults aged 20 and above, 67% of men and 57% of women were obese or overweight, with 25% of both sexes obese.

The National Sizing Survey in 2005 revealed that an average bra size of 34B had increased to 36C in ten years in Britain and Australasia. Since 1951, women's busts and hips had grown by 1½ inches (3.8 cm), while waists had increased by a huge 6½ inches (16.5 cm). Statistics from China in 2005 showed 18 million obese adults and 137 million overweight, out of 983 million adults, obesity rates doubling since 1980. Chinese soldiers have become taller and fatter and are outgrowing their equipment, aircraft cockpits and tanks designed 30 years ago.

Obtaining the correct bra size is complicated by diversity between a woman's breasts. Up to 25% of women have breasts displaying an asymmetry of at least one cup size. For 5% to 10% of women, their breasts are very different, with the left one usually being larger. Some people have feet of different sizes, requiring different-sized shoes.

3.2. Body fat and heat

The percentage of body fat is measured electronically and relates to body shape. Women tend to have more body fat than men, in their breasts and on their hips, for example. The American Council on Exercise gives these values for body fat. For men, essential minimum, 2 to 5%; athletes, 6 to 13%; fit, 14 to 17%; average, 18 to 24%; obese, 25+%, so my value of 17% is satisfactory. For women, essential minimum, 10 to 13%; athletes, 14 to 20%; fit, 21 to 24%; average, 25 to 31%; obese, 32+%.

Fat distribution affects temperature control. Fat around internal organs provides a reserve of nutrients but does not retain heat from skeletal muscle activity if heat retention is needed. Most fat is stored just below the skin, in the best position to act as an insulator.

In hot regions, heat dissipation through the skin is vital. A few groups in hot regions have evolved a visually striking pattern of fat distribution, where excess fat is stored in greatly enlarged regions of the buttocks, with special fibrous tissue to support it. It looks like a second layer of buttocks above the normal ones. The condition, steatopygia (Greek *stear*, suet, *pyge*, buttock), occurs in well-fed Hottentot and San women in South Africa, and Onge Pygmy women of the Andaman Islands. Strangely, the men are not affected. Locating so much fat in one region avoids the excessive insulation that it would provide if more generally distributed over the body in such hot climates.

The fat levels of models and actresses are 10 to 15%, as opposed to 22 to 26% for normal women. The unrealistic image of the female form being projected causes damage to girls trying to emulate it. Vulnerable adolescents and adults try for unachievable results, suffering eating disorders and low self-esteem.

3.3. Height

There are clear differences in height between populations. Average male adult heights were:

> 4 ft 8 (142 cm), Pygmies from the Congo;
> 4 ft 9 in (147 cm), Indian Veddas;
> 5 ft 0 in (152 cm), Bushmen of South Africa;
> 5 ft 2 in (157 cm), Lapps;
> 5 ft 4 in (162 cm), Japanese;
> 5 ft 6 in (167 cm), English;
> 5 ft 8 in (172 cm), Indian Sikhs;
> 5 ft 10 in (177 cm), Berbers from North Africa;
> 6 ft 0 in (183 cm), Northwest Europeans, Tutsi in Africa.

One very short group, the Pygmies of the Congo, live only a few hundred miles from one of the tallest groups, the Watusi.

A survey by the ONS showed that 15-year-old boys had an increase in average height between 1830 and 2001 of 9 inches (23 cm), compared with 4½ in (11.5 cm) for girls.

There is a diversity of heights in well-known people: Barbara Windsor, 4 ft 10 in (147 cm); Dolly Parton, 5 ft; Judi Dench, 5 ft 2 in; Sarah Jessica Parker, The Queen, Madonna, Britney Spears and Bernie Ecclestone, all 5 ft 4 in; Rod Stewart, 5 ft 5 in; Penelope Cruz, Joan Collins and Bob Dylan, all 5 ft 6 in; Sylvester Stallone, Prince Charles and Cameron Diaz, all 5 ft 9 in; Mick Jagger, Johnny Depp and Dave Bowie, all 5 ft 10 in; Uma Thurman, Leonardo DiCaprio, Iman and George W. Bush, all 6 ft; Tony Blair, 6 ft 0½ in; Boy George, 6 ft 2 in; Prince William, 6 ft 3 in; Jeremy Irons and Clint Eastwood, 6 ft 4 in; John Cleese, 6 ft 5 in (196 cm). Princess Eugenie, when 12, underwent an operation to correct spinal curvature, making her two inches (5 cm) taller.

Most Pygmies are shorter than 4 ft 9 in (144 cm). They occur in Equatorial Africa, the Andaman Islands, the Philippines, jungles in the Malay Peninsula and East Sumatra, and in New Guinea. There are about 150,000

Pygmies in small scattered tribal groups in tropical rain forests of central Africa.

Pygmies are not dwarfs (Chapter 18), although they also have a relatively long trunk but short limbs. Pygmies and dwarfs have overlapping heights. Unlike dwarfism, usually caused by an autosomal dominant allele, the genetics of Pygmies is less simple. Pygmies produce normal amounts of human growth hormone but are much less responsive to it.

There is an Annual Little People of America Convention, typically attended by 1,000 dwarfs and midgets, and 800 average-sized relatives and friends. The hotels put steps by the front desk, stools by the cash machine and pay-phone, and a rod in the lift for reaching the higher floor buttons. One dwarf said, "My average-sized friends can't know what it is like to be noticed every time you walk into a room. They don't know what it's like to drive a car with extensions, or to have to climb on an ATM machine like it was a jungle gym to get money out." Another male dwarf said, "It's good to dance with someone and look into their eyes. Don't get me wrong; head to boob is nice." About one in 10,000 births worldwide is of a dwarf.

There are also national tall persons' clubs and an annual International Tall Persons' Convention. An attendee said, "As a tall person, in a social situation, I find that people always want to talk about your height. But here, we're all the same, so we can just forget about it." As well as acting as dating clubs, they campaign for action, e.g., about leg room on planes. In Holland, the Tall Person's Club succeeded in raising the statutory height of doors to 7 ft 6 in (229 cm) and ceilings to 8 ft 2 in (249 cm) in new buildings.

Personal Account. *On being a six feet eight inches tall (204 cm) male*

I am 6 feet 8¼ inches (204 cm) tall and reached this at 16. Being tall runs in the family. Mother is 185 cm (6 ft 1 in), father is 195 cm (6 ft 5 in). My brother Patrick, 18, is even taller, 207 cm (6 ft 9½ in). My brother Sebastian, 13, is 5 ft 11 in (180 cm). When he is 14, he will start hormone treatment to reduce his growth. My eight-year-old sister is already as tall as most 11-year-olds. She is under observation for possible growth-reducing hormone treatment later.

My parents anticipated that I would be very tall and when I was 12 took me to a specialist with the option of injecting me with hormones to halt my growth. I enjoyed the attention and so my growth was not checked by injections. A German cousin started the hormone treatment when he was 14; now in his thirties, he is 198 cm (6 ft 6 in), with no lasting side-effects from the treatment. His brother, 21, did not receive treatment and is 204 cm (6 ft 8¼ in).

The standard height of British doors (6 ft 6 in, 198 cm) has not changed for 100 years despite an increase in average height of about one inch (2.5 cm) per generation. How often do I make violent and painful contact with the top of the door frame? Surprisingly infrequently. When I approach a door I automatically withdraw my head, a reflex laid down by reward and punishment.

If a bed is too short it causes problems if it has a foot board, meaning that my feet cannot hang over the end, when I put the mattress on the floor. There is little hope of finding clothes which fit me in normal British shops.

Seating on planes and trains usually causes no problems although I always ask for a seat with maximum leg room. If the opportunity arises I take two seats. If someone enquires whether the seat beside me is taken, I reply, "Yes, by my legs". One form of transport is unrelentingly punishing for tall people: the motor coach. The seats seem designed for women below the height of 5 feet (152 cm).

In the cinema or theatre I appreciate being much taller than everyone else, but have to ignore mumbled disapprovals from people behind. When I went to a rock concert, I enjoyed an unhindered view and received offers of money from girls wanting to sit on my shoulders. It has been proven that tall people are more likely to be promoted. I can see why: being six feet eight and standing a head above the crowd means you get noticed and remembered.

When I first meet people, my size offers an immediate talking point, but most people manage to find something else to talk about. In pubs, where inhibitions have been diminished by alcohol, comments are more forthcoming. Usually I'll get something like: "My God, you're tall", followed by a bit of good-natured banter. By now I've heard most of the comments that people are apt to make (including intrusive questions about my anatomy from drunken middle-aged women) and usually have some remark to fire *back. One disadvantage about standing at social functions is that conversation takes place at a lower altitude; I have to stoop to get involved.*

A few men can be aggressive because they want to prove to their friends that they could "have me" in a fight. That is true as I am a complete pacifist, who, unless severely provoked, will try everything to avoid a physical confrontation. Being very tall means that I am blessed with not a massive amount of co-ordination, which cancels out my height advantage.

I am asked, "Do you ever wish you were smaller?" The answer, even though there are situations in which I do want to be smaller, is overall no. Being this tall is inextricably part of me, and my whole psyche has been built up around it. The situations where I am uncomfortable or embarrassed are compensated for by those in which I bask in standing head and shoulders above the rest.

Casper von Wrede wrote this when an undergraduate, including information from his mother, Kristina.

For a female, being too tall has disadvantages. A 13-year-old died from a drug overdose after being bullied because she was 6 ft (183 cm) tall. Morgan Musson had been mocked and threatened by a gang of seven girls at her school in Nottingham. Susan Herbert, who is 6 ft 3 in (190.5 cm) tall, wrote that she sympathised with Morgan after her own experiences. A man she holidayed with was seven inches (18 cm) shorter than she was and in public referred to himself as "St George with his captured dragon". She was called "the leaning tower of Pisa", "lanky", "looby", etc. Being so tall was a misery during growing up and later.

Treatment with oestrogen (a female sex hormone) to reduce the adult height of tall girls has been available since the 1940s when it was used for girls suffering from acromegaly, where the pituitary gland produces excess growth hormone. Oestrogen therapy in adolescence reduces adult height by two to ten centimetres (0.8 to 3.9 inches). It increases self-esteem, social comfort and marriage prospects. Side effects included nausea, headaches and reduced later fertility.

There have been many studies on the effects of height on success. Judge and Cable (2004) made large-scale studies in Britain and the USA, and found that on average, each extra inch (2.54 cm) of height added about £493 a year in salary. Being tall improves self-confidence and therefore performance; it is psychologically associated with leadership, even though many leaders in history have been short.

In the rain forests of Brazil, anthropologists found that taller men attracted many more women than did shorter men. In 'lonely heart columns' many women specify that the man should be tall; men usually only give their height if tall and often exaggerate.

Mickey Rooney, the American actor who died aged 93 in 2014, was only 5 ft 2 in (157 cm) yet extremely successful in seducing women. The first of his eight wives, film star Ava Gardner, said that "He went through the ladies like a hot knife through fudge."

It seems reasonable to have height criteria for certain jobs. For the Fire Brigade the minimum height was 5 ft 6 in (168 cm) but the Home Office removed that limit in 1997. Katie Reid sued the Fire Brigade after being taken off active duty, claiming sex discrimination. She was only 5 ft 1 in (155 cm) and could not reach some equipment, pull out hoses or lift ladders. She had difficulty cleaning fire engines, using cutting equipment on large vehicles and could not reach the emergency keys in a lift. This was someone in the wrong job, not sex discrimination.

Vitamin deficiencies reduce height. The average Vietnamese man in his early twenties is only 5 ft 4 in (162.6 cm). The director of the National Institute of Nutrition in Hanoi hopes to increase that by two inches (5 cm) by giving all babies large doses of vitamin A. A Hanoi café worker commented, "All Vietnamese want to be tall because tall people are more pretty and handsome."

3.4. Weight and health

A Columbia University study in America published in 2013 showed that 18% of deaths among people aged 40 to 85 were linked to excessive weight, with women more affected than men. In America, the National Institute of Diabetes and Digestive and Kidney Diseases has linked medical conditions such as high blood pressure, strokes, heart disease and diabetes to obesity. Other health risks include asthma, cancer of the colon, rectum and prostate for men, and cancer of the gallbladder, breast, womb, cervix and ovaries for women. Being underweight can also have adverse effects. Some medics think that waist circumference is a better predictor than BMI for increased cardiovascular risk.

Ultra-heavy people cause problems for crematoria, hospitals, firemen, hotels, airlines and others. At 42, Patrick Deuel in Nebraska weighed 77 stone (1,078 lb, 489 kg). He had heart failure, thyroid problems, pulmonary hypertension and arthritis. When he became critically ill, no local hospital could take him so he was driven 200 miles to South Dakota, where they made a specially reinforced bed. Being put on a diet of 1,200 kilocalories a day, he lost 23 stone (322 lb, 146 kg) in eight weeks.

The second heaviest man recorded, Khalid Moshin Shaeri of Saudi Arabia, weighed 96 stone (1,344 lb, 610 kg) and needed hospital treatment in 2013. Khalid had been bed-bound for two and a half years and was unable to move. Part of his house was demolished to admit a fork-lift truck and he was transported by military plane.

Cosmetic surgery for weight reduction is popular in many countries, especially America. About one million people a year worldwide have liposuction, with excess fat sucked from under the skin, generally removing up to 2½ pints (1½ litres) of fat each time. Surgical removal of fat by knife is also done. Americans spend more than $2 billion (£1.4 billion) a year on trying to improve their body shapes.

One original idea for promoting weight loss is to undress for dinner. Taking your clothes off before you eat makes you more conscious of your size, and self-conscious about every fattening mouthful. Eating naked in front of a full-length mirror helps to curb cravings for large portions.

Many people desire to lose weight. The obvious ways are to eat less, or less calorific foods, and to exercise more. To lose one pound (0.45 kg) of fat, one could walk briskly for 37 miles (60 km). Table 3.2 shows the calories lost from various forms of exercise.

Some women opera singers were criticised for being overweight, even losing roles because of that. In 2004, American soprano Deborah Voigt was removed from a role at the Royal Opera House because she could not fit into a costume. Deborah had unsuccessfully tried various diets, then in 2004 had successful gastric-bypass surgery. She lost over 7 stone (100 lb, 45 kg), going from dress size 30 to 14.

Weight loss can be dramatic, takes only minutes and be permanent. Sounds ideal? It means having a tumour removed! Hungarian doctors

Table 3.2. Kilocalories used in exercises for half an hour, rough averages for people of different weights.*

Weight, stones and pounds/pounds/kilograms	9 st 4 lb/ 130 lb/59 kg	11 st 6 lb/ 160 lb/73 kg	14 st 4 lb/ 200 lb/91 kg
Cycling, vigorous, stationary	328	403	504
Football	250	307	384
Gardening	140	173	216
Golf, carrying clubs	172	211	264
Hiking	187	230	280
Housework	109	134	168
Ice skating	218	269	336
Running, 5 mph, 8 kmph	250	307	384
Running, 10 mph, 16 kmph	515	634	792
Swimming	187	230	288
Tennis	218	269	336
Walking, 3.5 mph, 5.6 kmph	125	154	192

*Adapted from *The Daily Telegraph*, 3/1/2003.

removed a benign tumour weighing 85 lb (6 st 1 lb, 38.6 kg) from a man's stomach after his weight had jumped from 12½ to 19 stone (175 to 266 lb, 79 to 121 kg). A Californian woman in 1979 had a benign ovarian tumour which had been growing for 15 years. She weighed 27 stones 2 pounds (380 lb, 172.4 kg) before the four-and-a-half hour operation, leaving her weighing only 12 stones 2 pounds (170 lb, 77 kg). The tumour weighed 14 st 4 lb (200 lb, 90.7 kg) and contained 20 US gallons (76 litres) of liquid!

There are two main hormones controlling appetite. Leptin is made by fat cells, and the fuller your fat cells are, the more appetite-suppressing leptin gets into the blood. Strangely, girls have three times as much leptin as boys. Ghrelin is released into the blood from the stomach when it is empty, increasing appetite. These two hormones have complicated interactions.

If you lose one to two kg (two to four pounds) a week, you largely lose fat, but faster weight loss involves losing more lean tissue. Unfortunately, continued low-calorie diets give lethargy and a reduced metabolic rate, reducing the diet's effectiveness. If your diet succeeds and your body fat reduces, you produce lower levels of leptin, which impels you to eat more. Surgical treatments for obesity, such as gastric bands and stomach stapling, have the highest rate of success for long-term weight reduction.

Extreme diversity for leptin production was discovered in Cambridge, England. Two children under the age of ten had insatiable appetites, giving enormous obesity. One child, aged nine, already weighed 94 kg (207 lb, 14 st 11 lb). They had no detectable leptin in their blood and carried a mutation knocking out their leptin gene. Daily injections of recombinant leptin stopped their compulsive overeating and they lost their excess weight. Leptin is extremely expensive, so is not generally available.

Girls overtake boys in average weight from ages 9 to 13, then boys catch up and overtake girls. A University of London study of 3,000 babies born in 1946 and followed for more than 40 years found that babies who were bigger at birth tended to do better at school and remained brainier in young adult life.

People's metabolic rates differ widely, affecting their weight. Harry Bullard was 6 ft 4 in (193 cm), weighed nearly 12 stone (168 lb, 75 kg) and had less than 10% body fat. He attributed being underweight for his height, in spite of eating large amounts, to having a very high metabolic rate.

Exercise is important for weight control and health. Dr Tuomilehto of Helsinki University said that four hours' exercise a week achieves an 80% reduction in the risk of type 2 diabetes. Eating a 50 kilocalorie biscuit a day can cause a weight gain of 5.5 lb (2.5 kg) a year; it takes a daily 20-minute walk to use up those calories. The British Heart Foundation recommends 30 minutes of moderate exercise, five days a week, for cardiovascular health.

3.5. Dissatisfaction with body image

In the Western World most females are dissatisfied with their body. *Bliss* magazine found that 90% of teenage girls in Britain were unhappy with their body shape. The desire to be slender and beautiful dominated the life of 14-year-olds. One fifth of the girls was overweight but three fifths thought they needed to diet, with 64% of under-13 girls having already been on a diet. Twenty per cent were already suffering from eating disorders such as anorexia or bulimia (Chapter 12). A study in 2001 in America found that 43% of men were dissatisfied with their bodies, three times as many as 25 years ago.

At the other extreme, in Calabar, Southern Nigeria, would-be-brides are sent to a fattening establishment. Arit AsuquoIbok was 35 and unmarried, so her family sent her for three months to a fattening house. As journalist Christine Lamb wrote, "Her thighs wobble like blancmange as she walks, her bottom is as round and squashy as two over-ripe pumpkins, and at least seven chins quiver when she swallows." From five in the morning she starts eating, spending the day reclining and stuffing herself. Arit said, "I must eat so I'll be fat and people don't laugh at my figure. It shows that my family has money and can afford to feed me properly and I will make a good bride." A local woman said, "Women who are not fattened are cursed. If you don't do it, the gods will be angry and terrible things will happen."

Sinisterly, the Nigerian fattening rooms are also used for female genital mutilation. A 15-year-old schoolgirl, Glory Ita Asuquo, said, "They will cut off part of my genitalia with a razor blade. It is painful but it's our tradition. It will be done in a hygienic way and they will put a mixture of gin and special herbs on the wound to stop the bleeding." (Information from *The Sunday Telegraph*, 25/3/2002.)

3.6. Shape, obesity and diet

In Europe, the OECD Fact Book gave figures for adult obesity (BMI greater than 30) in EU countries: 8% in Switzerland and Norway, 8.5% in Italy and Austria, 9% in France, Denmark and Sweden, 13% in Portugal, Germany, Ireland and Spain, 22% in Greece and the Slovak Republic, with the UK worst at 23%.

In 1999 Vasso Xyloyiannis studied the heights, weights and BMI of 358 Imperial College students of varying national origin, as shown in Table 3.3.

Male height exceeded female height by 5.1 inches (13 cm) on average, a difference of 7.8% of the average female height. Male weight exceeded female weight by 2 stone 0.6 lb (13 kg, 28.6 lb) on average, a difference of 22% of the average female weight. The fact that the sex difference is much greater in weight than in height is expected, as taller people are usually broader and deeper. Weights varied more than four times as much as heights. In females, the maximum height was only 28% greater than the minimum height, while the maximum weight was 120% greater than the minimum. In males, the maximum height was 28% greater than the minimum, while the maximum weight was 115% greater than the minimum.

Table 3.3.   Diversity in heights, weights and BMIs of university students.

	Females	Males
Height		
Average	5 ft 5.4 in (166 cm)	5 ft 10.5 in (179 cm)
Minimum	4 ft 9.1 in (145 cm)	5 ft 3.0 in (160 cm)
Maximum	6 ft 1.2 in (186 cm)	6 ft 8.3 in (204 cm)
Weight		
Average	9 st 6.3 lb 60 kg; 132.3 lb	11 st 6.9 lb 73 kg; 160.9 lb
Minimum	6 st 6.4 lb 41 kg; 90.4 lb	7 st 5.6 lb 47 kg; 103.6 lb
Maximum	14 st 2.4 lb 90 kg; 198.4 lb	15 st 12.7 lb 101 kg; 222.7 lb
BMI		
Average	21.7	22.8
Minimum	14.5	15.9
Maximum	32.0	37.1
Numbers and percentages of normal, overweight and obese students		
Normal	128 (84.8%)	138 (76.2%)
Overweight	18 (11.9%)	42 (23.2%)
Obese	5 (3.3%)	1 (0.6%)

The BMI allows for taller people being generally heavier, so there is less difference in average BMI between the sexes (5.1% of the female value) than for height or weight. Vegetarians had a greater tendency to be obese (6.9%) than people on normal diets (1.7%). This agrees with findings of Martins *et al.* (1999) that groups in Canada and Britain who eat little or no meat are those most likely to be overweight.

There is much diversity in the proportions of the legs to the trunk and head. This is measured by the ratio of the height when sitting (which omits the leg length) to the height when standing. It is the cormic index, or the sitting/standing height ratio. In most populations, it is about 50%, with the legs about half the total height. My sitting height is 35 inches and my standing height is 72 inches, giving a sitting/standing height ratio of $(35/72) \times 100 = 48.6\%$.

In many Chinese, Japanese and Native American groups, the legs are relatively short, giving sitting/standing ratios as high as 54%. In Australian

Table 3.4. Diversity for average height, weight and BMI.

Group	Height, metres	Weight, kg	BMI
Turkana (Kenya)	1.77	55.0	17.6
USA, European	1.76	69.0	22.3
USA, African	1.76	72.2	23.3
Batutsi (Africa)	1.76	57.0	18.4
Australia	1.73	60.0	20.0
Samoa	1.72	81.2	27.5
Finland	1.71	70.0	23.9
Berbers (N. Africa)	1.70	59.5	20.6
North China	1.68	61.0	21.6
England	1.66	64.5	23.3
Hong Kong	1.66	52.2	18.9
India	1.63	48.2	18.1
Japan	1.61	53.0	20.5
Inuit/Eskimo	1.61	62.9	24.2
Vietnam	1.58	49.1	19.8
Bushmen (S. Africa)	1.56	40.4	16.6
Pygmies	1.42	39.9	19.7

Aboriginals and tall Nilotic Africans of East Africa, the legs are relatively long, giving ratios of well under 50%.

The diversity in heights, weights and BMI between different groups is shown in Table 3.4, using data from various sources.

Average heights ranged from 4 ft 8 in (1.42 m) for Pygmies to 5 ft 9½ in (1.77 m) for Turkana. Average weights ranged from 6 st 4 lb (39.9 kg, 88 lb) for Pygmies to 12 st 11 lb (81.2 kg, 179 lb) for Samoans, with the largest value 104% greater than the smallest. The average BMI varied from 16.6 for Bushmen to a massive 27.5 for Samoans, well into the overweight category.

The Forbes organisation and the WHO in 2007 found that 80% of the most overweight populations were on South Pacific islands. The Pacific Islanders' diet used to consist of fish, coconuts, tropical fruits, vegetables and pigs. They had increasingly turned to eating mutton flaps (fatty, from New Zealand), Turkey tails (fatty skins from America), corned beef, hamburgers, fizzy drinks and crisps.

According to the UN Food and Agriculture Organisation, about 815 million people were chronically undernourished in 1999, with at least 180 million chronically undernourished children under 10 years of age in the developing world. Factors in malnutrition include high birth rates, improved survival rates and longevity, and poor agricultural and environmental practices. With medicine causing higher survival rates, it becomes even more important for groups to control birth rates. It has been estimated that the number of overweight people in the world exceeded the number of underweight people in the year 2000. Is that progress?

The effects of changes of nutrition can be seen in a classic study by Greulich which showed that American-born Japanese children were strikingly taller and heavier than a similar-aged group of Japanese children brought up in Japan. When they changed from their traditional 'natural' diets to ones high in refined carbohydrates and fats, many groups of Australian Aborigines, Pacific Islanders and Native Americans tended to obesity.

Of the selective forces on body shape, temperature and humidity are important. In colder climates, the body needs to prevent excess heat-loss and hypothermia, while in hotter climates, it needs to get rid of excess heat from muscular exertion. Fat is a good insulator, reducing heat loss and also reducing heat gain from hot environments. Sweating is a major method of heat loss but is retarded in humid environments. People from the humid Tropics tend to be slim, giving better heat dispersal. Tall, thin bodies have a larger surface area in relation to volume than do short, round bodies.

Although there are exceptions, people in colder climates tend to be heavier for their height than those from hotter climates, with higher BMIs for Inuit/Eskimos, and those from Iceland, Finland and England, than for Batutsi, Indians, Bushmen, Pygmies, and those from Hong Kong (Table 3.4). Samoans, Maoris and Hawaiians tend to be large yet live in warm regions. This has been explained by their history of long sea voyages between islands over perhaps 4,000 years. In open boats, heat is lost to winds and spray, so selection for being large, fat and heavy would give better heat insulation, better food reserves and perhaps better fighting ability to conquer the islands reached.

A range of adult female body shapes, from slim to plump, and short to tall, is shown in Photo 3.2. The mothers from Twickenham, near London,

Photo 3.2. Different female body shapes.

shown here at the rugby stadium, posed naked for a calendar for a local cancer charity.

People have a fascination with extremes, as exemplified by freak shows which continued in Britain until the 1960s. Two midgets said that they spent their lives being stared at so they might as well get paid for it. The oddities exhibited included midgets, Siamese twins, two-headed giants, bearded ladies, the Lion-faced Boy, the World's Fattest Couple and the Elephant Man.

Obesity is a major health concern, causing joint and mobility problems and increasing the risk of heart disease and strokes, doubling the risk of most cancers and increasing cancer of the womb five-fold. Obese people average two thirds the salary of the non-obese, have more depression and more divorce. While slim individuals tend to blame the obese for lack of self-discipline, for lack of moral (and dietary?) fibre, many obese people keep eating because their mechanisms for registering that they are full are faulty.

There are drugs such as Xenical and Reductil which are prescribed for obese patients who have managed to lose some weight. Both have unpleasant side-effects. There are bulking agents which, taken with water, swell in the stomach to reduce feelings of emptiness. Public pressure led to some food firms making their products less unhealthy, e.g., by reducing the levels of hydrogenated fats, salt and sugar, and by reducing portion sizes. Reducing fats but increasing the carbohydrates to maintain flavour has often increased calorific values.

Stomach staplings and gastric bypasses are used on seriously obese people to reduce calorie intake. Stapling makes the stomach smaller, so the people feel full earlier and eat less. Gastric bypass surgery makes the stomach smaller and causes food to bypass part of the small intestine, reducing the amount of food absorbed.

For ladies of America's National Association to Advance Fat Acceptance, fat is a virtue and beautiful. Members have an annual convention with like-minded and like-bodied women. One attender, weighing 23 stone (322 lb, 146 kg), said that she had no problem finding men. Her problem was getting rid of them.

According to the Health Development Agency, 8% of six-year-olds and 18% of 15-year-olds are now obese. It estimated that obesity kills about 34,000 a year in Britain and costs the National Health Service about £2.6 billion a year.

Very fat children are often bullied, ostracised, socially isolated, ridiculed and excluded from playground or organised games, which may encourage them to indulge in further 'comfort eating'. Only 5% of children now walk or cycle to school, compared with 80% 20 years ago. Being driven everywhere, watching TV, playing computer games, using the Internet and chatting on mobile phones do not consume many calories.

References

Judge, T. A., Cable, D. M. The effect of physical height on workplace success and income: preliminary test of a theoretical model. *J Appl Psychol* (2004) 89: 428–441.

Martins, Y., Pliner, P. and O'Connor, R. Restrained eating among vegetarians: does a vegetarian eating style mask concerns about weight? *Appetite* (1999) 32: 145–154.

Ng, M., Fleming, T., Robinson, M., *et al.* Global, regional, and national prevalence of overweight and obesity in children and adults during 1980–2013: a systematic analysis for the Global Burden of Disease Study 2013. *The Lancet* (2014) 384: 766–781.

Chapter 4

Differences Between Males and Females; Reproduction and its Production of Genetic Diversity

4.1. Introduction

Males develop facial hair, an Adam's apple, deeper voice, higher average height, a penis (for urination and sex) and testicles. Females have enlarged breasts (sex-attractants and baby-feeders), a womb, vagina and clitoris (the only organ in either sex solely for pleasure), the ability to give birth, menstruation, fluctuating hormone levels, and a longer life expectancy. Chapters 10 and 21 cover other aspects of reproduction.

4.2. Females

Females are sexually precocious. At five months' gestation, they have produced about six million germ cells, their final total. During the remaining months before birth, these germ cells start meiosis (two cell divisions which reduce the chromosome number from 46 to 23 for the eggs, with only one copy of each chromosome). Surprisingly, these primary oocytes do not complete meiosis. They arrest in the first division for 10 to 50 years. Many degenerate, so at birth each female has about two million egg follicles, about 200,000 at puberty, but only about 400 ever mature and release an egg.

Throughout childhood, these oocytes develop. Shortly after puberty, the eggs pass out of the ovaries, one about every 28 days, and into the womb through a Fallopian tube, above and partly around each ovary. The two plum-shaped ovaries are about 1.6 inches (4 cm) long and 0.9 inches (2.2 cm) wide. The vagina is a muscular tube about 3 inches (8 cm) long, expanding during mating and childbirth. It connects to the womb by the cervical canal. The womb is about 3 inches (7.5 cm) long, 2.4 inches (6 cm) wide at its widest, and 1 inch (2.5 cm) thick (front to back), narrowing down to the cervix. The womb expands hugely during pregnancy, and the cervix dilates enormously during childbirth. About one birth in seven requires an episiotomy, a cut between the vaginal and anal areas to widen the vaginal opening.

Before ovulation, the primary oocyte, whose nucleus has been arrested in the first division of meiosis for years, completes its first division, with most cell contents remaining in one daughter cell, the secondary oocyte. The other nucleus forms a small polar body. Sperm entry into the secondary oocyte triggers the second division of meiosis, with one set of 23 chromosomes going into the mature egg and the remaining set forming a second polar body. Once the egg has passed into the Fallopian tube, it dies unless fertilised within 26 hours. Females who never get an egg fertilised never reach the start of the second division of meiosis!

Menstrual cycles average 28 (23 to 35) days, with the egg released ready for fertilisation at around days 12 to 15, where day one is the start of bleeding. By the menopause, usually in the late forties or early fifties, the woman's reproductive life ceases, although child care may continue much longer. Women can have a sex life at an advanced age: in 2001 an 82-year-old prostitute nicknamed 'Grandma' was caught soliciting in Taipei, Taiwan.

Eggs, ovarian tissue, and embryos from *in vitro* fertilisation (IVF), can be frozen for years before being thawed for use. This is done for career reasons or for women who have their ovaries removed because of cancer.

Females are very conscious of breast size, often considering theirs to be too small or too big. Breast tissue is heavy and supporting that load can cause back, shoulder and neck pain. One to three kilograms (2.2 to 6.6 lb) may be removed in breast reduction operations. Laura Dunsby, 39, was large-chested but when breast-feeding her daughter, her breasts expanded to 38HH. She felt massive and uncomfortable, and resented men assuming that she was easy game because she had large breasts. Laura had more than 6 lb (2.7 kg) of breast removed, leaving her a C cup.

4.3. Males

Males start producing sperm (Plate 1B) at ages 10 to 13. Sperm production continues throughout life, but after 45 to 50 there is a reduction in testosterone levels and sperm production, although men of 70 or more may father children.

Sperm are produced in two testicles within the scrotum, collected in the epididymis on the rear surface of each testicle, then go through the 18-inch (45 cm) long vas deferens tube to the prostate gland. That has two storage seminal vesicles and an ejaculatory duct through which sperm pass to the penis at mating. The testicles contain about 800 feet (250 m) of seminiferous tubules, about 1,000 per testicle. Sperm take two months to mature. They cannot initially swim or fertilise, gaining these capabilities as they mature in the epididymis, which concentrates them a hundred-fold by absorbing fluid. Their fertilising ability is enhanced by secretions in the female reproductive tract, a process called capacitation.

The penis varies in size even in one individual in one day as activities such as exercise, swimming or bathing affect it. During erection, its three columns of erectile tissue are engorged with blood. It is not muscular contraction which produces erections, but relaxation of muscles in the walls of small arteries near the base of the penis, allowing blood to flow into the erectile tissue.

At the climax of mating, smooth muscles contract in the prostate, vas deferens and seminal vesicles, delivering fluids and sperm (together making semen) into the tube (urethra) within the penis. This emission of semen into the urethra triggers rhythmic contractions at about 0.8 second intervals in muscles at the base of the penis, increasing pressure within the penis until the semen is ejaculated.

Men ejaculate about two to six ml (0.07 to 0.2 fluid ounces) of semen, containing about 200 million to 600 million spermatozoa. Sperm can be treated and frozen, remaining viable for many years at minus 196°C. The volume ejaculated in ml is: man, 2 to 6; turkey, 0.2 to 0.8; boar, a massive 150 to 500; bull, 2 to 10; ram, 0.7 to 2; stallion, 30 to 300; dog, 2 to 14.

Sperm concentrations in millions per ml are: man, 100; turkey, 7,000; boar, 100; bull, 1,00 to 1,800; ram, 3,000; stallion, 100; dog, 3,000. There are aids for men with erectile dysfunction, difficulty in achieving or sustaining an erection, e.g., Viagra. Anxiety about penis size can cause psychological and physical problems.

4.4. Fertilisation

The baby's inherited characteristics are determined by what genes the egg and the successful sperm carry. Its genes could result in a genius, an average person or an idiot.

The English writer Aldous Huxley (1894–1963) put it beautifully in *The Fifth Philosopher*:

> *A million million spermatozoa,*
> *All of them alive:*
> *Out of their cataclysm but one poor Noah*
> *Dare hope to survive.*

> *And amongst that billion minus one*
> *Might have chanced to be*
> *Shakespeare, another Newton, a new Donne -*
> *But the One was Me.*

The egg is about 1/175th of an inch (0.14 mm) in diameter and weighs about one 20-millionth of an ounce (0.0014 mg). It is 85,000 times the volume of a sperm. Eggs are usually produced singly from alternate ovaries, although more than one may be released, when non-identical twins may be produced from two fertilised eggs.

Sperm have a long whip-like tail (Plate 1B) and swim like tadpoles. Sperm are about 1/500th of an inch (0.056 mm) long and swim about 20 inches (51 cm) per hour (0.00032 miles an hour, 0.00051 kmph). The immobile egg is swept into the Fallopian tubes from the body cavity by finger-like fimbriae. Although sperm are motile, their main movement up the female reproductive tract comes from wall contractions, so dead sperm travel as fast as live ones.

The cervical canal is penetrable by sperm for only about three days per menstrual cycle, and sperm are dispersed into the womb by its muscular contractions. The egg uses chemical attractants for the swimming sperm. It takes sperm 1 to 3 minutes from ejaculation to reach the cervical canal, 10 to 20 minutes to reach the womb, and 30 to 60 minutes to reach the fertilisation site, usually the upper third of a Fallopian tube. About 3% of sperm get through the cervical canal, 0.1% reach the womb, and 0.001% reach the fertilisation site. Perhaps 100 sperm get close to the egg.

Each sperm head has an enzyme-filled acrosome around it, and when the sperm meets the zona pellucida barrier around the egg at fertilisation, the acrosomal membrane is disrupted, releasing enzymes which attack the zona pellucida, allowing sperm to tunnel through it to the plasma membrane around the egg. Although only one sperm fertilises the egg, large numbers are required around it so that collectively they have enough enzymes to make holes in the zona pellucida. The need for many sperm to reach the egg is why men are considered clinically infertile if their sperm concentration is below 20 million per ml. About 2% of men have very low sperm counts and are effectively infertile. They are now able to become fathers through ICSI, intracellular sperm injection, where a single sperm is injected into an egg's centre.

Within an hour of the first sperm penetrating it, the egg has undergone the second division of meiosis and the sperm and egg nuclei have fused. The fertilised egg now has 46 chromosomes and is a zygote. It stays in the Fallopian tube for three to four days, then descends into the womb where it floats for three to four days before implanting. The egg is moved by fine hairs in the reproductive tract. The week's delay between fertilisation and implantation allows the zygote to undergo a series of cell divisions to reach the blastocyst stage, and for the womb lining to prepare itself with nutrients and more blood vessels.

The womb readies itself for implantation of a fertilised egg in the second half of the menstrual cycle. Its innermost layer, the endometrium, increases in thickness two- to three-fold, reaching 0.16 inches to 0.2 inches (4 to 5 mm). If there is no implantation, hormonal changes at the end of the cycle trigger menstruation. Most of the womb's inner lining sloughs off. The resulting bleeding and mild contractions of the womb help to expel the debris out through the vagina, with typical blood losses of 1.1 to 4.5 fluid ounces (30 to 130 ml). Menstruation generally lasts three to six days, with the first day of bleeding as day one. An embryo, if produced, implants at about days 20 to 28.

At birth, the baby has about 200 billion (200 thousand million) cells, while the adult has about 100 trillion cells (100 million million). It takes more than 46 successive cell divisions from egg to adult. The fertilised egg weighs about 1.5 micrograms; just over one 20-millionth of an ounce), growing to about 7 pounds (3.2 kg) at birth, 2.1 billion times as much.

At every cell division there is a chance of errors in the distribution of chromosomes to daughter cells, and at every DNA replication there is a chance of errors arising (mutations from a change in base pairs, or additions or removals of base pairs, or larger additions, deletions or structural changes). Only errors in germ-line cells, those that will form eggs or sperm, will be passed on, and those in somatic (non-germ-line) tissues will not, although they may cause cancers from unregulated divisions.

4.5. Diversity from reproduction

If the chromosomes passed unchanged through meiosis, 46 chromosomes in 23 pairs could give 2^{23}, 8,388,608, different combinations. With the same number in the egg, that makes 7×10^{13} possible types of offspring, 70 trillion (70 million million), from one couple. Chromosomes are not passed on intact as the members of a chromosome pair have reciprocal exchanges (crossovers) in meiosis. Each couple therefore has a virtually infinite number of genetic combinations in the offspring. All children of the same parents will differ genetically unless they are identical twins, triplets, etc.

4.6. Physical variations

Some men are born without a penis, or with one or no testicles, or the testicles fail to descend into the scrotum from the body cavity, usually causing infertility. Some women have the rare genetic Mayer-Rokitansky-Kuster-Hauser syndrome, where the vagina and womb are absent or seriously underdeveloped. As reported in *The Lancet* (2014), four girls aged

13 to 18 were treated with laboratory-grown vaginas engineered on scaffolds with the patients' own cells.

4.7. Conception through the oral route, a true virgin birth

Verkuyl (1988) reported the case of a girl born without a vagina. The 15-year-old unmarried barmaid in Lesotho in southern Africa was found to be nine months pregnant and in labour! An immediate Caesarean operation delivered a healthy boy, but how had she become pregnant in the absence of a vagina?

Nine months earlier, she had been interrupted by a former lover while having oral sex with her new boyfriend. In the ensuing fight she was stabbed in the upper abdomen. In hospital, a puncture hole of the stomach was explored and her abdominal cavity was washed. Sperm must have leaked out of the stomach and been flushed into the Fallopian tubes where fertilisation occurred. The child grew to resemble the new boyfriend, ruling out an even more miraculous conception.

References

Raya-Rivera, A. M. *et al.*, Tissue-engineered autologous vaginal organs in patients: a pilot cohort study. *The Lancet* (2014) **384**: 329–336.

Verkuyl, D. A. Oral conception. Impregnation via the proximal gastrointestinal tract in a patient with an aplastic distal vagina. Case report. *Br J Obstet Gynaecol* (1988) **95**: 933–934.

Recommended reading

Lamb, B. C., *The Applied Genetics of Humans, Animals, Plants and Fungi*, 2nd ed. (2007) Imperial College Press, London.

Wolpert, L., *Why Can't a Woman Be More Like a Man? The Evolution of Sex and Gender*, (2014) Faber & Faber.

Chapter 5

Personal Choice, Cosmetic and Preventative Surgery, Clothing and Make-Up

5.1. Introduction

Life is a series of choices, perhaps including career, partner, location, beliefs, clothing, entertainment, food and drink. Individuals face different choices. There is a classic cartoon of a man in an igloo, with a frozen whale outside, asking his wife, "What's for supper?" We often make life-or-death choices, as when crossing the road or driving. Our dietary choices can aid health or promote disease. A couple from Zimbabwe came to England and for religious reasons chose a vegetarian diet. Their son died at the age of five months of acute rickets, caused by deficiencies of vitamin D, phosphorus or calcium.

We limit the effort of choosing by establishing routines, e.g., for what we do upon waking up. A choice requiring thought might be what to wear. One major choice may restrict further choices. Choosing to become a Buddhist monk restricts one's choice of clothing, diet, lifestyle, possessions, habitation, etc. In the 18th century, Italian parents wanting their boy to be a singer with a high, powerful voice, a castrato, had many future choices cut off as well as his testicles. A choice for one person may be trivial, yet crucial for another. Whether to eat peanuts is of little consequence to most, but for people with an allergy, one bite could kill them.

Often we fail to achieve our chosen objective because it depends on other people. The person from whom we desire friendship, a kiss, marriage, a job, may rebuff us. We sometimes make wrong choices.

Choices are constrained by circumstances, especially money, as when buying a car. Sometimes, the young like to rebel against tradition. Many young Japanese wear odd make-up (Plate 12A), dye their hair (Plate 12C) or wear strange clothes.

Choice is restricted by monopolies. Because they lack the stimulus of competition, they are often bad for the consumer, with high prices and poor service. The break-ups of state monopolies for telephones, gas and electricity in Britain were very successful.

Strict political regimes hugely restrict choice, including freedom of expression and travel. For many years, Russian geneticists were not allowed to visit the West. Four senior Russian geneticists were eventually allowed to visit America in 1967 when I was at the California Institute of Technology. The head of biology invited two of them to dinner and, to my surprise, invited me. When someone rang the doorbell, which had Westminster Chimes, the Russians were fascinated. They tried it several times, like children. They remarked that such wonderful things must only be available to high officials, as they would be in Russia. Our host showed them a Sears Roebuck catalogue which included this doorbell, which anyone could choose to buy.

Singapore has a reputation for being strictly regulated. Only after a partial liberalisation in 2003 were its inhabitants allowed to buy *Cosmopolitan Magazine* (if wrapped in plastic) and to watch *Sex and the City*. Singaporeans were allowed to buy chewing gum 'for therapeutic purposes' on production of a note from a doctor or dentist.

Liberalisation of choice in Britain was shown by the proliferation of sex shops in the 2000s, and by the wider social spectrum of their customers. Staff comments included "Anything willy-shaped is very popular," and "It shows how far the British have come in their attitude towards sex. Once, women were embarrassed to pick up a vibrator. Now they ask, 'How many speeds has it got?'"

One's choice of job is limited by education, talents and circumstances. In 18th century Britain, aristocratic families usually wanted at least two boys, 'an heir and a spare'. One's choice of food and drink is limited by money, availability, season, allergies and likes and dislikes. Religious food laws were covered in Chapter 2.5.

In a language there is a choice of tone of voice, words, grammatical constructions, and even some choice within English spelling: hair drier or dryer, connection or connexion, focused or focussed, specialise or specialize. For choice of names, see Chapter 7. A chemistry lecturer used to tell first-year students that he was known as 'piggy', which suited his face.

With the availability of contraception and assisted fertility, many modern couples can choose whether to have children, which has an enormous effect on their lives, leisure time, sleep, finances, accommodation and transport needs. In Britain, many parents choose to pay higher house prices to live in the catchment area of a good state school, or pay for private education.

Some choices are difficult, with a big diversity of opinion. Should terminally ill people be able to choose to have their lives ended by a medical practitioner (euthanasia) or with the help of a friend (assisted suicide)? Should pregnant women be able to choose to have an abortion?

Many Western youngsters choose to have body piercings, visible in ears, eyebrows, nose, lips (Plate 19B) or tongue, or generally invisible as in the nipples, penis or vaginal labia. They were popularised by pop stars: Scary Spice had her tongue stud and she and Robbie Williams had pierced navels. The Queen's granddaughter, Zara Phillips, aged 17, flaunted her tongue piercing and stud at Prince Charles's 50th birthday party. A Greek student had her tongue pierced 'for no particular reason'.

5.2. Cosmetic surgery and other treatments

Cosmetic surgery is for aesthetic purposes. Nose surgery (rhinoplasty) is mentioned in ancient Egyptian texts from 3000 to 2500 BC, in ancient India and the Roman Empire. In Britain in 2013, rhinoplasty was the fifth most popular cosmetic surgery operation, with 4,878 nose jobs.

Members of the British Association of Aesthetic Plastic Surgeons carried out 38,274 procedures in 2010, with the number increasing each year. Breast surgery on women was the most frequent operation with about 9,500 cases. The biggest rise was in breast reduction for men, with 700 operations. For men, the choices in order of popularity were: nose jobs, breast reduction, eyelid surgery, ear correction, liposuction, face/neck lift, brow lift and tummy tuck. For women, the order was: breast enlargement, eyelid surgery, face/neck lift, breast reduction, nose job, tummy tuck, liposuction, brow lift and ear correction.

Other procedures include removal of fat from the waist and abdomen and of 'bags under the eyes'. *OK!* (4/3/2014) reported that former Spice Girl Victoria Beckham had had breast implants, enlarging her from 34A bust size to 34DD, but had the operation reversed in 2009. A face lift may involve an incision near the hairline, removal of skin to tighten the face to remove wrinkles and sags, trimming the eyelids, tightening face and neck muscles, and lip-filling.

The commonest cosmetic operation for men and women is liposuction, sucking out fat, particularly from the chin, 'stomach', bottom, thighs and upper arms. As reported in *The Daily Telegraph* (29/1/2007), Lorraine Midwinter put on weight after reaching 40, going from dress size 10 to 16. Dieting and exercising did not work so she spent £4,400 on liposuction. She said, "The bruising and swelling wasn't too bad and after three weeks I was really good."

There are many non-surgical procedures such as injection of botulinum nerve toxin. Botox® relaxes the muscles which cause frowns, crow's feet and wrinkles, giving smoother skin. Wrinkles can be injected with biological fillers such as collagen to smooth them out. 'Botoxing' only takes minutes and lasts several months. An American attorney in his fifties thought that his professional success depended on looking imperturbable. He paid about $10,000 a year for six Botox injections to give him a really smooth face. Botox injections around the bladder have proved successful for women who have 'urgent bladder' problems, suddenly having to urinate.

There are laser methods used medically and cosmetically to remove blemishes such as scars, birthmarks and tattoos. Argon lasers affect haemoglobin so are used to seal or destroy blood vessels to treat thread veins and port wine birthmarks. Thread veins, varicose veins and piles can be treated with injections to shrivel them. Ruby lasers can treat melanin problems as they target that pigment. For skin abrasion or cutting, carbon dioxide lasers are used as their wavelength vaporises tissue. Surface skin removal (dermabrasion) and chemical peels are used to remove acne scars, enlarged pores, warts, liver spots and sun damage. Medical tattooing can fill in missing areas of pigmentation. For a woman with breast reconstruction, it can put pigment around the new nipple.

Sometimes the desire for surgery is psychological. In 2000, a NHS surgeon removed healthy limbs from two private patients, at their request. They suffered from Body Dysmorphic Disorder, in which they became convinced that those limbs were defective and had to come off. They paid £3,000 each.

The American Society of Plastic Surgeons has more than 7,000 members, carrying out more than a million procedures a year on men, especially face fillers. In 2006 there were 5,000,000 procedures on people aged 40 to 54, and 2,800,000 on those aged 55 or older, including lifts to the thigh, lower body, upper arms, face, breasts, and tummy tucks.

In China, many middle-class women choose cosmetic surgery to look 'more modern'. One businesswoman asked the surgeon for "modern eyes, big eyes", with having more open eyes being a common desire. Many Korean girls have eye-opening operations (107,000 in 2014), and want to look like film stars. Yu-Hoi, a Korean, told me that 90% of South Korean women wanted plastic surgery to have 'tall noses and big eyes, like Europeans'. Korean women thought that Europeans were much more attractive. She said that the Japanese often wanted longer noses, bigger eyes and thicker lips. Lots of her friends had had plastic surgery and looked really nice, but kept it secret as men did not like women to have had it as the children might not resemble their mother!

Many girls want to be taller, and high heels are the non-surgical solution. Josephine from Hong Kong is 5 feet 5 inches (165 cm) tall but would like to be 6 feet (183 cm). She told me that many Chinese girls had leg-extension operations, which were painful and prolonged. One of her friends was a dwarf and had lower limb extensions. The skin, nerves, blood vessels and muscles must also be stretched.

According to *The Daily Telegraph* (23/9/2006), Li Ping, an attractive 23-year-old from Beijing, paid doctors £1,600 to break her legs because she felt too short at 5 ft 1 in (155 cm) and wanted to grow two and a half inches (6.4 cm). In the weeks after both legs had been broken, the fractured ends of her tibias were braced, pinned and gradually screwed apart in splints to extend the bones as they fused together again. Her bones failed to knit so she was unable to walk properly and had a corrective operation.

In China this became a popular way to boost height for the short and even the not-so-short. There is status attached to being taller in China: height is seen as a sign of wealth and Westernisation. Employers consider looks and height important. The Foreign Ministry has a minimum height requirement as it does not want Chinese diplomats to be looked down on.

Brazil is obsessed with body image. One cosmetic surgeon in Rio performed more than 60,000 operations over 40 years. Even girls aged twelve wanting to be models undergo cosmetic surgery. Dr Ailthon Takishima operates on about eight would-be models a day in São Paulo. In his opinion, in Germany they like big breasts; in Japan, they don't like breasts at all; the Japanese want women with a white skin, like a doll, with a small nose; in France, they like the larger curved nose. He treats the models according to where they hope to work.

The Brazilian wax, using wax to remove hair from the female pubic region, often leaves a small strip of pubic hair at the front. Hollywood waxing removes all pubic hair. The beauty of many Brazilian women was noted by 16th century Portuguese colonisers such as Pedro Alvares Cabral. He commented on their good faces and wrote of their private parts that 'the hair on them was well shaved and arranged'.

Art critic John Ruskin failed to consummate his marriage to Effie Gray as he was put off by her having pubic hair. In the West, most women shave their legs and armpits, and deal with dark hair on the upper lip by bleaching, shaving or depilation. Roman women used razors and pumice stones to remove facial hair, and Queen Elizabeth I used walnut oil and ammonia. Women in Tudor England plucked hair from their eyebrows and temples.

While many people think that their bottoms are too large, 2,361 women in America had buttock augmentation surgery in 2006, typically costing about $20,000. Fat from the stomach and thigh areas is injected into the buttocks. A surgeon said that Asian patients want small, shapely buttocks, while whites are after 'voluptuous Playboy bottoms' or an 'athletic tiny rear'. Hispanics want full round buttocks, while African-Americans and Afro-Caribbeans often ask for 'as big a buttock as will fit their bodies'.

As people put on weight, their skins stretch. If they later lose weight, the skin does not contract so much and folds of loose skin can result. An enormous American woman had gastric-bypass surgery, losing 280 pounds (20 stones, 127 kg). Her skin hung down in yellowish folds, with two-foot (61 cm) long overhangs from her upper and mid-torso, with one flap draping her pubic area like a miniskirt.

The *Daily Mail* (23/11/2001) reported the case of a divorced mother of two, Christine Terry, 48, of Worcestershire. She attracted many younger men after an addictive series of operations: breast implants, £1,500; nose job (bump removed), £1,700; nose job (tip shortened), £1,950; further breast implants, £2,500; eye lift, £2,500; neck and jawline, £3,000; liposuction on stomach, £2,000; full facelift, £5,000; Botox, £600; total cost £20,750. She loved the admiring looks she got from men and women. Further operations were planned.

5.3. Preventative surgery

People choose preventative surgery if their chance of developing a serious disease is high because relatives have had it (Chapter 11.11). About 6% of breast cancer cases in women are associated with faulty genes *BRCA1* and/or *BRCA2*, rising to 15% in men. Women with those faulty genes have a roughly 50% chance of getting breast cancer. In Britain, about 41,000 new cases a year are of breast cancer in women, and 300 cases in men. Not having a faulty cancer gene reduces the risk of breast cancer in men 100-fold.

The Daily Telegraph (20/3/2006) reported that Ian Bentley underwent a voluntary double mastectomy (breast removal) at age 39. His younger brother, mother, grandmother and an aunt had died of breast cancer and his sister had a preventative mastectomy after finding a breast lump. It took Ian a year to recover from the surgery which removed pectoral muscles, fatty tissue and lymph nodes, leaving his chest slightly concave.

There are many cases of women having preventative surgery, normally a double mastectomy to reduce the risk of breast cancer. Actress Angelina Jolie chose a double mastectomy in 2013 after finding that she had a mutated *BRCA1* gene. She later had her ovaries removed, as *BRCA1* and *2* also increase the risk of ovarian cancer. In Greek mythology, the

Amazons were a race of female archers who had their right breast burnt off to prevent damage to it from their powerful bowstrings.

There was a case of preventative surgery in Canada in 2001, when five members of a family had apparently healthy stomachs removed. Natasha Benn was found to have the same rare gene which had caused her sister, mother, grandmother, great-aunt and great-grandfather to die of virulent diffuse stomach cancer. Surgeons removed Natasha's stomach and small intestine and made a smaller stomach from a loop in her large intestine. Four relatives with the faulty gene chose preventative surgery.

5.4. Clothing, hats, shoes and make-up

Clothing and make-up are influenced by environment, custom, religion and personal taste. Wearing purple was the Roman emperor's prerogative. In China, Mao Tse-tung (1893–1976) imposed a dull national uniform and banned the wearing of skirts. In 1974, however, he had a visit from Imelda Marcos, wife of the Philippines president. Mao and his wife were impressed by her beautiful dresses. Skirts were permitted in China again by 1975.

In England, Henry VIII passed sumptuary laws which marked a person's status, specifying what they could wear, eat and what furniture they could have. He wanted to ensure that social strata were maintained, with no one looking or living better that his superiors. He passed the 1509 Act against Wearing of Costly Apparel, where only the king and his immediate family could wear cloth of purple silk or gold. Dukes and marquises could use cloth of gold woven into their coats and doublets. Only Tudor Royalty were permitted clothes trimmed with ermine. Lesser nobles were allowed fox and otter fur. Other imported furs could be worn by graduates, yeomen grooms and pages of the king's and queen's chambers and by men having land yielding an income of eleven pounds a year.

Clothing (Plate 2 onwards) has appeared in diverse forms throughout history, from nakedness and near-nakedness (Plate 9G) to the total enclosure of a deep-sea diver. Picture the work clothing of surgeons, policemen, judges, Bunny Girls, strippers, pearl divers and air hostesses. In 2014, Skymark Airlines in Japan was criticised for putting female flight attendants into a uniform of daringly short minidresses. It was accused of inviting sexual harassment; the job involves a lot of reaching up and bending down, and passengers might use mobile phones to take pictures up the short skirts.

Protective clothing is worn by policemen, workmen and Formula 1 race car drivers. A retired radiologist told me that she had to wear an 'OP', a heavy lead-containing overall 'ovary protector' as a radiation shield for females.

While many women now wear trousers, some males wear skirt-like garments such as the Scottish kilt and the fustanella, worn by Greek Evzoni soldiers. Saris in Sri Lanka (Plate 5D) and India vary from rich materials for elegant ladies to plain cotton for women working in the fields (Plate 17D). A friend told me that it took 12 pins to hold her sari's folds in place. Saris are great in that ladies can wear each other's saris as their size does not matter. The material is wrapped around the waist, with one end draped over the shoulder, leaving a bare midriff if the blouse is cropped.

Muslim communities often insist on women covering their heads and wearing all-enveloping chadors, or burkas which cover the body and head, with small eye-holes. In London one sees Middle Eastern women in black robes and black head-dresses, and with face veils hardly even showing the eyes, while the men are casually dressed.

The Muslim head coverings for females are diverse, depending on the culture and choice. As shown in Plate 10, the head scarf may be plain black, coloured or even multi-coloured. The Malaysian lady (Plate 10A) with the pink-red headscarf, matching outfit and eye make-up looked charming and well dressed, not all repressed. Even a plain black headscarf can frame the face attractively (Plate 10C). The niqab, however, leaving only a slit for the eyes, looks repressive, hiding its wearer's personality and expression (Plate 10B). It inhibits communication as facial expressions are an important part of human interactions. Moderate Muslims do not regard the wearing of the niqab as a requirement, just a voluntary expression of belief. The French banned its wearing in public.

The wearing by immigrants and minorities of distinctive clothing can be a source of pride and group cohesion. However, it marks them out as different, non-assimilated, with alien customs. It can raise inter-group tensions and mark out members of minorities for attack. In parts of Europe in the Middle Ages, Jews had to wear a yellow badge. In Nazi Germany,

Jews had to display a yellow patch or star. Hasidic Jews wear distinguishing dress, with diversity between sects. Members of one sect wear a long black coat even in summer, a black hat, a white shirt, and have long hair, often with curls beside their faces.

Hat-wearing by British men is less common than in the past, except in really cold weather. Some older men wear trilbies and homburgs. Religious Jews often wear the skull cap (yarmulke) and Muslim men may wear special caps. Women often wear hats for weddings.

Head coverings vary enormously, as shown in the Plates. There are crowns, tiaras, feathered hats, hats with artificial fruits (Plate 9B), fascinators, helmets, boaters, ten-gallon hats, Alice bands, scarves, swimming hats, toques, scrum, baseball, cricket, school and flat caps, deerstalkers, fedoras, Panamas, sun hats, pith helmets, Cardinals' red hats, babies' bonnets, etc.

In some countries, women dress in duller clothes after marriage, and widows may permanently dress in black. In Britain 50 years ago, people in mourning wore a black armband for several months, but that has largely died out.

The French are famous for Haute Couture. One French friend who worked as a fashion buyer in England eagerly scanned Paris fashions to see what to wear. English people usually treat Haute Couture as a joke. Most modern fashions look hideous, yet when I visited exhibitions of clothes by Versace, Armani and Givenchy, I was impressed by some pieces of genius and imagination. Françoise Sagan wrote that 'a dress makes no sense unless it inspires men to want to take it off you'.

Clothes are usually advertised using pictures of attractive models with almost perfect hair, faces and bodies, often the result of photographic manipulation: slimming of legs, arms and waist, removing creases, lines or stray hairs; whitening teeth and brightening eyes; increasing shadows to emphasise chest cleavage; flattening the stomach, etc. They make people unhappy with their own faces and bodies, e.g., crushing girls' self-esteem, leading to eating disorders, unnecessary dieting and general unhappiness.

Many items carry words. These may be brand names, where Chinese tourists love to buy things with famous names such as Gucci. There are many fakes: in India I bought a leather wallet bearing the Gucci name for about £1, knowing it was fake. In London I saw an innocent-looking young Japanese girl wearing a hat bearing the word *Shagged*. It seemed incongruous and vulgar. I wondered if she knew what it meant.

Uniforms restrict choice. Branches of the armed services wear different uniforms and ranks may be distinguished by chevrons, rings, badges, braid, epaulets, stripes, etc. One practical aspect of human diversity is how to tell friend from foe in conflicts. Different uniforms help but many combatants do not wear uniforms, especially in guerrilla warfare and terrorist attacks.

What you choose to wear conveys a message, whether of respect, indifference or contempt. I like to dress smartly for the opera as it adds to the sense of occasion. At Imperial College, a former Head of Department dressed almost like a tramp even for staff meetings, giving a bad impression of the department and suggesting that we were not worth making an effort for.

Shoes are a great passion and expense for many women. Christian Louboutin's are famous for red-lacquered soles and can cost thousands of pounds. He helped to bring stilettos back into fashion in the 1990s and 2000s, designing many styles with heels of 4.7 inches (12 cm) and higher. His goal has been to 'make a woman look sexy, beautiful, to make her legs look as long as he can'. The former First Lady of the Philippines, Imelda Marcos, was said to have had at one time 15 mink coats, 508 gowns, 1,000 handbags and 3,000 pairs of shoes.

The Sunday Telegraph Magazine (28/12/2003) reported English singer Sophie Ellis-Bextor's views on high heels. As a little girl, she lusted after 'the full-on, trampy vertiginous versions'. Sophie said, "I really think that first pair of heels [at 14] changed my life. ... I've got around 200 pairs of shoes now and only about five are flat. ... I love the way heels make your legs look more elegant; I love the fact that they take my height over six foot; and I love the poise and discipline they give you."

Journalist Sarah Smith wrote: 'High heels do wonders by lengthening legs, lifting bottoms and defining calves, but they also kill your feet and knacker your back'. Camilla Morton's book, *A Girl for All Seasons: the Year in High Heels* (Hodder & Stoughton, 2007) includes: 'In my six-inch stilettos I keep my head held high and my eye on the goal. They are my shot of confidence and secret weapon against any rivals. It's true that as a heels devotee, plasters, pedicures, paracetamol and taxis have become an integral part of life. ... Live a little — and let this precarious mode of transport bring out a fabulous new you'.

Models in absurdly high heels often fall when parading. Naomi Campbell's stumble and fall in 1993 during Vivienne Westwood's Paris Fashion Week show are famous. The shoes had very thick soles and nine-inch (23 cm) heels. Shoes and clogs with very thick soles have been used sensibly to keep feet above mud and water, especially in ancient Japan and modern Holland. Some women drive wearing high heels. A 2013 study by Sheilas' Wheels showed that driving in flip-flops was even more dangerous than in high heels.

The 15th century crackow shoes had extremely long toes. Crackows were worn by men and women, but men's were longer. Laws allowed the nobility two-foot-lengths, merchants one, and peasants one-half. At the Battle of Nicopolis in 1396, French Crusaders cut off the tips of their crackows to facilitate running away.

Short men often choose shoes with higher heels, especially if they have taller girl friends or wives. Short politicians are frequently unhappy beside taller ones. There are jokes about the shortness of French politicians Nicolas Sarkozy (5 ft 5 in, 165 cm) and François Hollande (5 ft 7 in, 170 cm, as is Vladmir Putin). David Cameron and Barack Obama are 6ft 1 in (185 cm), while General Charles de Gaulle was 6 ft 5 in (195 cm). Sir Winston Churchill was 5ft 6 in (168 cm) when old.

Make-up (Plate 12) goes back thousands of years. Girls feel more grown up when allowed to wear it, but parents are often horrified by their choices. Some male politicians wear discreet make-up. Finger and toe nails may be painted exotically. In 2013, it was reported that 58% of males and females in Liverpool were fake-tanned.

Primitive tribes often wear body or facial colouring. Some early British tribes wore woad, a blue dye from leaves. Julius Caesar commented on the Britanni using blue on their bodies. The Northern tribes, Picts, used body colour; 'Picti' comes from the Latin for 'painted ones'. Most tribal cultures use clay and natural pigments for body and face painting. Tattoos are covered in Chapter 13.

A 2,000-year-old pot of Roman face cream was dug up in London. It was based on cows' fat, the whitener tin oxide (healthier than lead acetate) and vegetable starch. In the West, there is an expectation that most women will wear make-up. As journalist Bryony Gordon wrote, 'It seems that the 21ˢᵗ-century woman can juggle a high-powered job with running a family — just as long as she does it in her Lancôme Définicils mascara, Yves Saint Laurent Radiant Touch concealer, Estée Lauder Lucidity Foundation, Bobbi Brown lipstick and Clinique pressed powder'.

Many harmful chemicals have been used in make-up. Ancient Egyptians used eye shadow containing copper, lead sulphide and kohl, and applied a white lead face cream to clear complexions. Roman women used white lead to lighten the face, with red lead for a rosy glow. From the 15th to 18th centuries in Europe, white lead and vinegar were used to achieve the popular 'dead white' look. White lead was mixed with mercuric chloride for a skin peel. In 1760, the Irish beauty Marie Gunning died 'a victim of cosmetics'. As late as the 1930s, many unsafe products were still used.

'Goths' wear strange make-up, concentrating on white and black.

Isolde, a Goth, wrote: A Goth's make-up is second only to her (or his) clothes in defining her (or his) image. The staple item is the chalk-white foundation known as 'corpse paint', perfect for making one look like some kind of vampire. ... Next in importance is the eyeliner. This can take many forms: smoky, smudgy, subtle or intricate. On special occasions I have been known to cover large areas of my face in fiddly little patterns. ... Others go for the Alice Cooper/The Crow caught-in-the-rain look, with streaks running down the face. ... Last but not least is lipstick. The preferred colour is black but anything 'weird' will do. Blue is good (especially for the heart-failure dead chic thing).... A man in make-up is, apparently, a complete freak, and probably gay.

Shaving body or facial hair is another choice. Many men shave daily but some fashionable types prefer stubble. Others grow beards and/or moustaches (Plate 11). Educator and politician Sir Rhodes Boyson's mutton-chop sideburns made him look Dickensian and old-fashioned, which he liked.

Our genes determine our hair characteristics, and hair is an important part of our appearance. Unless we go bald, we can choose to control its length by cutting, its cleanness by washing, its curliness with curlers or straighteners, its colour by bleaching or dyeing, and its stiffness, thickness and sheen by using appropriate products. With age, the hair changes, becoming sparser and losing colour, an unwanted sign of ageing. Hundreds of millions of people dye their hair (Plates 12C, 13F), from teenagers who want to stand out with green, crimson or blue hair, to the elderly with a light blue rinse or black dye.

For older males, there are characteristic patterns of hair loss, from thinning (Plate 7D) to a bald patch to total baldness. Chemicals such as minoxidil are claimed to promote hair regrowth or to restrict further loss. Hair transplants are expensive. Wigs in many styles and colours may be worn over the normal hair, or used to disguise hair loss from age, chemo-therapy or disease, e.g., the autoimmune condition, alopecia areata.

Chosen hair styles are diverse, including waist-length hair, all-cut-off skinhead-style, gelled into small spikes (Plate 13C), tall, high-crested Mohican styles (Plate 13E), chignons or French pleats. Plaits are common for schoolgirls. Buns are sometimes associated with old ladies, especially strict disciplinarians. Bobs, fringes, 'kiss curls' and pony tails come and go in fashion. A 'short back and sides' was traditional for British males. In the 1950s, the DA (duck's arse, with greased hair combed to a vertical ridge at the back) was popular with Teddy Boys and Rockers, often with a quiff at the front. Dreadlocks and 'corn rows' are mainly worn by Negroids. Sikh men are not supposed to have their hair cut, while Buddhist monks often have shaven heads (Plate 8B). Many Christian monks have a tonsure, with the crown of the head shaved.

Chapter 6

Languages — A Rich but Frustrating Diversity

Our language is an important part of our identity. Unfortunately, language barriers restrict friendships and getting to know people from other nations and cultures. It is sad to want to communicate with someone but not have a common language. In Sri Lanka I would love to chat and joke with friends' children but do not know enough Sinhala. It is difficult to believe that what two Japanese people say to each other in Japanese makes perfect sense to them while meaning nothing to me. The incomprehensibility of unfamiliar languages was used by the Allies on D-Day in 1944. A team of 14 Comanche Indians took part in the landings, relaying messages in their native language with no fear of being understood by any Germans. As their language did not have the terms, the Indians used *turtle* for *tank* and *pregnant aeroplane* for *dive bomber*. The US forces recruited from several different Native American tribes for passing messages, including Navajos whose language was unwritten, tonal and had a complicated grammar.

There is enormous diversity between languages in vocabulary, grammar and punctuation, and in metaphors and similes. Take expressions for heavy rain. In Britain, we say that it is raining cats and dogs, or stair rods. According to Charlie Connelly's book, *Bring Me Sunshine* (Abacus, 2013), in Portugal it rains toads' beards; in Germany, cobblers' boys; shoe-makers' apprentices in Denmark; pipe stems in Holland; wheelbarrows in the Czech Republic; tractors in Slovakia; chair legs in Greece; female trolls in Norway, and Afrikaans-speakers say that it rains old women with knobkerries. In Mandarin Chinese, they say "qing pan da yu", the rain is like someone overturning a large bowl of water. In Cantonese, they say "lokgou xi", raining dog poo.

Punctuation differences include Spanish putting an upside-down question mark before questions as well as a normal one at the end: *¿Y Inés?*, and German doing that with quotation marks. Correct punctuation is essential for clarity. John Richards of the Apostrophe Protection Society gave an example of the importance of apostrophes. For a notice outside a block of flats, compare *Residents refuse to be put in bins*, and *Residents' refuse to be put in bins*. That apostrophe changes the meaning from residents declining to be put in bins, to an instruction for residents to put rubbish in bins. It changes the pronunciation and part of speech of *refuse*.

The human voice expresses a diversity of moods and emotions through tone. We can sound loving, aggressive, pleading, accusatory, bored, interested or angry. Voices have different pitches. Surprisingly, the gene giving high voice in females gives low voice in males.

One thing, such as a hand, is expressed by thousands of different words in different languages. Conversely, many words have several meanings in the same language, the source of **puns**. In Afrikaans, *haas* means *rabbit* and *haste*. Some words have many meanings (about 180 for *set* in the Oxford English Dictionary) and can be several parts of speech: *still* is a noun, verb, adjective, adverb and conjunction. *Nick* (verb) means to make a notch, arrest or steal. *Nick* (noun) means a notch, printer's groove or condition (in good nick), or the last moment (nick of time). As a proper noun, it is a name, a nickname, a shortened version of Nicholas, or the devil (Old Nick).

Many languages have regional dialects, varying in vocabulary and grammar, and spoken with different accents. English has the Queen's English (a traditionally 'pure' form, not Queen Elizabeth II's own English), American English, Australian English ('Strine'), 'Singlish' (in Singapore), etc. British accents include Birmingham, Welsh, Geordie and Scottish.

Accents can be attractive and catching, but can make understanding difficult. At meetings of the National Guild of Wine and Beer Judges, we enjoy and joke about regional accents. At one meeting, there were amusing speeches by a Yorkshireman and a Welshman. Because of their accents, I missed 10% of what they said but my wife missed 50%. Announcements on public transport in London and elsewhere are often incomprehensible when made with a strong foreign accent, which can be dangerous when safety or destinations are involved.

In Scotland, people brought up only 20 miles (32 km) apart may have difficulty with each other's accents. The different accents in Northern and Southern Ireland helped the Northern Ireland security forces identify possible terrorists from the south. Northern Ireland's regional accents include Armagh, Ballymena, Fermanagh, Londonderry and Strabane. A Belfast lady said that she had problems understanding people from Ballymena and Strabane.

There are **different usages** in different countries. In America, a man can fall on his fanny (bottom), but in Britain that is slang for the female genitals. While 'bumbag' is known in America and UK, its American alternative of 'fanny pack' would shock many Britons. Americans say, "I'll write him", while we write *to* someone. Indians often say, "Please do the needful." In Sri Lanka, one hears, "Please off the light."

Words' meanings vary with context. A dry rock has no water but a dry wine is mainly water. In 1976 the Russians called Margaret Thatcher 'The Iron Lady', meaning stubborn and inflexible. Her supporters took it as a sign of strength, transforming her image and probably helping her win three General Elections. Earlier 'Iron people' were Germany's 'Iron Chancellor', Otto von Bismarck, and British/Irish Arthur Wellington, 'the Iron Duke'.

There have been occasional cases of foreign accent syndrome. Sarah Colwill spoke with a West Country accent, then an acute migraine left her speaking with a Chinese accent. In 2014, Roy Curtis woke from a coma after a car crash and suddenly spoke fluent French.

The observation that Britain and America are 'two countries separated by a common language' has been made by Oscar Wilde, George Bernard Shaw, Bertrand Russell, Dylan Thomas and Sir Winston Churchill. In California *pipette* is pronounced as in *pipe*, instead of as in *pip*. *Route* is pronounced *rowt* instead of *root*, and *vase* as *vays*, not *varse*. My request for *drawing pins* baffled the Americans. It took a sketch and much discussion before one of them said, "Gee, you mean a thumb tack!" What is dangerous is Americans' use of *pavement* for the road part for vehicles, while for Britons it is the part for pedestrians. 'Gas' (US, petrol; UK, gas) can also be confusing.

George Bernard Shaw wrote in the Preface to his 1912 play *Pygmalion*, 'It is impossible for an Englishman to open his mouth, without making some other Englishman despise him'. Accents can show regional origins, class, attitude and education, so a man may feel that the speaker is *above or below him* in such aspects, and therefore despises him or her. An Edinburgh engineer said that when he detects from the accent that a fellow Scot is from Aberdeen, he expects him to be "miserable, tight, mean with money." Some accents are so strong that others find them hard to understand, as with Glaswegian (Glasgow, Scotland). In 2014, a woman's call to the police was misinterpreted as a murder confession because of her strong Liverpool accent.

In India, in Delhi, street signs are in Hindi, English, Urdu and Punjabi. The official language is Hindi, in Devanagari script (देवनागरी), with English as the official second language. In the 2001 census, 30 languages were recorded as spoken by more than a million native speakers and 122 languages were spoken by more than 10,000 individuals. Most of those languages are mutually unintelligible. Indians usually understand Hindi and at least one other language.

According to Ethnologue, mainland China has 292 living languages. Most have dialects and accents. Mandarin Chinese has three written forms: simplified or traditional symbols, or in Romanised form, pinyin.

Simplified Chinese: 汉语

Traditional Chinese: 漢語

Pinyin is the system for transcribing Chinese characters into Latin script. Compare the pinyin and symbol forms for friend: *péngyou* and 朋友. Pinyin has 25 Romanised letters from the Latin alphabet, excluding 'v'. Chinese, like Bantu languages and Vietnamese, is tonal, where the same sound in different tones (different pitch or pitch sequence) has different meanings. Thus *mā* (媽/妈) means *mum*; *má* (麻/麻) means *hemp*; *mǎ* (馬/马) means *horse*; *mà* (罵/骂) means *scold*, and *ma* (嗎/吗) is an interrogative particle. The same symbol is used for *man* in Mandarin, Cantonese and in Japanese kanji scripts, although the spoken languages are very different. Vietnamese (*tiếng Việt*) has abundant diacritics, above and below vowels. It is tonal with six tones in the north, five elsewhere. With no diacritics, *ma* in Vietnamese means *ghost*; *mà* means *but*; *má* means *cheek* (or *mother*, in the south); *mả* means *tomb* or *grave*; *mã* means *horse; mạ*, with the dot underneath, means *rice seedling*, with little correspondence in meaning with the Chinese words. Other marks are shown in *thẽ* (thus) and *răng* (how).

By a lake in Hanoi, there is a small memorial. Omitting the diacritics here, it is headed *NGAY 26-10-1967*, and begins *TAI HO TRUC BACH, QUAN VA DAN THU DO HA NOI BAT SONG PHI CONG, JOHN SIDNEY McCAIN …* This shows the shortness of most Vietnamese words, and that *Ha Noi* should be two words, as is *Viet Nam*. It commemorates the shooting down of John McCain (later an American presidential candidate) when he tried bombing the Hanoi waterworks.

Vietnamese formerly used a modified set of Chinese characters called *chữnôm*, but today uses the Vietnamese alphabet (*quõcngữ*), which is a Roman alphabet with diacritics and some different letters, such as *đ*. It does not show case, gender, number or tense, and has no finite/nonfinite distinction. Vietnamese has a strange vocabulary omission, not distinguishing between *blue* and *green*, having to add "like the sky" or "like a leaf" after the word *xanh*.

Lao or Laotian (ພາສາລາວ) is also tonal and has diacritical marks, as in ຂອບໃຈຫຼາຍໆເດີ (thank you very much). *Lao lao* means rice wine, with tones distinguishing the two words, which are written differently in Lao script, an abugida script, as are Cambodian and Thai. In Lao and Cambodian there are no inflections, conjugations or case endings, with particles and auxiliary words being used to indicate grammatical relationships. Lao is partly mutually understandable with Thai (ภาษาไทย) and is spoken by many in northeastern Thailand. In Laos, I saw a wonderfully pithy notice in English outside a bar: *DRINK TRIPLE. SEE DOUBLE. ACT SINGLE.*

Khmer or Cambodian (ភាសាខ្មែរ) is not tonal, unlike Chinese, Thai, Burmese, Lao and Vietnamese. The script looks strange to a European, as in សៀវភៅនេះថ្លៃណាស់, which means *This book is expensive!*

Some languages are phonetic, with spelling and pronunciation matching closely, as in Finnish, Albanian, Georgian and Italian. It is not so in English where the same sound may be represented by several different letter combinations, and where one set of letters may be pronounced in several ways: *ough* can be pronounced *up* (hiccough), *oh* (dough), *off* (cough), *ow* (plough), *er* (borough), *oo* (through), *ock* (lough) or *uff* (enough).

Languages vary greatly in written forms. Some use the 26 letters of the English/Latin alphabet. Others have **diacritical marks**, including accents, umlauts, cedillas and 'strike-throughs', e.g., Norwegian, *Bokmål, Østnorsk*, and Welsh, *dŵr* (water). They modify pronunciation or indicate vowel length or stress. A grave accent can indicate that a word-combination is pronounced, not silent, as in Andrew Marvell's poem, 'To his coy mistress': *Time's wingèd chariot hurrying near*, where the pronunciation is 'wing-ed', not 'wingd'.

Table 6.1 shows different scripts, including joined up ones, ones with separate letters, pictorial ones, linear Cuneiform and beautifully rounded Sinhala letters. Cuneiform was used more than 5,000 years ago by Sumerians

Table 6.1. Examples of different scripts.

Language or script	Country or area	Script
Hindi	India	**देवनागरी**
Sanskrit	Indian literature, especially Hindu texts	**संस्कृतम्**
Japanese	Japan	風立ちぬ
Korean	Korea	한국어/조선말
Cyrillic	Russia/Eastern Europe	**БернардЛамб**
Georgian	Georgia	ქეთევან "ქეთი" მელუა
Greek	Greece	Ελληνικό αλφάβητο
Tamil	India, Sri Lanka	**தமிழ் அரிச்சுவடி**
Sinhala	Sri Lanka	ශුද්ධසිංහල
Arabic	Middle East	**أبْجَدِيّة عَرَبِيّة**
Hebrew	Israel	**עִבְרִיאָלֶף־בֵּית**
Old German	Germany	**AaÄäBb**
Farsi	Iran	**فارسی**
Cuneiform	Ancient Sumeria	
Runic, Anglo-Saxon	Ancient Northern Europe	**ᚠᚢᚦᚨᚱᚲᚷᚹᚺᚾᛁᛃᛇᛈᛉ**
Hieroglyphs	Ancient Egypt	

in what is now Iraq. They used sharpened, wedge-shaped reeds to inscribe clay tablets which were then baked hard.

Having different languages in one area causes expense and difficulties. In Britain, some authorities have their leaflets translated in up to 20 languages. A 2014 NHS leaflet, *The GP Patient Survey,* is in Arabic, Bengali, Czech, English, French, Gujarati, Mandarin, Polish, Portuguese, Punjabi, Slovak, Somali, Turkish and Urdu. In 2014 a gang of Roma and Iraqi Kurds was strongly criticised by the judge for their appalling crimes and for failing to learn our language, burdening the British taxpayer with a bill for £40,000 for translators at their trial.

Printing offers different fonts, such as Times New Roman, `Courier`, Calibri, *Lucida Handwriting* and 𝔒𝔩𝔡 𝔈𝔫𝔤𝔩𝔦𝔰𝔥 𝔗𝔢𝔵𝔱, as well as *Italic* and **Bold** forms. One can also have different sized print. Text may be left-justified (aligned at the left margin but with the ends of lines not aligned), or left- and right-justified, or centred.

One huge blessing is a fixed alphabetical order, a, b, c, d… This enormously speeds up finding words in a dictionary or index, or a name in an address book. A fixed numerical order is also invaluable. Finding a house in a road where the houses have names but no numbers is hard.

According to the 2011 census, England and Wales have 800,000 people living here who cannot speak English, and 3.7 million for whom English or Welsh is not their main language. Large-scale immigration in many countries causes enormous language problems in schools. The Department for Education found that in 2013, English was *not* the first language for the majority of pupils in one in nine UK schools. At Maidenhall Primary School in Luton, 98.9% of pupils do not speak English as their first language. Between them, they speak Pahari, Urdu, Bengali, Punjabi, Somali, Polish, Hindi, Gujarati, Tamil, Portuguese, Arabic, Spanish, Pashto and English. Many translators are needed at parents' evenings. Much teaching time has to be devoted to English before other subjects can be understood. It is very hard on teachers and pupils.

The average number of words understood is about three at 12 months, 270 at two years, 1,540 at four, 6,000 at five, then about 3,000 more each year up to 18, making about 45,000 by then. Usually one has a general vocabulary plus specialist words for work or hobbies. In America, Theodore

Roosevelt said in 1907, "We have room for but one language in this country, and that is the English language. For we intend to see that the crucible turns our people out as Americans, of American nationality, and not as dwellers in a polyglot boarding-house." In spite of that, America has no official language, although American English is the *de facto* national language.

Some states have only English as the official language; others add Spanish or local languages, while Native Americans have very diverse languages. In Israel, the official languages are Hebrew and Arabic, although English is widely used.

There are six official languages at the United Nations, Arabic, Chinese, English, French, Russian and Spanish, used in all meetings and documents. The costs of 30 translations between all these languages are very high. The situation in the European Union is much worse as it has 24 official and working languages: Bulgarian, Croatian, Czech, Danish, Dutch, English, Estonian, Finnish, French, German, Greek, Hungarian, Irish, Italian, Latvian, Lithuanian, Maltese, Polish, Portuguese, Romanian, Slovak, Slovene, Spanish and Swedish. The number of translations is 552! The costs are appalling.

While most of us want our language clear and unambiguous, that does not apply to politics. George Orwell wrote that 'Political language is designed to make lies sound truthful and murder respectable, and to give an appearance of solidity to pure wind'.

Political aims, especially independence, autonomy or devolution of powers, are often linked to language demands. There has been devolution of authority from Parliament to the Welsh Assembly and the teaching of Welsh is compulsory up to age 16 in Welsh schools. In spite of that, only 19% of the Welsh population claim to be able to speak Welsh, although being able to speak it is mandatory for many public service jobs and in some private firms. In 2012, a Royal Marine died in Snowdonia because the map he was given was in Welsh. He fell down a gorge labelled '*ceunant*' on the map, and none of the organisers realised what it meant.

Many Muslims learn Arabic for reading the Koran. Hebrew is the language of Judaism and Latin was associated with Roman Catholicism. In some religions only the priesthood were taught the language of their holy book, giving them extra power over devotees.

The Bible has verses relating to language and human dispersal. Genesis 11:1: *And the whole earth was of one language, and of one speech.* In Verse 4, men plan to build a tower *whose top may reach unto heaven.* Their presumption did not please the Lord. In Verse 7, the Lord says: *Go to, let us go down, and there confound their language, that they may not understand one another's speech.* By Verse 9, the men have stopped building the Tower of Babel, their language has been confounded and the Lord has scattered them abroad upon the face of all the earth.

There are claims of modern 'speaking in tongues' (glossolalia). William Samarin from Toronto University found that Pentecostal Church glossolalia was a 'meaningless but phonologically structured human utterance, believed by the speaker to be a real language but bearing no systematic resemblance to any natural language, living or dead'.

Being able to speak more than one language has advantages; bilingualism and multilingualism are thought to aid intelligence in the young and to delay dementia in the old. Exposure to other languages is a motivating factor. Our Swiss cousins, in a country in which several neighbouring-country languages are spoken, are fluent in Swiss German, German, French, English, Spanish and Italian.

Exposure to different languages can be confusing. Eriko, a Japanese lady working in Singapore, wrote that her 11-month-old daughter is confused by hearing several languages. Eriko and her mother speak to her in Japanese while the husband uses English and a helper uses Tagalog. Bilingual individuals often switch languages even in mid-sentence, and mix words of one language in with those of another. Most languages use words from other languages. In Britain we might administer the *coup de grâce* or commit a *faux pas*.

In 2005, Taiwan ended the practice of writing from *top to bottom* and *right to left*. All government documents were to be written horizontally, from *left to right*, as in the West. A spokesman said that the old style was confusing, especially when numbers and alphabetically spelled English words were included. The Chinese changed to horizontal writing in the 1950s when Mao introduced simplified characters to aid literacy. The initial list of 230 simplified characters in 1956 was later expanded to 2,000.

In most languages, writing is read from left to right, with gaps between words. Arabic and Hebrew go from right to left. Some ancient Greek inscriptions used **boustrophedon**, with lines running alternately from left to right and from right to left, turning like a plough-pulling oxen at each end of the furrow, from *bous*, ox, *strephein*, to turn. The Etruscans had writing from right to left, with no spaces between words. Reading involved deciphering which symbols belonged to which word. Spaces really are important. Note the difference between *The rapists are evil* and *Therapists are evil.*

As an example of language diversity, take **Sinhala,** සිංහල, spoken by 17 million people in Sri Lanka. It is Indo-European but looks different. It has 54 letters, with 18 vowels and 36 consonants, but only 12 vowels and 24 consonants are needed for writing colloquial Sinhala. The consonants are written with letters but vowels are indicated by diacritical marks on, under or around those consonants. If no diacritic is indicated, a following short *a* or *u* sound is understood. Thus the letter ක (*k*) on its own indicates *ka*, either /ka/ or /kə/ as in the last syllable of *baker*. That consonant with different following vowels is written thus: කා (*kā*, long *a*, as in *car*), කැ (*kä* as in *cat*), කෑ (*kā̈* as in *cash*); those marks are after the consonant. Some marks are above it, කි (*ki* as in *kit*), කී (*kī* as in *keep*) or below it, කු (*ku* as in *cook*), කූ (*kū* as in *cool*), before it, කෙ (*ke* as in *kept*), කේ (*kē* as in *cake*), or surrounding it, කො (*ko* as in *cot*), කෝ (*kō* as in *coke*). For *k* without a following vowel, a vowel-cancelling diacritic like a small superscript *p* is used: ක්.

Kataragama, a pilgrimage site, is written කතරගම, with consonants *k*, *t*, *r*, *g* and *m*, but no diacritics as all the vowel sounds are short *a* or *u*, so Sinhala needs five letters when English needs ten. A vowel at the beginning of a word has its own symbol. A city of historical importance is Anuradhapura, අනුරාධපුරය. Here the first letter, අ, is for *a*.

It is impossible to get accurate numbers for speakers of the world's approximately 7,000 languages, and what degree of proficiency is needed to count as a speaker? Table 6.2 shows rough estimates for the top ten languages. Mandarin Chinese comes top as a first language and English as a second language. Even someone fluent in all ten languages could be understood by only half the world's population.

Many languages do not distinguish between *he, she* and *it*, including Maori, Vietnamese, Swahili, Turkish, Hindi and Japanese. English, French and German use **initial capital letters** for proper names, *Thames, Fritz*. German uses initial capitals for common nouns, *Baum* (tree), *Seele* (soul).

Table 6.2. Estimates for numbers of speakers of the top ten languages (figures from Ethnologue).

Language	Total for first and second languages, millions	Total first language, millions	Total second language, millions
1. Mandarin Chinese	1,026	848	178
2. English	765	335	430
3. Spanish	466	406	60
4. Hindi	380	260	120
5. Arabic	354	Difficult to estimate	
6. Russian	272	162	110
7. Bengali	250	193	57
8. Portuguese	217	202	15
9. Indonesian	163	23	140
10. Japanese	123	122	1

English uses capital initial letters for adjectives of nationality, *French, Malaysian,* but many languages do not; the French language body is the Académie française.

Titles vary between cultures. In Korea, the painter-monk Lee Man-Bong (1909–2006) was given the title of *Important Intangible Cultural Property.* At Geisenheim in Germany, there is a bust of the wine-grape breeder, captioned *Prof. Dr.Dr.hc.H. Müller-Thurgau,* where *hc* shows an honorary doctorate. Strange British titles include *Groom of the Stole, Black Rod,* and *First Chief Gentleman and Groom of the Stool* (to Henry VIII).

Repetition features in many languages. Poetry can have alliteration (repetition of initial letters), rhyme, eye-rhyme (the word endings look alike but are pronounced differently), and repeated patterns of syllable stress or syllable length. One can have repetition with variation, as in Shakespeare's *Hamlet:* "To be, or not to be: that is the question." A modern example is in the film *Carry on Cleo* (1964), where Kenneth Williams as Julius Caesar says as he is stabbed, "Infamy! Infamy! They've all got it in for me!"

All written languages have the possibility of errors in grammar, vocabulary, spelling and punctuation. Errors in a foreign language can be lethal if someone is working in enemy-held territory and pretending to be local, with mistakes showing that they are not natives. In World War 2, a German agent in Britain was caught when he wrote a letter in perfect English but used an exclamation mark after the salutation beginning 'Dear ...', a German practice. Another was caught after handing over five pounds and six shillings (more than many workers' weekly wage) when told that the rail fare was 'five and six'. While camping in France, my brother-in-law asked a farmer for six *boeufs* (cattle) instead of six *oeufs* (eggs).

Errors by foreigners in one's own language are a source of innocent amusement, including notices in foreign hotels. Here are some from India:

- *Laal Maas, a deadly meat dish.*
- *WINE & BEAR SHOP, COLD BEARS.*
- *CAUTION SURFACE NEAR FOUNTAIN MAY BE SLIPPER AVOID PHOTOGRAPHY BY CLIMBING ON IT.*

As external examiner, I saved a Sri Lankan university from setting an essay on 'Fermented diary products'. It was not about obtaining alcohol from dates.

Errors in one's own language are frequent in many countries, even among educated groups. A UK health circular about blood donation used *expatriots* instead of *expatriates*, with a quite different meaning. Barclays Bank had an advertisement about *excepting* cheques, when it meant *accepting*.

Imperial College London has very good students. In my research on their standards of English, the home undergraduates made three times as many errors in English as did the overseas ones, who had had more grammar teaching, more correction of errors and more emphasis on correctness.

In poetry, deviations from correct usage may be accepted as '*poetic licence*'. There may be omitted syllables to help with metre or odd pronunciations to help with rhyme. Poems may exploit odd pronunciations, such as Leicester being pronounced 'Lester'. Here is a classic limerick about diversity between Oxford colleges. 'Magdalen' and 'dawdling' do not seem like rhymes, but Magdalen is pronounced 'Mordlin'; the apostrophe on 'dawdlin' indicates the omitted 'g' to get the rhyme.

> *A dozy old don of Divinity*
> *Made boast of his daughter's virginity.*
> *They must have been dawdlin'*
> *Down at old Magdalen —*
> *It couldn't have happened at Trinity.*

Jokes involve quirks of language or behaviour, or playing with expectations. People have diverse senses of humour so that a joke that one person finds hilarious leaves another cold or disgusted. Mozart wrote exquisite music but liked lavatory humour.

Some jokes involve taking figures of speech literally, as in: *How do you stop your mouth from freezing on an icy day? You grit your teeth.* Other jokes involve dodgy definitions (*coffee,* a person coughed upon), errors, puns (Where did Noah keep his bees? In the archives!), mishearings, misunderstandings (1st lady, "Joan had to answer the door in her nightdress." 2nd lady, "What a strange place for a door."), paradoxes and purported behaviour, as in jokes about blondes' alleged lack of intelligence: *Two blondes walked into a bank. They should have looked where they were going.*

Anagrams include *realfun* for *funeral; Flit on, cheering angel,* for Florence Nightingale; *I am a weakish speller* for *William Shakespeare; That great charmer* for Margaret Thatcher.

Mixing languages can be funny, as in **Franglais**, mixing French and English. In *The Sunday Telegraph* (19/1/2014), Michael Deacon imagined a French journalist reporting to friends that former British prime minister Gordon Brown did not have mistresses, unlike French politicians:

> *"Mon ami! Vous will jamais believe le rumour j'ai heard au Royaume-Uni!"*
> *"Quoi, quoi?"*
> *"Leur prime minister ne cheat pas sur sa femme!"*
> *"Non! Ce ne peut pas be vrai! Pourquoi would il behave in cette bizarre way?"*

Causes of misunderstandings include **homographs**, words with the same spelling but different meanings, and **homophones**, words with the same sound but different meanings. For example, *row* can mean to propel a boat with oars, or a line of objects, both pronounced with a long *o*, but can mean a quarrel, pronounced like *cow*. Homophones include *meet/ meat, bier/beer, lead* (metal)/*led, stair/stare, flee/flea, herd/heard.*

Linguistic hazards include idioms, euphemisms, double-meanings, slang, jargon, 'false friends' and ambiguity. In **idioms**, the words' meaning differs from their face-value, as in *He kicked the bucket*, meaning *He died*. Many expressions are not meant literally, as in *I laughed my head off*. If one driver *cut up* another motorist, he drove dangerously but did not perform a dissection.

In **euphemisms**, inoffensive terms are substituted for ones considered rude or offensive, particularly where excretion, sex, death or religion are involved. *They slept together* sounds peaceful and passive, whereas it usually means that they had sexual intercourse.

Slang and **jargon** are common and sometimes used to puzzle and exclude strangers. There are dictionaries devoted to slang, including Cockney rhyming slang, such as 'trouble and strife' for 'wife', and 'Adam and Eve' for 'believe'. Jargon has its uses within groups if it leads to brevity, e.g., medical jargon. **'False friends'** are foreign words which seem like ones in one's own language but with different meanings. The Spanish *embarazada* resembles *embarrassed* but means *pregnant*. When dancer Ivan Nagy called Margot Fonteyn "My lady", she called him "Sir". He replied that in Hungarian, *Sir* means *pubic hair*.

Not everyone can speak, hear, see or write. Many blind people use **Braille**, with raised groups of dots representing letters, words and punctuation. Deaf people can use sign language, aided by lip-reading. A dumb friend in Sri Lanka can make only a few squeaky sounds but works as an eagle-eyed guide on jeep safaris, using gestures, facial expressions and pointing to what he wants to show the tourists. It must be terrible not being able to talk to your own children, or anyone.

Less bad than being dumb is having a stammer, a stutter, repeating sounds, as in 'st-st-st-stammer', prolonging sounds as in 'mmmummy' or pausing when a word gets stuck. There are many famous stutterers or ex-stutterers, including the young Winston Churchill, Hugh Grant, Marilyn Monroe, Emma Blunt, Tiger Woods, Charles Darwin and King George VI.

On screens, **subtitles** can be invaluable for the deaf or if the work is in a foreign language. In the opera house, **surtitles** can transform one's understanding even if the opera is in one's own language.

Dyslexia is common and is an impaired reading ability in the presence of normal vision and letter recognition. The main features are apparent in this originally handwritten account.

Personal account. *I am suverly Dyslexic*

I am a 20 year old man and my currently at Imperial collage london. I am suverly Dyslexic. There is a significont differnce Between my Verbal and performance IQ. I have a greatly impaird ability to sequence code and this means that I had great difficulty learning to read and still have a very pore spell age.

My lastest educational siecologest Report said that my Beading/word decoding was at the 53rd percentile and that my Spelling/writing Endoding was at the 2nd persentile, so I can Read as well or Better than 53% of the population and that I can Spell as well or Better than 2% of the population. My Verbal IQ is at the 90th percentile.

Although I have the Intelligence to acheive accademically I have needed Surton aids. For every puplic exam I have used a Reader to read me the questoins and an amonuensis (scriber) to write the Ansers that I have dictated to thim.

Has it Effected My life? Yes, on many occations. I went on a train from Reading to twickenham I have done this journey many times, but one day I passed through Richmond and knew that on the next stop I had to get of the train, so the next stop come and I jumped off and saw to my horrer that I had got of at the wrong stop "Gentalmen" so befor the doors closed again I jumped Back on only to see a sign that said "Twickenham" so again I ponced of the train and walked Back to inspect the origernol Sign. It was the Sign for the male water closset. So yes there are situeation that because of my difficulty I find more confusing than most.

Do I see myself as dissabled? Only when It Suites me. I have an odd disablity I have been fortunate to have a good enough long term memory to get though A-levels without any legable organised notes.

And would I change It? No, not for the world. Dyslexia is a term used to disribe many specific learning difficultys. It absolutly Is not a p.c*. Sinonim for Stuppid.

I use a persons Ability to articulate their arguments as a gage of intellegence Because I see myself as an Intellegent person, and that is what I am good at. I have had to exept that even inarticulate people can be quit

Brillent and therefore I think that sociaty as hole shouldn't right of children who struggle with the 3 R's.

It is possible that If I not have Dyslexia then I could be an ausome mind capeable of anything. But If my "Disability" was taken away I would not be me, like or not Dyslexia is an integural part of who I am.

Alex Baines-Buffery (hand-written, not from his usual dictation into a computer)
*p.c., politically correct

Languages have more uses than conveying information. Speaking to one another links humans socially in the way that mutual grooming does in animals. 'Talking grooming' is a valid concept. Silences can seem awkward or hostile.

Although people say that our language is changing, changes in English are slow and minor. We can easily read books written three hundred years ago as we have the same basic grammar and vocabulary. The Authorized Version of the Bible was published in 1611 yet is easily understood more than 400 years later. A big dictionary with 100,000 headwords might each year add 100 new words, many of them for new technologies. A change of one thousandth a year is not rapid.

English was enriched by the Authorized Bible, the Book of Common Prayer and by Shakespeare's works. It is an apocryphal story that the theatre critic from a tabloid newspaper went to see a Shakespeare play and slated it for being full of clichés. Many phrases that Shakespeare coined (or pinched) are used today, including: 'Green-eyed monster', 'The world is your oyster', 'Love is blind', 'Wild goose chase', 'A heart of gold', 'Wear your heart on your sleeve', and 'Let slip the dogs of war'.

The august members of the Académie française try to stop foreign words coming into French, wanting *ordinateur* instead of *computer*, but terms such as *le weekend* are used widely. One can understand them objecting to *switcher* (to switch) and *forwarder* (to forward). Groups such as Défense de la langue française and Verien Deutche Sprache (German-speaking association) rightly try to maintain their language, deploring the use of English and American words when there are good French or German words.

In Britain we have the Queen's English Society. We believe that 'Good English matters', and campaign for better teaching of English in schools, more enjoyment of our wonderfully rich language, and better standards of language in public life and the media. We do not object to new words or usages unless they lead to a loss of clarity and precision. We object to dictionaries listing *imply* as a meaning of *infer*. *To infer* requires a mind to make an inference, deduction or conclusion, whereas words, gestures or findings can all imply something. While some children use *wicked* to mean *very good*, dictionaries should mark such uses as incorrect. It is the end of clarity if opposites are listed as synonyms.

Changes in languages can spread rapidly through modern media. If a broadcaster or journalist comes up with some new expression or usage, even if wrong linguistically, it can be quickly adopted by others. The use has spread from America of calling young females 'guys', when the word was for men only, as in 'Guys and Dolls'.

In 2014 there was a competition for the best punning shop name in Britain. 'Junk & Disorderly', a second-hand store, won, followed by 'Pane in the Glass', for double-glazing, 'Barber Blacksheep', for hair dressing, and 'AbraKebabra', for take-away food. Radio listeners came up with 'Tree Wise Men', tree surgeons, 'Cirrhosis of the River', for a dredger boat, and a clever variant of Shakespeare's 'Now is the winter of our discontent' in a camping shop: 'Now is the winter of our discount tent'. Some speech is meaningless, as in saying, "Guchiguchi goo" to a baby, but it conveys the desire of the speaker to communicate ('grooming talking') and may be appreciated by the recipient. In our **conscious minds**, we think in our native language and conjure up visual images, but I do not know how or in what language my **subconscious mind** operates. Often it is cleverer than my conscious mind, arriving at solutions to problems which defeated my conscious brain. We communicate non-verbally by facial expressions, body language, the meaningful glance, the hand on the knee, the kiss and the hug. In what language do cats and dogs think?

Different languages have diverse grammatical rules, such as those about word order even in closely related languages such as English and German. In Indonesian, tenses are not used. Instead, the present tense is used with a description of when the event took place, as in, "Yesterday, the man go to the shop." Sinhala has no definite or indefinite article.

In English there are subtleties of pronunciation. In *entrance*, stress on the first syllable distinguishes the noun (as in door) from the verb with a totally different meaning (to enchant), which has stress on the last syllable and a longer *a*. Although this is seldom taught in Britain, we pronounce **the** as *ther*, with a short vowel, before a noun beginning with a consonant, as in *the weather*, but as **thee**, with a longer vowel before a noun beginning with a vowel, as in *the arrow*. We also have liaisons, so *the bow and arrow* is usually said as *the boa narrow*.

Many languages have silent letters. In English there is the *k* of *knee*, the *g* of *gnat*, the *p* of *pneumonia*, the *l* of *could* and the *w* of *wrapped*. Although the second consonant of doubled consonants is unsounded, it has profound effects on meaning and on pronunciation. Consider *mating* (long *a*, sexual intercourse)/*matting* (short *a*, carpeting), *raping*/*rapping*, *bared*/*barred*, *caning*/*canning* and *hoping*/*hopping*. In *canon*/*cannon*, there is no difference in pronunciation.

Recommended reading

Burgess A. *A Mouthful of Air: Language and Languages, especially English.* (1992) Hutchinson, London.

Crystal, D., *The Cambridge Encyclopedia of the English Language*, 2nd ed. (2003) Cambridge University Press, New York.

Lamb, B. C., *The Queen's English and How to Use It.* (2010) Michael O'Mara Books, London.

Chapter 7

Names and Identity

7.1. Introduction

Amongst the vast diversity of names, our own is dear to us, part of our identity. Names can be intriguing. Will Tallulah and Gloria be exotic, Honey sweet, Grace graceful, Sonny sunny or Priti pretty?

Christian names give no indication of relationship but surnames can, especially if rare. In Britain there are about 560,000 Smiths but only about 50 Feakins, so two Feakins have a much greater chance of being related than two Smiths. In the 1881 census, there were 440,000 different surnames in Britain; widespread immigration has added others.

Having a common name can cause trouble. Mohamoud Ali is common amongst Muslims. In 2014, a man of that name died in prison in Wales. His death was announced to the wrong family, which also had a Mohamoud Ali, but not in prison. An extraordinary concentration of surnames is that of Baj, where in the Italian village of Stoccareddo, 370 out of 380 (97.4%) inhabitants share that surname which is rare elsewhere. Many Bajs share the same Christian name, with seven Luigi Bajs and 20 Maria Bajs. As the streets did not have names until 30 years ago, the village was a nightmare for the postman, with love letters and bills going to the wrong Baj, causing confusion and fights. Nearly all the world's 8,000 Bajs can be traced to that village.

The usefulness of names for distinguishing people diminishes when they are too common. In 1967, in a largely Pakistani district of Leicester, England, about 45% of borrowers from the public library were called Patel and about 45% were Khan. Many had the same forenames and there was much multi-occupancy of houses, so even forenames and addresses did not uniquely identify individuals. In Britain there are many people with the same name, such as Janet Smith, Ann Jones, Robert Williams and Tom Brown. Second or third forenames can help distinguish between them, but each has a unique government-provided identifier, the National Insurance Number, such as YA 14 51 39 B. In any country, passport numbers and/or national identity card numbers should be unique.

Names can be a matter of life or death. During violence in Sri Lanka between the Buddhist Sinhalese majority and the Hindu Tamil minority, the names revealed a person's origins.

Foreign names can cause confusion. The admissions staff at Imperial College failed to identify an honorific term, Hj. Ilzam Zainulabidin from Brunei went through college listed under the surname Hj. Her father had been on the pilgrimage to Mecca, so he had the honorific Hj, short for Hadji, after his surname, Zainulabidin.

In Britain, we put Christian or forenames before the surname. In Hungary and China, they put the surname first, which caused a lot of trouble with Chinese students at Imperial College. Which out of 'Sisu Pan' was the surname? Chairman Mao's name was Mao Tse-tung, where Mao was his surname.

Larger-than-life English personality **Clarissa Theresa Philomena Aileen Mary Josephine Agnes Elsie Trilby Louise Esmeralda Dickson Wright** (1947–2014) said that her parents had problems in deciding on her names. They got drunk on the way to the Christening, blindfolded her mother and turned her loose in a library, where the first book she pulled out was Richardson's *Clarissa*. An **aptonym** is a name fitting the person's job and is closely related to nominative determinism, in which someone's name influences their choice of career. A supreme example is Igor Judge (I. Judge), who became a judge and Lord Chief Justice of England and Wales.

Many forenames are shortened to seem friendlier, less formal, or to save space. Taking British politicians, David Cameron is Dave (his wife Samantha is Sam), Nicholas Clegg is Nick, John Vincent Cable is Vince and Edward Miliband and Edward Balls are both Ed. When the Prime Minister's wife became pregnant, one headline was 'Wham bam! Sam Cam to be mam'.

There is a strong tendency to shorten Christian names, especially by young females, to sexually ambiguous forms. Alex could be Alexander or Alexandra; Sam could be Samuel or Samantha; Chris could be Christopher or Christine; Pat could be Patrick or Patricia; Terry could be Terence or Theresa. Names used for both sexes include Ashley, Averil, Beverley, Dee, Gay, Jo, Kerry, Kim, Paddy, Robin, Sacha, Tracy and Valentine.

7.2. The history of surnames

Several thousand years ago, the Chinese imposed a system of surnames. In a survey of AD 627, there were 593 different surnames there. Today, 40% of Chinese have one of the 10 commonest surnames. In Britain only 20% have one of the 75 most common surnames. Li is the commonest in the world, often transliterated as Lee. In China, there are 93 million Wangs and 92 million Lis, with the top ten surnames being Wang, Li, Zhang, Liu, Chen, Yang, Huang, Zhao, Zhou and Wu. Most can be spelled in different ways in Chinese characters, with different meanings. Wang could be king, ruler, vast or pool; Li could be dawn, plum, logic, jasmine or strength; Chen/Chan could be great, ancient or official. Jingjing Su did not like her forename "because those double names are for pandas". In Chinese the two Jings have different characters and meanings. ChiChi and AnAn were giant pandas.

In Japan, surnames were not allowed except for the governing classes, so the man-in-the-street or paddy field only acquired a surname about 150 years ago. After imperial rule was re-established in 1868, surnames became mandatory for all in 1875. There are now more than 120,000 surnames (usually consisting of two or three Chinese characters) for the 100 million Japanese.

In Britain, the Norman conquerors introduced detailed written records as in the *Domesday Book* of 1086. Names were needed for property rights and taxation. Some knights who came with William the Conqueror had surnames such as FitzOsbern. Even in the 14th century many commoners were called 'John' or 'William', with no surname. In Anglo-Norman times, Fitz was a prefix for surnames in noble families, so FitzOsbern indicated Osbern's son. Fitz became used for illegitimate children of the nobility as in Fitzjames, son of James II of England, and FitzClarence, son of the Duke of Clarence. The names of former president John Fitzgerald Kennedy mean 'Gracious gift of God, bastard son of the ruler with the spear, ugly head'.

Many surnames came from the father's Christian names, as in Williams, Jones (John's son), Johnson, Richardson, O'Neill and MacDonald. **Toponyms** came from place names or their derivatives, e.g., Marie-France, Windsor and Scotland. In England, Bottom, as in Ramsbottom, is common up north; there is Thorp in the north east, as in Mablethorpe, and Combe in the south west, as in Widdicombe.

The use of connectives such as 'of', 'de', 'da' and 'von' before place names is ancient, e.g., John of Gaunt, Madame de Pompadour, Leonardo da Vinci and Otto von Bismarck. Richard the Lionheart (1157–1199) fought Saladin — in full — Ṣalāḥ ad-Dīn Yūsuf ibn Ayyūb (Arabic: ح صلاح الدين يوسف بن أيوب). Here, ad-Dīn means 'of the religion', not a place name.

Geographical features or buildings gave Hill, Woods and Monteverdi (Italian, green mountain), or Castle and Church. Materials gave Cotton and Leather. **Metonyms** came from occupations: Thatcher, Gardner, Miller, Cartwright, Cook, and Mahler (German, painter).

Some surnames were descriptive: White, Armstrong, Longfellow and Schwarzkopf (German, dark head), but what about Peabody? There are surnames from animals: Lamb, Fox, and plants: Moss, Vine. A couple with an animal/plant surname, Cocksedge, had problems with emails because spam filters picked up 'cock' and treated Cocksedge as a dirty word!

Surnames can be used as Christian names. The 41st and 43rd US presidents used the surname Walker as a Christian name: George Herbert Walker Bush and his son, George Walker Bush. In Spain and Spanish America, there are usually two forenames, followed by the fathers' surname, then the mother's. In Portugal and Brazil, the forenames are followed by the mother's surname, then the father's.

Foreigners have some very short surnames, such as Ng and Lu in China, while those in India, Sri Lanka and Greece can be very long. The Sinhalese name Wickramaratne means 'picture of the jewel of the king'. Long Tamil names include Theivendarampillai and Venayakamoorthy. In Britain, we can identify some surnames as foreign even if their possessors were born here. We would recognise Balasubramanian as Indian and Wrzyszczynski as Polish.

Here we have **double-barrelled surnames** such as Paterson-Fox, or triple-barrelled as in Anstruther-Gough-Calthorpe or Cave-Brown-Cave. The surname of the Dukes of Buckingham and Chandos was **quintuple-barrelled**: Temple-Nugent-Brydges-Chandos-Greville. Vaughan Williams is a double-barrelled surname with no hyphen. Multiple surnames are often associated with the upper social echelons. The French often use double-barrelled Christian names, as in Jean-Paul Sartre and Marie-Claire Alain.

7.3. DNA sequencing and surnames

The Y chromosome confers maleness and goes straight down the male line, father to son. The X goes through both sexes. Mitochondrial DNA is inherited only through the mother, so it goes down the maternal line to males and females. With DNA sequencing, one can distinguish between particular Ys, particular Xs and particular mitochondrial DNA sequences, which can be very useful in determining ancestries.

At Leicester University, Dr Jobling found a strong link between surnames and particular Y chromosome DNA sequences, as one might expect where both go down the male line. Professor Sykes of Oxford University investigated sequences on the Y chromosomes of 61 men called Sykes. About half had an identical DNA fingerprint not seen elsewhere.

7.4. Name changes

In Britain and the West, surnames usually go down the male line, with the wife and children taking the man's surname. Sometimes the wife keeps her maiden name (her father's surname) for all purposes or for work. Princess Anne and Mark Phillip's daughter keeps her maiden name, Zara Phillips, when competing in equestrian events, instead of her married name, Mrs Michael Tindall. In Sri Lanka, the wife often keeps her maiden name and the children take the father's surname, so the wife and children have different surnames. In China, the wife keeps her maiden name but children can choose either parent's surname, usually taking the father's.

The biggest cause of surname change is marriage. A wife may have the British surname Smith but could come from anywhere, with a maiden surname such as Wozniaki, Muguruza, Hsieh, Wittenhof or Li. Adopted babies usually take the adoptive parents' surname. Singer Dame Kiri Te Kanawa was born Claire Rawstron but was adopted by Thomas TeKanawa. Actress Kate Winslet did not take their surnames on marrying Jim Threapleton or Sam Mendes or Ned Rocknroll. A 2013 Facebook survey found that 62% of married women in their 20s took their spouse's surname, compared with 74% in their 30s and 88% in their 60s.

People can change their names by deed poll if they do not like their name, for sentimental reasons, to get a legacy, or sound less foreign. A newspaper tycoon changed his name twice, starting as Jan Hoch, a Jewish boy in eastern Europe. Later, he changed to Ivan du Maurier and then Robert Maxwell. At Imperial College London, the 1987/8 student union president Ian Howgate changed his name by deed poll to 'Sydney Harbour Bridge' in aid of the Comic Relief charity.

In Sri Lanka the Portuguese ruled from 1505 to 1658 and many people have surnames such as Fereira, De Silva and Pereira. They do not necessarily indicate Portuguese descent because the voluntary adoption of Portuguese names was fashionable.

Hitler's father was Alois Schicklgruber but at 37, before Adolf was born, Alois changed his surname to Hitler. 'Heil Hitler!' has more resonance than 'Heil Schicklgruber!'.

Audrey Hepburn was a great and brave actress, born Audrey Ruston in 1929 in Belgium. Her father, Joseph Ruston, double-barrelled his surname to the more aristocratic Hepburn-Ruston, wrongly believing himself descended from James Hepburn, third husband of Mary Queen of Scots. Audrey's Dutch mother was Baroness Ella van Heemstra. After the Germans invaded the Netherlands in 1940, Audrey adopted the pseudonym Edda van Heemstra because having an English name was dangerous. She ran errands for the Resistance, then resumed the name Hepburn after the liberation.

During World War 1, our Royal Family changed its name from the House of Saxe-Coburg and Gotha to the (non-Germanic) House of Windsor. One of the rarest causes of name change is becoming king. Albert Frederick Arthur George Windsor chose the last of four Christian names to become King George VI. His great grandson, Prince Harry, does not have Harry amongst his names, Henry Charles Albert David.

A lady in London has the surname Viles; her father's surname 'Vile' upset her mother who added the 's'. Vile dates from the 12th century in various forms, and first appeared in writing as Reginald Le Viel in 1173 in pipe rolls of the City of London, financial records by the English Treasury.

Young people who have not changed potentially embarrassing surnames include Japanese violinist Ryota Ichinose and Rosie Shufflebotham. Fartman was common but has disappeared. In Glasgow in the 1881 census there were 1,300 Smellies, which was down to 424 by 1998, with a rise in Smillies. In Britain, Ed Balls, a Labour politician, married Yvette Cooper, who kept her maiden name. Their children are called Cooper as Ed thought his surname (slang for testicles) might lead to them being bullied. Many entertainers have stage names, and the actors' guilds of Britain and America decree that no two actors may have the same name. Some changed their names because of that, or to sound more British or American, or to sound more appealing. Archibald Alexander Leach seems less romantic than Cary Grant. Dino Paul Crocetti became Dean Martin, while as Russian ballet dancers were supreme, Lilian Alicia Marks 'Russified' her name to Alicia Markova. Dirk Bogarde sounds more dashing than Derek Niven Van den Bogaerde.

Many authors use 'pen names', sometimes with different ones for different styles of book such as crime and romance. Harry Potter's female author J. K. Rowling wrote 'The Cuckoo's Calling' under the masculine pen name Robert Galbraith.

7.5. Local associations of surnames

Hashimoto is Japanese, Schmitt is German. Goldberg, Levi and Cohen have Jewish associations. Jewishness passes down the mother's line but the tribal affiliation, such as Kohen, Levite or Israelite, depends on the father. Even though ancestors may have come to Britain more than a thousand years ago, feelings and identifications can still run strongly. A lady from South Pembrokeshire is of Viking or Saxon descent, a tall, blue-eyed blonde. Her opinion of the people of North Pembrokeshire was: "They are small, dark, hairy and primitive - they speak Welsh! And I hate them."

7.6. Russian names

Russians use **patronyms**. The father's first forename is given an extra ending and used as his children's second forename. The endings *-ovich*,

-yevich, -yich are for sons, *-yevna, -ovna, –ichna* for daughters. A man named Ivan with a father named Nikolay would be known as Ivan Nikolayevich. The patronymic is an official part of the name, used in official documents and when addressing somebody formally and among friends. The Russian president is Vladimir Vladimirovich Putin. So many Russian lady tennis players had surnames ending in *–ova* that they were collectively called 'the ovas', e.g., Sharapova, Makarova and Vaidisova.

7.7. Iceland's unusual (dottier?) way with surnames

In Iceland, the child's surname is based on the father's (sometimes the mother's) forename, so is patronymic (or matronymic). If Jón Einarsson had a son named Ólafur, his surname is not Einarsson but Jónsson, son of Jón. For a daughter, dóttir is added, so Jón Einarsson's daughter would have the surname Jónsdóttir. An **Icelandic Naming Committee** controls which forenames can be used. They must contain only Icelandic letters, be declinable according to its case system, and gender-appropriate. Carolina would be banned as C is not in their alphabet.

7.8. The frequency and geography of names

In 1998, out of 37 million surnames in a British database, there were 18,465 Lambs, the 292nd most frequent surname, 496 per million, with concentrations in Sunderland, Newcastle, Edinburgh and Glasgow, and in Australia, New Zealand, USA and Canada. Widgery, with only 162 cases, ranked 19,178th, while Smith came first with 514,898 cases. Jones is hugely concentrated in Wales and Thorne in the west of England.

At the start of Queen Elizabeth 1's reign (1558), 70% of Englishmen were named John, Thomas, William, Richard or Robert. When the Geneva Bible appeared, Puritan names began, e.g., Be-courteous Cole, Safety-on-high Snat and Fight-the-good-fight-of-faith White. Congregationalists in New England picked names at random from the Bible, such as Mene Mene Tekel Upharsin Pond and Notwithstanding Griswold.

French, Spanish, Portuguese, British and Italian names were exported throughout their empires. Britain has had massive immigration in the last 30 years, so that in the 2011 census, only a minority of Londoners, 47%, were 'white British'. There are now many Hussains, Mohammads, Patels, Khans, etc., in England. Very non-British surnames are common here, including tennis players Anne Keothavong and Elena Baltach, and runner Christine Ohuruogu. Although born in Somalia, champion runner Mohamed (known as Mo, which sounds less foreign and lacks the religious connotations of Mohamed) Farah is popular as a British athlete.

Some British surnames listed on the website Ancestry.co.uk are extremely rare, such as Rummage, Hatman, Chips, Silly, Strongitharm and Harred. About 200,000 surnames disappeared in the century after the 1901 census, including Raynott, Woodbead and Nithercott, but thousands of new ones have come in. Stoke-on-Trent has the greatest concentration of people surnamed Salt and also of Pepper.

Some names have particular associations in fact, fiction and jokes: Father Seamus O'Flaherty, an Irish priest; Fritz, German; Olaf, Scandinavian; Giuseppe, Italian. Blodwen, Bronwen and Angharad are Welsh females. Jean is an English female name, pronounced 'gene', but in France it is masculine (Jean-Paul Sartre), pronounced a nasal 'Jon', with Jeanne as the feminine (Jeanne d'Arc, Joan of Arc). In jokes, Mick and Paddy are dim Irish men.

In Singapore, English names once carried a colonial stigma, with Chinese names preferred. Their first prime minister was known as Harry Lee at Cambridge but in Singapore was Lee Kuan Yew (Bright Shining). Now English forenames are common, such as Deborah, Ann, Thomas and John. Even young people with poetic-sounding Chinese names are choosing to be called by English ones. Low Eng Kee's sons prefer to be called Daniel and Donald rather than Hao Yu (Grand Universe) and Hao Dian (Grand Grace). Siew Hong (Kind Eagle) changed to using Catherine, and her brother Chun Pay (Cultivate Handsomeness) preferred the less romantic Kevin.

7.9. Choosing forenames

Many Christians choose Biblical or saints' names for their children. In an ONS (Office for National Statistics) survey in Britain in 2013, there were 10,681 different first names given to boys and girls. The top 10 boys' names were: first, Harry (popular because of Prince Harry and the Harry Potter books), then Oliver, Jack, Charlie, Jacob, Thomas, Alfie, Riley, William and

James. For girls, it was Amelia, Olivia, Emily, Lily, Ava, Mia, Isla, Sophie and Isabella. George, the royal baby's name, was 12th and expected to rise quickly. Muhammad was 19th and two of its variants were also in the top 100. Taking all variants of Muhammad, the name came a close second to Harry. Similarly, Lily plus Lilly came second amongst girls.

In that survey, some the oddest names inflicted by parents on children were Lexi-Jade, Riley- (with the hyphen but nothing hyphenated with it) and Demi-Mai. 18 boys were Baby, 22 were Zi, and three were just C. TV is influential. Names from the American series *Breaking Bad* were favoured, with a 70% increase in Skyler. *Homeland* led to increases of 40% for Brody, 200% for Carrie and 66% for Dana.

President Harry S Truman had a 'dummy' second name, as the 'S' did not stand for anything. Many people have been known by their second forename, as with (Henry) Warren Beatty, (Richard) Buckminster Fuller, (Adeline) Virginia Woolf and (Ahmed) Salman Rushdie.

7.10. Unusual names

Zimbabwean banker Gideon Gono called his daughters Pride and Praise, and his son Passion. Sir Paul Getty and his wife Talitha Pol called their son Tara Gabriel Galaxy Gramophone. David Bowie used Zowie for his son. Frank Zappa chose Moon Unit, Dweezil, Diva and Ahmet. Other names for celebrities' children include Jermajesty, Zuma Nesta Rock, Fox, Fifi Trixibelle and Heavenly Hiraani Tiger Lily.

Peaches Honeyblossom Geldof hated her Christian names. This celebrity daughter of singer Bob Geldof died aged 25 in 2014, leaving two strangely named boys, Astala Dylan Willow and Phaedra Bloom Forever. In mythology Phaedra was a tragic female. Curiously, Peaches had written, 'Why do the rich and famous give their children such ridiculous names? Mine has haunted me all of my life. … I get a lot of lascivious comments: "Ooh, you're a juicy piece of fruit. … I'd like to take bite out of that peach … Look at those peaches."… "Oi, Peaches, are your parents bananas?"'

Foreign names may be misused. The Hausa name Tanisha means 'born on Monday'. Non-Hausas have used it for girls born any day of the week. Tiponya is from the Miwok American Indians and has been frequently used by outsiders. It is said to mean 'great horned owl sticking her head under her body and poking an egg that is hatching', a lot for one word to convey.

In *Quest* 116, 2014, Terry Wilton gave examples of Christian names from the Cayman Islands, including sisters Anique, Banique and Canique, and identical twin brothers who shared their father's Christian name, Jahmal. Other local names were Rozzard, his brother Rezzard, Jelice and Gerewarifucha, who when asked how to pronounce her name, replied, "I'm always called Gerry. I don't really know my full name. . It's long … I don't even know how to spell it." Kamaljit gave her name's pronunciation as 'Camelshit'. Mr Wilton mentioned a Turkish boy, Ufuk, in the same school as a Japanese girl, Fuku, and the names of Birmingham University students taking name-appropriate courses: Sue, law; Beryl, geology; Carol and Melody, music; Mark, teacher training; Oscar, drama; Esther, chemistry, and Nick, applied criminology!

An unusual name helped catch a mugger in 2013. Having had his bike seized, the boy tracked down his attacker through Facebook after hearing her accomplice shout her unusual name, 'Shermya Pyatt'. He passed her details to the police.

7.11. The meanings of some Christian names

These examples show diversity of origin and meaning. They come from Hebrew, French, Old French, German, Germanic, Greek, Celtic, Gaelic, Latin, Welsh and Scottish but are usually regarded as English.

Aaron, lofty, mountaineer; Abigail, father rejoiced; Absalom, father of peace; Adelaide, noble; Agatha, good; Alan, harmony; Albert, nobly bright; Alexander, defender of men; Alfred, elf counsel; Algernon, moustached; Alma, nourishing; Amadeus, love God; Amy, beloved; Andrew, manly; Angela, angel; Annabel, lovable; Anselm, God-helmet; Anthea, flowery; Archibald, genuine and bold; Arnold, eagle strength; Arthur, bear; Audrey, noble power; Augustus, venerable; Averil, boar-favour.

7.12. Names of Native Americans

Native American Indian leaders showed a wonderful diversity of names, some in English, others in native languages: Opchanacanough,

Rain-in-the-Face, Red Cloud, Cornstalk, Crazy Horse, King Philip, Chief Pontiac, Sitting Bull, Wovoka, Sakayengwaraton.

7.13. Initials, combinations and titles

The British Association of Dermatologists has a good name but a BAD acronym. My surname Lamb made me concerned about the initials of ladies whom I might marry. Amanda Jones would become A. Lamb, and Barbara Ann Smith and Mary Angela Thomas would become B. A. (baa!) or M. A. (maa!) Lamb. For university, I chose science, as arts subjects would have put BA or MA after my name.

Children enjoy jokey combinations of made-up book titles and authors' names, such as 'The Lamppost', by Eileen Onnit; 'The Mystery Murder' by Huw Diddit, and '50 Years in the Saddle' by Major Bumsore.

The record office in Cornwall has some odd combinations of surnames in marriage: Bone/Skin, Swine/Ham, Mutton/Veale and Dinner/Cook. In the Baptist Register in Truro, the names included Susan Booze, Edward Evil, Charity Chilly, Gentle Fudge, Obedience Ginger and Offspring Gurney.

One can have odd combinations of titles and names, such as Miss Hurd, Mrs Target or Master Bates. Titles may include place names, as in Prince Philip, Duke of Edinburgh. A holder of many titles was a Spanish lady, Victoria Eugenia Fernández de Córdoba y Fernández de Henestrosa, the 18th Duchess of Medinaceli, who died in 2013 at the age of 96. She was 9 times a duchess, 18 times a marchioness, 19 times a countess, 4 times a viscountess and 14 times a grandee of Spain, and head of a family whose members included three saints and two Popes. She could not remember how many castles she owned, perhaps 90 to 100.

7.14. Permitted names and ugly names

Some countries such as Germany, Sweden, China, Japan and Iceland ban humiliating names. A New Zealand child called 'Tallulah Does The Hula From Hawaii' was made a ward of court so that she could change her hated name. In that country, the Wheaton's saw an ultrasound scan of their baby, realised it was 'for real', and when he was born, called him

'4Real'. The Registrar did not accept that. Denmark has about 3,000 permitted forenames for boys and 4,000 for girls.

In America, some families like to keep the same forename and surname, using Senior (Sr.) and Junior (Jr.), as in the black civil rights activist, Martin Luther King Jr., son of Martin Luther King Sr. Over more generations, Latin numerals are used, as in John Paul Getty III.

There was an East Asian phenomenon of ugly names such as the Mongolian Muunohoy (nasty dog) and the Hokkien Chinese Ah-ti (pig), chosen to prevent spirits becoming jealous of a child.

7.15. Nicknames

Nicknames are pet, derisory, shortened or jokey versions of the full name, such as Bob for Robert. They were common among the Vikings and have been used for thousands of years. Sandy is a nickname for males or females with red hair and a shortened version of Alexander/Alexandra. Insulting nicknames include fatso, beanpole, curly and train tracks (a wearer of braces on the teeth, Photo 15.2). 'Four eyes' and 'carrot top' are school nicknames for those with glasses or red hair. Some occupations attract nicknames such as Sparks for electricians or Sawbones for surgeons. 'Quack' is a derogatory term for doctors.

7.16. Scripts, diacritical marks and punctuation in names

Names can differ in their scripts. Aglaia in Greek is Αγλαια. Novelist Fyodor Mikhailovich Dostoyevsky was Фёдор Михайлович Достоевский in Cyrillic. Mao Tse-tung in Chinese is 毛泽东. Many foreign names have **accents** and other **diacritical marks** (Chapter 6), especially French, Hungarian, Norwegian and German, e.g., Thérèse, Zoltán and surnames Øybjørg and Müller.

Punctuation includes hyphens in Christian names and surnames, e.g., Marie-Therese Rolly-Gassmann. Apostrophes occur in abbreviations of words like 'de', as in Jeanne d'Arc. A Cameroonian footballer is Samuel Eto'o. A missing apostrophe in a woman's name led to her death in 2012. Marje Piner's maiden name was L'aimable, but her medical records had that as Laimable, so staff at a mental health unit were unaware of her history of severe depression; they discharged her and she killed herself.

7.17. Illegitimacy and surnames

Illegitimacy, adoption and surrogate parenting bedevil work on surnames, especially on DNA sequence-matching to surnames. Around 1960, genetics classes at the California Institute of Technology involved pedigree analysis of the students' families for inherited characteristics. Many students found out that the man they thought was their father could not be their father, causing anguish, court cases, divorces, etc. Those practicals were abandoned.

The frequencies of children born outside wedlock are increasing. In a National Vital Statistics Report, 41% of babies born in the United States in 2009 were born to unmarried mothers (up from only 5% fifty years ago), including 73% of non-Hispanic black children, 53% of Hispanic children and 29% of non-Hispanic white children. From English parish registers and census data it has been calculated that illegitimacy frequencies were 4.4% in 1540, 1% in 1650, 6% in 1850, 7% in 1964, 11% in 1979, up sharply to 38% in 1998, and a record 48% in 2011.

7.18. Consequences of name confusions

Charles Joseph Frederick from Calgary had his home seized, his credit rating shattered and was compelled to make unjustified child-support payments because the Canadian government confused him with a man with the same name but from Edmonton. Their birthdays were close, 10 June 1958 and 12 June 1958, and both were divorced in 1993. The Calgary man endured 14 terrible years, with agencies pursuing him for money owed by the other man.

Mrs Patricia Kem of Aurora, Colorado, was sent a tax bill for $3,000 for a job in Oregon, a state she had never visited. The bill should have gone to Mrs Patricia DiBiasi of Hines, Oregon. Both were born on March 13, 1941, shared the maiden name of Patricia Ann Campbell, been given the same social security number and had fathers called Robert. Both had married military men within 11 days of each other, worked as bookkeepers, and had children aged 21 and 18. In Italy, Andrea Ruga, 47, had the same name, date and place of birth as a Mafia godfather accused of kidnapping and terrorism. He had his house raided several times, was stopped at road-blocks, and had hotel staff calling the police when he tried to book in. After phoning three senators, three magistrates, a newspaper editor and the local police chief to say he could not stand the harassment, he committed suicide.

Chapter 8

Albinos, Colour Blindness and Height: How Human Characteristics are Inherited

8.1. Introduction

For understanding human diversity, it is very useful to have an understanding of inheritance and of how genes interact with environments. One needs to know about dominance versus recessiveness, sex-linked versus autosomal genes, and genes of major and of minor effect.

We can understand how human characteristics are inherited by considering albinos, red-green colour blindness and normal heights because they are inherited in different ways. See Plate 1A and C for human chromosomes, showing the sex chromosomes: two Xs in females, one X and a smaller Y in males. The genes on the X which are absent from Y are called **sex-linked**, including ones controlling red-green colour blindness. The 44 non-sex chromosomes are **autosomes**, with the gene controlling albinism on an autosome. Genes have other genes on the same chromosome, with a tendency to be inherited together, especially if close: they are **linked**.

Meiosis is process which reduces the two copies of each chromosome per cell to one copy of each in sperm and eggs. At meiosis, parental genes can show **segregation** and **recombination**, and so different embryos from the same parents can come from genetically very different eggs and sperm, giving much genetic variation among their children (Chapter 4).

Some characteristics show **clear-cut variation**, such as normal skin versus white albino skin, or colour-blind versus normal colour vision. They are often controlled by **single genes of major effect**. Other characteristics such as height show **continuous variation** and are controlled by many genes, some with small effects, and often with effects of the environment. Those characteristics are **polygenic** (*poly*, many; *genic*, of genes).

Patterns of inheritance are demonstrated using genetic symbols, reflecting whether a particular gene is dominant, recessive, incompletely dominant or sex-linked. Crossing diagrams help one understand what is going on. Symbols for a gene usually consist of one to four letters, in italics, with an **initial capital letter for the dominant** and **lower-case letters for the recessive**, such as *A* for the dominant normal skin colour gene and *a* for the recessive albino gene. Sex-linked genes are shown as **superscripts on the X**, with nothing on the Y, for example, X^{Cb} for dominant sex-linked normal colour vision and X^{cb} for the recessive colour-blindness gene.

The term 'gene' is used in **two ways.** In the sense of an **allele** (short for *allelomorph*, from Greek, *allel*, one another; *morphe*, form), a gene is of one of a number of alternative forms which can occur at one place (locus) on a chromosome, such as albino or non-albino alleles. 'Gene' is also used to mean a **locus**, the place of the gene on the chromosome, such as the albinism locus. A locus may have hundreds of different alleles. With the **normal allele** having one particular base sequence, there are a vast number of **possible mutant genes** with DNA sequences differing at one or more base pairs. There can be many different mutations at a locus, some having different effects from each other, including being neutral.

For **autosomal** genes, we have two copies of each locus per cell, one on each pair of chromosomes (one from mother, one from father). If the two copies of the allele are identical, the individual is **homozygous** (*homo*, alike; *zygous*, yoked together); if they are different, the individual is **heterozygous** (*hetero*, unlike). If the two copies are alike then they can be expressed, so an individual with two copies of the normal skin colour gene (the *A* allele), will have normal skin colour, while an individual with two copies of the allele for albinism (*a*) will have white skin. The homozygote produces a single kind of **sex cell**, so *AA* has *A* eggs or *A* sperm, but the heterozygote produces equal numbers of the two kinds, so *Aa* gives a ratio of 1 *A* to 1 *a* eggs or sperm.

If the individual is **heterozygous**, two things can happen. One copy (the **dominant** allele, e.g., the one for normal skin colour) may be expressed and the other copy (the **recessive**, e.g., albinism) may not be expressed, so heterozygotes for normal and albino alleles will express the dominant, normal skin. Alternatively, both the different alleles may be expressed, giving an intermediate manifestation, from **incomplete dominance**.

Sex-linked genes show different patterns of inheritance from autosomal genes because they are present on the X but not Y. For sex-linked genes, females have two copies of the gene (one on each X) but males have only one copy (on their one X). Red-green colour blindness, a recessive sex-linked characteristic, is much more common in males than in females, as are other sex-linked recessives, such as haemophilia.

For simplicity, I shall use '**genetic make-up**' for the genes carried instead of 'genotype', and '**manifestation**' instead of 'phenotype' for the characteristics expressed.

8.2. How an autosomal characteristic is inherited, albinism

Albinism has a dominant **normal** skin colour allele, (**A**) and recessive allele **albino** (**a**, white skin), with only the homozygote *aa* showing the albino characteristic (Plate 15E). There are several different types of albinism (Chapter 15). Here we consider oculocutaneous (eye and skin) **albinism type 1**, caused by a mutation in a gene on chromosome 11. Sufferers have milky white skin, white hair and pale blue eyes.

Most people are normal, **AA**, but a minority are unaffected carriers, heterozygotes **Aa**, with A dominating a. In Europe, about 1 in 20,000 individuals is albino, so the calculated frequency of heterozygous symptomless carriers is 280 in 20,000 (1 in 71), and of homozygous normal individuals is 19,719 in 20,000.

When adults with two copies of the albino locus produce eggs or sperm, these **carry one copy of the gene**, as shown in these crossing diagrams of all possible combinations of parents, where 'x' means 'mated with'. Females are shown first.

Albino (*aa*) x albino (*aa*) gives all albino children (*aa*):		
aa woman	x	*aa* man
Ⓐ Ⓐ	x	[sperm with *a* and *a*]

All eggs are *a* and all sperm are *a*, so all offspring are *aa*, albinos, from fusion of two sex cells carrying *a*.

Normal (*AA*) x normal (*AA*) gives all normal children (*AA*):		
AA woman	x	*AA* man
Ⓐ Ⓐ	x	[sperm with *A* and *A*]

All eggs are *A* and all sperm are *A*, so all offspring are *AA*, normal, from fusion of two sex cells carrying *A*.

Normal (*AA*) x albino (*aa*) gives all heterozygous carriers, normal children (*Aa*):		
AA woman	x	*aa* man
Ⓐ Ⓐ	x	[sperm with *a* and *a*]

All eggs are *A* and all sperm are *a*, so all offspring are *Aa*, normal in appearance, from fusion of one sex cell carrying *A* and another carrying *a*.

Normal (*AA*) x heterozygous carrier (*Aa*) gives a ratio of one normal child (*AA*), to one heterozygous carrier, normal child (*Aa*):		
AA woman	x	*Aa* man
Ⓐ Ⓐ	x	[sperm with *A* and *a*]

All eggs are *A*, while half the sperm are *A* and the other half *a*. All offspring look normal but half carry the unexpressed recessive albino allele.

Heterozygous carrier (*Aa*) x heterozygous carrier (*Aa*) gives an expected ratio of 1 (*AA*) normal: 2 (*Aa*) heterozygous carriers and 1 (*aa*) albino:		
Aa woman	x	*Aa* man
	x	

The ratio is 1*A*: 1*a* in eggs and sperm. The expected ratio of **observed** types is ¾ normal to ¼ albino. Plate 15E shows two carrier parents with affected boy albino twins and two unaffected children. **By chance, observed ratios may differ somewhat from expected ones** .

Heterozygous carrier (*Aa*) x albino (*aa*) gives an expected ratio of 1 (*Aa*) heterozygous carrier : 1 (*aa*) albino:		
Aa woman	x	*aa* man
	x	

The heterozygote gives a ratio of 1 *A*: 1*a* in the egg, while the albino's sperm are all *a*, so the expected ratio in children is 1*Aa* unaffected heterozygous carrier: 1 albino, *aa*.

We can make generalisations about the inheritance of autosomal recessives. Examples include albinism, sickle-cell anaemia and cystic fibrosis.

- Females and males are affected equally often.
- Affected individuals must be homozygous because the characteristic is recessive.
- If an affected individual marries a heterozygous carrier, about half the children will be sufferers and about half will be symptomless carriers.
- If an affected individual marries a normal homozygote, all children will be carriers.
- If a pair of outwardly normal individuals has an affected child, they must both be heterozygous carriers, and any further child has a chance of one quarter of being normal, a chance of a half of being a carrier, and a chance of one quarter of being a sufferer (Plate 15E).

We can make generalisations about the inheritance of autosomal dominants. This could be for the normal gene for skin colour (non-albino), or for a disorder specified by a dominant gene, such as the ordinary form of **polydactyly**, where sufferers have extra fingers and/or extra toes.

- Males and females should be affected equally often.
- If the dominant characteristic is very common, as for non-albino, most individuals in a population will be homozygous dominant and will have children with the dominant manifestation whomever they marry.
- If homozygous dominants marry a heterozygote, with the dominant manifestation, one expects all children to show the dominant manifestation, but half to be homozygous and half to be heterozygous.
- For a rare dominant characteristic such as polydactyly, most sufferers are heterozygous and usually marry normal recessive homozygotes, so about half the children will be sufferers and half will be normal recessive homozygotes.
- If the sufferer is a dominant homozygote, all their children will be affected, whomever they marry.

Variable expressivity is where a characteristic is expressed to different extents in different individuals, as in polydactyly having small effects such as a little lump (Chapter 18, Photo 18.2) or a major effect such as two extra toes. The term is about *how much* a characteristic is expressed when it is expressed. Some genetic characteristics are sometimes expressed fully and sometimes not expressed at all, even when the appropriate genes are present: they show **incomplete penetrance**. A dagger either penetrates the skin or it doesn't; similarly a gene may either penetrate into being manifested or not. A gene which always shows when the appropriate genes are present has 100% penetrance, but penetrance can range from low to high. There is a form of inherited cancer of the colon caused by a dominant allele with incomplete penetrance, so some people with that allele get the cancer and some do not. Incomplete penetrance is the main reason why a dominant gene sometimes appears to 'skip a generation', with an individual passing the gene on although not showing its effect. Recessive genes can easily 'skip a generation' through being hidden in heterozygotes.

8.3. How a sex-linked characteristic such as red-green colour blindness is inherited

Red-green colour blindness (Chapter 15) is a common condition which affects males much more often than females. The allele for colour blindness is recessive to that for normal colour vision and is carried on the X chromosome in the part missing from the Y. About 1 in 12 men is affected and 1 in 167 women, with 1 in 7 women being **unaffected heterozygous carriers**.

Women can be **normal**, with two normal Xs, $X^{Cb}X^{Cb}$, or **heterozygous carriers** $X^{Cb}X^{cb}$, or colour blind, $X^{cb}X^{cb}$. Men have only two types, normal, $X^{Cb}Y$, or colour blind, $X^{cb}Y$. We have to specify each individual's sex.

Colour blind woman ($X^{cb}X^{cb}$) **x colour blind man** ($X^{cb}Y$). All children will be colour blind. All her eggs carry X^{cb} and his sperm are in a ratio of 1 X^{cb}: 1 Y, giving a ratio of 1 colour blind female, $X^{cb}X^{cb}$, to 1 colour blind male, $X^{cb}Y$.

Normal woman ($X^{Cb}X^{Cb}$) **x normal man** ($X^{Cb}Y$). All children will be normal as the eggs all carry X^{Cb} and sperm carry X^{Cb} or Y.

Colour blind woman ($X^{cb}X^{cb}$) **x normal man** ($X^{Cb}Y$). All eggs carry X^{cb} and the sperm carry X^{Cb} or Y. All daughters will have normal colour vision but will be heterozygous carriers, $X^{Cb}X^{cb}$. All sons will be colour blind, $X^{cb}Y$.

Normal woman ($X^{Cb}X^{Cb}$) **x colour blind man** ($X^{cb}Y$). All eggs carry X^{Cb} and the sperm carry X^{cb} or Y in a 1:1 ratio. All daughters will be normal but will be heterozygous carriers, $X^{Cb}X^{cb}$. All sons will be normal, $X^{Cb}Y$.

Carrier woman ($X^{Cb}X^{cb}$) **x normal man** ($X^{Cb}Y$). The eggs will be in a 1:1 ratio for X^{Cb} and X^{cb}, and the sperm carry X^{Cb} or Y in a 1:1 ratio, so the offspring will be in ratio 1 normal female ($X^{Cb}X^{Cb}$), 1 carrier female with normal vision ($X^{Cb}X^{cb}$), 1 normal male ($X^{Cb}Y$) and 1 colour blind male ($X^{cb}Y$).

Carrier woman ($X^{Cb}X^{cb}$) **x colour blind man** ($X^{cb}Y$). The eggs will be in a 1:1 ratio for X^{Cb} and X^{cb}, and the sperm carry X^{cb} or Y in a 1:1 ratio, so the offspring will be in ratio 1 carrier female with normal vision ($X^{Cb}X^{cb}$), 1 colour blind female ($X^{cb}X^{cb}$), 1 normal male ($X^{Cb}Y$) and 1 colour blind male ($X^{cb}Y$).

We can make **generalisations about the inheritance of any sex-linked recessive**, such as red-green colour blindness or haemophilia.

- Males are affected more often than females; for very rare traits, nearly all sufferers are male. The rarer the condition, the more the imbalance between the sexes.
- Affected males usually have normal sons, as sons receive the father's Y, not his defective X.
- All daughters of affected fathers and normal mothers are heterozygous carriers, and one expects half their sons to be affected and half unaffected.
- Sex-linked recessives cannot be passed from fathers to sons, because sons get their father's Y, not his X.
- Affected females have an affected father and a mother who is affected or a carrier.

We can make generalisations about the inheritance of any **sex-linked dominant**. These include normal traits such as non-colour-blind and non-haemophiliac. There are few known **sex-linked dominant disadvantageous traits**. They include inherited vitamin D-resistant rickets. For a sex-linked dominant harmful characteristic:

- One expects more female sufferers (X^DX^D and X^DX^d) than male sufferers ($X^D Y$).
- For affected males marrying normal women, all daughters are affected but none of the sons.
- Affected women can have affected sons and affected daughters.
- If affected women are homozygous, all children will be affected, but if they are heterozygous, half their sons and half their daughters will be affected, on average.

8.4. How characteristics with continuous variation, such as normal height, are inherited

Characteristics with **continuous variation** include height, weight, intelligence, blood pressure and longevity. Many are influenced by the environment, making predicting the children's characteristics from their parents' characteristics difficult.

For characteristics showing continuous variation, there are often **many loci involved**. If each locus has two alleles, B and b at one locus, C

and *c* at the next, etc., with three genetic possibilities per locus, e.g., *BB*, *Bb* and *bb*, then there are many possible combinations. For example, if 10 loci control a characteristic, with two alleles at each locus, there are $3 \times 3 \times 3 \times 3 \times 3 \times 3 \times 3 \times 3 \times 3 \times 3 = 59,049$ genetic combinations. If there is complete dominance at each locus, with two manifestations each, dominant and recessive, there are 1,024 combinations; with incomplete dominance there are 59,049 combinations. Characteristics determined by several to many loci, and perhaps with environmental effects, are called **multifactorial**.

Multifactorial disorders include club foot, cleft lip, coronary artery disease, diabetes and schizophrenia. Multifactorial disorders are much more common than are single gene disorders. Compare in Britain (population about 60 million) the approximate numbers for multifactorial traits: Alzheimer's, 400,000; diabetes, 2.5 million; heart disease, 1.25 million; cancer, 1.5 million, and high blood pressure, 5 million, with those for single-gene disorders: Huntington disease, 2,500; Duchenne muscular dystrophy, 3,000, and cystic fibrosis, 7,000. Multifactorial conditions are extremely important medically and in human diversity.

Their inheritance is complicated because a single genetic type may have many possible manifestations from environmental effects, and any one manifestation may have many different possible genetic types. A heavy man may be heavy because his genes predispose him to heaviness, or because he eats huge meals, or from a combination of factors. If he is heavy largely because of his genes, then some of those genes for heaviness will probably be passed to his children, although there will also be the influence of his wife's genes, and the genes may be dominant, recessive or incompletely dominant. If he is heavy largely because he eats too much, then he will not tend to pass on genes for heaviness. If his example of overeating leads his children to overeat, they may be heavy from the parental environmental influence.

Take **height in non-dwarfs** as an example of a **multifactorial continuous characteristic**. Let us initially ignore environmental effects and that males tend to be taller than females. The controls proposed are hypothetical. Height means adult height.

If we have incomplete dominance, we use a capital letter for the locus, with a **prime symbol (′)** for the allele **increasing** the effect and one **without the prime** for the allele **not increasing** it. Suppose that height is controlled by four loci, each with **incomplete dominance**, each locus having an **equal effect**, with a base height of 160 cm (5 ft 2 in) for no primes, and a maximum height of 200 cm (6 ft 7 in), with each primed allele adding **5 cm** (2 in). The shortest people would be *A A, B B, C C, D D*, and the tallest would be *A´ A´, B´ B´, C´ C´, D´ D´*, where having eight primes, two at each locus, gives them an extra 8×5 cm of height $= 40$ cm, or 160 (base height) $+ 40 = $ **200 cm**.

The most extreme types are homozygous at each of the four loci, and we can predict the outcomes of various matings. With a very short man (160 cm) mated to a very short woman (160 cm), both parents and children would be *A A, B B, C C, D D*, so all children should be very short, 160 cm, like their parents. Similarly, with a very tall man (200 cm) mated to a very tall woman (200 cm), both parents and children would be *A´ A´, B´ B´, C´ C´, D´ D´*, giving only very tall children, 200 cm.

For a mating between a very tall man, 200 cm, *A´ A´, B´ B´, C´ C´, D´ D´*, and a very short woman, 160 cm, *A A, B B, C C, D D*, all children will be heterozygous at each locus, *A´ A, B´ B, C´ C, D´ D*, and **should be of the same height**, the base height of 160 cm plus the effect of four primes, each giving 5 cm, so **they will be of intermediate height between their parents**, at $160 + 20 = $ **180 cm**.

When one has **extreme parents**, like very tall × very short, there is a tendency for the children to be less extreme, to be nearer the population average; this tendency is called **regression**. If one has dominance or effects of the environment, regression often occurs when two individuals of similar manifestation mate, such as tall × tall (giving offspring who are on average shorter than the parents, though they may still be fairly tall) or short × short (giving offspring who are on average less short than their parents, although they may still be fairly short) if the parents are heterozygous.

If the parents were tall because of an exceptionally favourable environment, the offspring may well have a more average environment and so be less extreme than their parents, showing regression. Similarly, if the two parents were short because of a very unfavourable environment, the offspring may well have a more average, less unfavourable, environment and be a bit taller than their parents, again showing regression. Regression is always towards a more average, less extreme type.

Predictions become harder for other genetic types because there will be segregation at egg and sperm formation in heterozygotes, and recombination between the loci. One usually does not know a person's genetic type, only the manifested characteristic. Thus a man who is 180 cm tall, with four primed genes, could have a number of possible genetic types: homozygous at each locus, such as *A′ A′, B′ B′, C C, D D*, or *A A, B B, C′ C′, D′ D′*, or *A A, B′ B′, C′ C′, D D*, or heterozygous at all loci, *A′ A, B′ B, C′ C, D′ D*, or homozygous at some loci and heterozygous at others, as in *A A, B′ B′, C′ C, D′ D*. All these genetic types have the same medium height manifestation, **180 cm**, because all have **four primed alleles.**

Some matings, such as those between intermediate height men and women, 180 cm, who are homozygous at all loci and who mate with individuals of the same genetic type for height, will have children identical to each other and the same as their parents, with no regression, e.g., *A′ A′, B′ B′, C C, D D × A′ A′, B′ B′, C C, D D*, or *A A, B′ B′, C′ C′, D D × A A, B′ B′, C′ C′, D D*.

Individuals heterozygous at all four loci, *A′ A, B′ B, C′ C, D′ D*, 180 cm, intermediate height, can have eggs or sperm with no, one, two, three or four primed alleles, depending on how the genes segregate at meiosis, so if two such individuals mate, their children can have no, one, two, three, four, five, six, seven or eight primed genes, giving heights of 160, 165, 170, 175, 180, 185, 190, 195 or 200 cm, respectively, with some children (with four primed genes) resembling the parents but most are taller (with five to eight primes) or shorter (with none to three primes) than their parents, with large differences between children of the same parents. Those children who are taller or shorter than their parents show **transgression.** In genetics, transgression means exceeding the limits of both parents, e.g., being taller or shorter than either parent; it does not carry the implication of sinning! In the skin colour limerick in Chapter 13.5, white × black giving khaki would show regression, while khaki × khaki giving some white and black would show some transgression.

By considering the genetic types of eggs and sperm from the parents in any mating, we can work out the children's range of expected adult heights, and the frequency of children of any given height. In practice, we would have to allow for the effects of environment and the fact that males are on average taller than females.

This example shows how one can get distributions of nine different heights (160, 165, 170, 175, 180, 185, 190, 195, 200 cm) from four loci, with two alleles each, with incomplete dominance and equal effects of each locus, if there is no effect of the environment. This is not a continuous distribution but **nine discontinuous values**, 5 cm apart.

There are two factors which can change a **discontinuous set of manifestations** into a **continuous set**. The first is **effects of the environment**. People whose genetic types would lead you to expect a certain manifestation, say eight primes giving 200 cm adult height, might eat more or less well than average, with additional effects of exercise, disease, climate, etc., changing their height upwards or downwards by varying amounts, perhaps from 0.01 cm to several centimetres. Each genetic type could therefore give a whole range of continuous manifestations, giving a truly continuous manifestation distribution, although there is an underlying discontinuity of genetic types.

The second factor giving continuous manifestations from discontinuous genetic types is the existence of many segregating **polygenes** for a characteristic, each with very small effects. While alleles at the A, B, C and D loci in the example above gave differences in height of 5 cm each, and can be called **genes of major effect**, the effect of one polygene might only be 0.01 cm or less, making it very hard to distinguish their effects from minor environmental effects.

A man with genetic type $A'\ A'$, $B'\ B'$, $C'\ C'$, $D'\ D'$ and an average environment might thus have a height of say 195, 195.01, 195.02, 195.03 etc., to 204.98, 204.99 or 205 cm, if there were many polygenes segregating, with a maximum effect of minus 5 cm to plus 5 cm. At the maximum value, 205 cm, all the polygenic loci have their height-increasing allele present, and at the low extreme, 195 cm, all polygenic loci have their height-decreasing allele present. Intermediate values come from mixtures of height-increasing and height-decreasing alleles at the polygene loci.

In summary, multifactorial characteristics are determined by a number of loci, perhaps including several genes of major effect and many polygenes with minor individual effects, and by the environment. **Continuous distributions** have underlying **discontinuous genetic types**, but with blurring effects of environment and polygenes.

8.5. Threshold characteristics

There are some characteristics which one either does not have or one has, like cleft lip or club foot, being born with them or without them, or characteristics that may or may not develop with time, such as schizophrenia or diabetes. They are **discontinuous characteristics**, although people with them get them to different extents. Cleft lip and club foot are multifactorial, so **multifactorial characteristics can be discontinuous.** In ones developing with time, such as schizophrenia, there may be environmental triggers needed for expression as well as underlying genetic predispositions.

A theory to explain such conditions is one of a distribution of the genetic risk among the population, with only individuals above a certain threshold of risk being able to get the disorder. Even then, those at or above that threshold have a probability, not a certainty, of showing the disorder.

Here is a theoretical example. Suppose that there are **polygenes predisposing one to cleft lip**, with four being the average number of such predisposing polygenes per person, and that 10 such polygenes is the **threshold** above which people are likely to get cleft lip, and that the more they are over the threshold of 10 predisposing polygenes, the more likely they are to get cleft lip on one side of the mouth. It might take an average of 15 such predisposing polygenes to cause cleft lip on **both sides** of the mouth, and perhaps an average of 20 such polygenes to cause a **cleft palate** as well as cleft lip.

The more severely affected an individual is, the more likely he or she is to carry a high number of predisposing polygenes, and hence to have affected children and affected brothers or sisters. Switching from theory to practice, if a child has cleft lip on one side of the mouth, the chance of a brother or sister being affected by cleft lip is 1 in 40, but if a child has the more serious condition of cleft palate as well as cleft lip, indicating more predisposing polygenes, the chance of a brother or sister being affected rises to 1 in 17.

Congenital pyloric stenosis is a narrowing of the pylorus, the opening valve from the stomach into the small intestine. It results in 'projectile vomiting' from birth but babies grow out of this habit. It occurs in 1 male in 200 and 1 female in 1,000. The fact that it is five times more frequent in boys than girls suggests that the threshold for manifestation is higher in girls than in boys, so girls need more predisposing polygenes than do boys to develop it. The relatives of affected girls are therefore likely to carry more predisposing polygenes than are relatives of affected boys, and in practice the children of affected females are three times more likely to be affected than are children of affected males.

8.6. The human genome

The draft human genome was published in *Nature* 409 in February 2001. The researchers found 30,000 to 40,000 protein-coding genes, only twice as many as in the fly or worm, although our genes are more complex. Later work gave lower estimates, nearer to 24,000 genes. Hundreds of our genes have come from bacteria by horizontal transfer. More than 1.4 million single base pair differences were identified and this number has vastly increased. Many DNA sequences have been largely conserved in evolution, even over millions of years, and occur in very different types of organism, from microorganisms to primates. We share about 60% of our gene sequences with the kiwi fruit!

One megabase (Mb) is one million DNA base pairs. The smallest human chromosome, number 21, is 45 Mb long, compared with 48 Mb for number 22, 51 Mb for Y, 163 Mb for X, and 279 Mb for the largest, number 1 (Plate 1A). The total genome length was estimated at 3,289 Mb, and we have two genomes per cell, or about 6,578 million base pairs in the nucleus, with substantially fewer in DNA outside the nucleus, mainly in the mitochondria.

Less than 2% of our sequence is coding DNA, with repeated sequences making more than 50%. Much of our DNA is of unknown function. A report by Rands *et al.,* (2014) stated that about 8.2% of human DNA was likely to be functional for **coding** or **gene switching**. The average heterozygosity rate in humans is about 1 in 1,300 base pairs. The average nucleotide diversity in gene-containing regions has been estimated as 8 differences in 10,000 base pairs, and only half (4.7) that in the X chromosome and even less (1.5) in the Y. In any two individuals, about 99.9% of their DNA sequences are identical, with all their genetic differences between them coming from that **0.1% difference**!

References

Lander, E. S. *et al.*, Initial sequencing and analysis of the human genome. *Nature* (2001) **409**: 860–921.

Rand, C. M. *et al.*, 8.2% of the human genome is constrained: variation in rates of turnover across functional element classes in the human lineage. *PLoS Genet* (2014) **10**: e1004525.

Chapter 9

The Brain, Intelligence, Mind, Personality, Mental Problems, Learning, Memory, Creativity, Happiness

9.1. The brain and nerves

There is a huge diversity in intelligence, personality, behaviour and creativity, and brain afflictions such as schizophrenia, autism and dementia. See Chapter 1.3 for the heritability of some mental conditions. The brain is roughly 8 inches long and wide and 6 inches high (20, 15 cm), weighing 46 to 49 oz (1,300 to 1,400 g) in males, and 42 to 46 oz (1,200 to 1,300 g) in females. It is about 1.7% of one's weight but consumes 13 to 20% of the blood supply. We have about a hundred thousand million nerve cells. The number of synapses — the connections between nerve cells — is about a thousand million million. Humans form about one million new nerve connections a second.

The spinal cord is about 18 inches (45 cm) long and 1.5 inches (3.8 cm) wide at its maximum. From it come 31 pairs of spinal nerves. Peripheral nerves become extremely thin as they spread out into smaller bundles of fibres. Their total length is about 93,000 miles (150,000 km). Although most signals to and from the brain go through the spinal cord, messages from the head's sense organs, and to head and face muscles, do not. Instead, they go through 12 pairs of cranial nerves which pass through holes in the skull, like the optic nerve through the eye socket.

Neck injury can result in paralysis. A young cousin dived into the sea, struck his head on a rock, and has had to use a wheelchair for 50 years since. That has not stopped him from being an excellent cook, having a job, a long-term girlfriend, and driving his adapted car at speed. People react very differently to adversity, some managing splendidly, some feeling useless and miserable.

After age 20, about 10,000 to 100,000 brain cells are lost each day and not replaced, but it is a small fraction. The annual brain loss is about 1/30 to 1/10th oz (1 to 3 g) between ages 20 and 60, and 1/10 to 1/7th oz (3 to 4 g) for the over 60s, about 0.15% and 0.27% respectively.

The right side of the brain controls the left of the body, and vice versa. The left temporal lobe is associated with piecing sounds into words, and the right lobe with processing melody and intonation. Brain scans showed left temporal activity when English speakers listened to English but activity in both lobes when Chinese speakers listened to Mandarin, a tonal language.

The two large, grooved cerebral hemispheres are topped by a folded sheet of **grey matter** only 1/8 to 1/5th inch (3 to 5 mm) thick, the seat of consciousness and thought. It contains billions of nerve cell bodies which have many connections to other nerve cells. Below the grey matter is the much thicker **white matter**, mainly nerve fibres to and from nerve cell bodies in the grey matter. The brain is an important source of hormones which act on the body via the blood stream, e.g., from the pituitary gland.

Motor nerves run from the brain to muscles while sensory nerves come to the brain from sense organs. Nerves in the stomach wall tell the brain how full it is and allow the brain to control when it empties. Nerve fibres carry signals in only one direction. In the brain, a single nerve cell may have 50,000 branches (dendrites) and be in communication with a quarter of a million other nerve cells, making for tremendous complexity.

Consider a sprinter's reactions. When the starting pistol is fired, the sound takes about 0.03 seconds to reach the sprinter's ears. In 0.01 seconds, sensory cells in the ears have sent a message to the brain and in another 0.01 seconds the brain has processed the signal and identified its importance and purpose. The brain sends a signal down the spinal cord and out through the peripheral nervous system to the legs, arms, chest, etc. The muscles start propelling a sprinter from the starting blocks about 0.1 second from the pistol firing.

Nerve impulses in large insulated fibres travel at up to 400 feet (122 m) per second, dropping to 2.5 feet (0.7 m) per second in very small uninsulated fibres. Their maximum voltage is about 100 millivolts and the number of impulses varies from about 300 per second in large insulated fibres to 50 per second in small ones. Some nerve cells are more than 3 feet (91 cm) long. The main stem of a nerve cell, its long axon, has an insulating fatty myelin sheath around it, damage to which causes plaques which may be associated with multiple sclerosis. See Chapter 15.3 for head and spinal cord defects.

If a brain region is used more, it functions better. A violinist's left hand uses the thumb just for support, but the other fingers are used intensively. The brain regions for those fingers are much more active than that for the thumb. Damage to one part of the brain may result in compensatory improvements in others. Nadia had extensive brain damage and could barely talk, yet she drew vivid pictures of horses as the relevant part of her brain was intact.

9.2. Intelligence

Some people are more intelligent than others. Definitions of intelligence include: intellectual skill or knowledge; mental brightness; quickness of understanding; wisdom; having a lively mind; the ability to make connections between things and adapt to circumstances. Intelligence differs from having knowledge, although involving the ability to use it. Intelligent individuals may have poor memories.

Some people postulate different types of intelligence, such as general, abstract, concrete, social, practical, emotional, or topic-based intelligences such as mathematical, musical or linguistic. Emotional intelligence is the ability to understand and empathise with others, to react appropriately, being sensitive to one's own and other people's emotions and feelings.

A list of European geniuses might include Aristotle, Mozart, Handel, Bach, Leonardo da Vinci, Newton, Shakespeare, Goethe, Molière, Gregor Mendel, Robert Adam, Brunel, Einstein and cartoonist Matt Pritchett. What links them is originality, creativity and an ability to handle complex ideas. At the California Institute of Technology, I had lunch each weekday with Nobel Prize winners. They were very bright indeed.

At the other extreme are people with serious mental defects. In a Californian institute for severely handicapped children, most could not feed themselves, keep themselves clean, play or interact with others. Some lay or sat inertly; others hit their bandaged heads against the wall to get some sensation. There is much more human diversity than one generally encounters because some is hidden away in institutes and hospitals.

Causes of low intelligence include problems before or around birth, such as maternal malnourishment, oxygen starvation, chromosomal disorders such as Down syndrome (Chapter 23.3) and genetic defects such as phenylketonuria (Chapter 12.5) in which most untreated babies have IQs below 20.

Intelligence can measured as an **Intelligence Quotient (IQ)**. Although such tests have been criticised as inaccurate or culturally biased, they were devised so that people who were regarded as very intelligent scored highly, and dim people had low scores. They are useful for rough comparisons. **Particular IQ ranges** are: 0–20, profoundly disabled; 21–50, severely disabled; 51–70, moderate disability; 71–80, borderline disability; 81–90, low average; 91–110, average; 111–120, high average; 121–130, superior; 130–150, very superior; over 150, near-genius to genius. IQ scores give a symmetrical bell-shaped normal distribution curve with a peak and average of 100.

Geneticists have standard ways of determining what proportion of variation for a character in a population is due to genetic variation and what proportion is due to environmental variation, that is, the **heritability of a character**, by studying relatives brought up together or separately, such as identical and non-identical twins, brothers, sisters, cousins. The calculation is objective, not subjective.

60 to 80% of variation in intelligence is due to genetic differences, and 40 to 20% to environmental differences, such as nutrition, school, parental and cultural influences, as well as climate. Heritabilities were found to be about 63% for verbal ability and 76% for numerical ability. There are thousands of genes of small individual effect involved. A child's intelligence potential is genetically determined, but how far that potential is achieved depends on environment, especially education and effort.

Having a higher degree is evidence of high intelligence. Someone with no qualifications could be as intelligent or of low intelligence. One cannot assume that people in lowly jobs are of low intelligence. Students often take mundane jobs to pay their way through university; one brilliant Sri Lankan PhD student worked part-time as a supermarket cashier and was more intelligent than almost all the customers she served. Intelligence is not always combined with common sense, with university students getting up to stupid pranks and some Nobel Prize winners making fatuous remarks about topics outside their subject, especially religion, society and politics.

In most British schools there is a pernicious attitude that to be intelligent or to work hard is bad. Children hate being called swots. Marathon runner Paula Radcliffe was very bright at school but never put her hand up when she knew the answers as she feared being picked on by classmates. She wasn't worried about excelling at sport as that didn't get anyone picked on.

Some abilities 'run in families', such as musical talents in the very large Bach family over several generations. There was a strong genetic predisposition, reinforced by growing up in a musical environment, performing in family groups, being taught from an early age by parents and relatives, and having musical instruments and music to hand.

Although men's brains are on average 7% larger than women's, the two sexes have roughly equal IQs. At school, girls usually out-perform boys. There are sex and race differences in achievement at different ages and in different subjects. According to the Department of Education and Skills in 2007, girls made up 60% of England's brightest pupils at 16. Girls at 11 occupied 65% of the top places in English, but boys were better at maths. In English state schools, the top 10% of GCSE students included only 10% of white British pupils, 31% Chinese and 18% Indians, way out of their proportions in the population.

9.3. Stress

Some stress is normal and can be stimulating. Overcoming challenges is satisfying. People differ greatly in what they find stressful, so that an audition for a part in a play could be highly stressful, boring or enjoyable, depending on personality, attitudes and abilities. Personalities range from easy-going to nervous stressed types. According to the Samaritans, 20% of

Britons suffer undue stress every day, especially from concerns over jobs and money, leaving them feeling depressed and isolated. High stress levels are associated with increased risks of heart attacks and strokes. Stressed individuals may turn to drugs, 'comfort eating', excess alcohol or cigarettes in efforts to relax.

Stress, depression and anxiety often lower the concentration of blood antibodies and may reduce the activity of infection-fighting lymphocytes. Prolonged stress can increase adrenaline levels, increase heart action (possibly with palpitations or raised blood pressure), and increase secretion of stomach acids, increasing the chance of ulcers.

Many sufferers from stress and anxiety are prescribed tranquillisers such as Valium. Most come off them after a few weeks and never need them again. Others become dependent on and addicted to tranquillisers, antidepressants or sleeping tablets, especially benzodiazepams such as Valium. They were widely prescribed in the 1960s and 1970s, with Americans taking about two billion Valium tablets in 1978, an average of more than 90 each.

Feeling different is a cause of stress. A headmistress told me about a difficult nine-year-old boy who threw frequent tantrums. When asked about it, he explained that everyone else in his class had two daddies and he had only one: he wanted to be like everyone else! His parents had not separated, unlike those of most other children, who were spoiled by their father and by their mother's new partner.

Bereavement is a common cause of stress and strong emotions. The Yale Bereavement Study found that the typical symptoms of disbelief, yearning, anger and depression peaked by six months after the death, except in cases of prolonged grief disorder, from which Queen Victoria suffered. Anger against God or fate is common.

9.4. Mind-altering drugs

Mind-altering drugs can be **medical**, such as tranquillisers, or harmful **'recreational'**, such as heroin and ecstasy. Prozac is an antidepressant. The tranquilliser Diazepam is used for anxiety and Temazepam for insomnia; both increase the activity of a compound (GABA) inhibiting activity of the brain's emotion-handling areas. Manic-depression mood swings were often treated with lithium compounds, but less so now as anti-convulsants are usually prescribed as mood stabilisers.

Writers who used drugs to enhance creativity include Aldous Huxley and Samuel Taylor Coleridge, who wrote *Kubla Khan* under the influence of opium. Hilary Mantel found inspiration for a book while on morphine for endometriosis. A survey in Britain found that half of young prisoners suffered from drug-induced memory loss, with amphetamine misuse before entering prison a frequent cause.

9.5. Personality; self-esteem

Whether you like or dislike a person is a function of their personality and yours. Factors in personality include genetic predisposition, education, culture, upbringing, environments at formative times, experience, choice and chance. Events can change someone's personality temporarily or in the long term, such as betrayal in love or business making a person less trusting, more suspicious and more introverted. It is said that long dark winters make Scandinavians more gloomy and that bright sunshine makes West Indians more sunny in personality.

Personalities are sometimes described in terms of different dimensions, such as **extraversion/introversion**. Extroverts tend to be outward looking, sociable, cheerful, talkative, sure of themselves, while introverts tend to be introspective, withdrawn, quiet, unsure of themselves. A question to allocate people between these categories is "Would you rather spend an evening with friends in a pub or at home with a book?" For neuroticism, the emotionally stable are carefree, relaxed and adaptable while the neurotic are anxious, tense, emotionally unstable and easily upset. In **openness/closed mindedness**, some like new experiences, new people, new feelings, new challenges, while others prefer the old and familiar, the simple and unchallenging. In **conscientiousness/irresponsibility**, one has orderly, responsible, reliable, dependable, hard-working types as opposed to the feckless, impulsive, unpredictable and aimless. In **agreeableness/antagonism** one has those who are co-operative, modest, easy to get on with, and those who are uncooperative, quarrelsome and anti-social.

Aggression is unpleasant but excessive passivity can make societies prone to being conquered. There is diversity in aggression between races, cultures and the sexes. In Britain, men are 24 times more likely to kill or assault someone, and 263 times more likely to commit a sexual offence, than are women. Only 3% of same-sex murders in Britain, America and Canada are committed by women. The Yanomamö tribe in the Amazon Basin are renowned for aggression; young men undergo a ritual purification which involves killing. A third of their young men die violently.

Others do not see us as we see ourselves. I consider myself to be mild, friendly, helpful, approachable, agreeable, reasonably modest and well-balanced, but one colleague described me as aggressive, arrogant and paranoid, while my Head of Department called me intransigent!

People engaged in a range of activities usually find life interesting, making plenty of friends, especially from group activities such as clubs and societies. A sense of comradeship, of shared dangers or unsocial hours, leads some people mainly to socialise with colleagues, as with the police and soldiers.

People show different emotionality. Some like showing and talking about their feelings — 'wearing their hearts on their sleeve' — while others hide their feelings. Women are generally more outwardly emotional than men. British men are expected to refrain from crying at funerals, even if deeply moved, and in many societies women do the wailing and weeping at funerals.

We convey emotions through facial expressions, body language and speech. Similar facial expressions are used in most societies to express emotions such as anger, disgust, fear, happiness, irritation, sadness and surprise. Making friends and finding people to love normally involve a prolonged, complicated set of exchanges of expressions, as well as of words.

Self-esteem involves feeling accepted, valued, useful, that one has some talents and friends. **Low self-esteem** is common in children who are bullied or feel like failures, socially, academically or in sports. It is a feature of depression and many mental disorders, tending to make people sad, unadventurous, unsociable, unmotivated. They tend to project their self-rejection onto others, feeling unwelcome in company, becoming preoccupied with negative, unhappy thoughts and ignoring positive events. Low self-esteem can be part of a vicious circle, leading to unfulfilling lives and depression.

Unrealistic parental expectations often lead to low self-esteem, anxiety and depression. In 2013 NICE reported a steep rise in the number of children suffering from severe depression, 80,000 in Britain, including many below the age of ten and some below five. Exam stress is increasing. **Excessive self-esteem**, vanity and feeling one is absolutely wonderful, are features of the manic phase of manic-depressiveness.

9.6. Mental disorders; Alzheimer's; depression; manic-depressiveness; autism; schizophrenia; dyspraxia; dyslexia; recognition disorders; self-harming; epilepsy; obsessive-compulsive disorder

Mental problems vary from mildly inconvenient and temporary to permanent and life-quality-destroying. Some are treatable by drugs or psychotherapy and some clear up spontaneously. Some develop after traumatic events; some are inherited but may be expressed at different times.

Triggering factors include marital break-up or other relationship problems, bereavement, losing one's job, moving house, overwork, trying to cope with too many problems, financial losses and suffering humiliation. Physical accidents and illnesses can precipitate mental illness, for example, severe depression and loss of will to live on losing the use of one's legs, becoming paralysed or blind.

In 2014, Public Health England announced that children who spent too much time on the internet, social media and computers were developing mental health problems, including loneliness, depression, anxiety, low self-esteem and heightened aggression. One in ten children now has a mental health problem; about 750,000 teenagers in England think they have nothing to live for.

The approximate frequencies of mental problems in Britain are: acute anxiety, 1 in 7; depression, 1 in 8; eating disorders such as anorexia, bulimia and compulsive overeating, 1 in 33; obsessive-compulsive disorder, 1 in 85; phobias, 1 in 85; schizophrenia, 1 in 100. Dementia affects 1 in 90 generally, 3% of people aged 65 to 74 and 47% of the over 85s. At least one person in four has a mental disorder at some stage, more if one includes alcohol, drug or nicotine addiction.

Although serious mental disorders need medical treatment, many respond to simple remedies such as increased exercise, healthier eating, better sleep and relaxation techniques. Hypnosis (for phobias) and faith-healing sometimes work. Helping those who have worse afflictions is therapeutic as sufferers become less obsessed with their own problems. Laughter is a traditional medicine for mental problems, and having pets can be beneficial. Supportive friends, family and colleagues can make a big difference to recovery rates in mental disorders.

Frontal lobotomy, removing or disconnecting part of the brain's frontal lobes, was a frequent treatment for mental illnesses such as severe depression and schizophrenia until the 1950s; it is rarely used today. Complementary therapies of aromatherapy, reflexology, acupuncture and Reiki have their advocates for treating mental disorders, especially if not too severe.

A report in 2003 by the Royal College of Psychiatrists found that 10% of women at one UK university had eating disorders and found excessive drinking by 50% of medical students. About 3% of university students used the counselling services each year, with two thirds being women.

Although most people welcome retirement, with more free time and none of the aggravation of commuting, deadlines, fractious bosses or colleagues, it often triggers depression and loss of focus. There is a lack of the organised day, companionship of fellow workers, loss of status and self-esteem. There is more time to spend, with less money. The retired person usually adapts, often claiming to be busier than ever, but retirement can lead to prolonged depression.

Addictions can be physical, e.g., to nicotine, heroin, cocaine, cannabis or amphetamines, where withdrawal gives symptoms such as sickness, sweating and trembling. Estimated deaths in Britain each year from substance abuse are: nicotine, 120,000; alcohol, 35,000, and illegal drugs at least 2,000 but probably much higher. Addictions can be psychological, as with stealing, gambling, shopping, work or sex. Addictions can ruin the lives of sufferers and their families. Alcoholism has a genetic element and an environmental one from growing up in circumstances where heavy drinking is normal.

Dementia refers to any form of insanity characterised by loss of mental powers, with deterioration of intelligence, memory, behaviour and orientation. The commonest form is Alzheimer's disease. For people over 65, 10% suffer from Alzheimer's; for the over-65s, the incidence doubles every five years, making it very common amongst those in their 80s or 90s. Dementia can come from lack of blood to the brain (clots, tumours and narrowed arteries), HIV, Huntington's disease, Creutzfeldt-Jakob Disease or head injury. Symptoms may include very poor short-term memory, disorientation and getting lost, loss of identity and failure to recognise friends and family. Simple domestic tasks become difficult and the abilities to reason, plan, write or even talk may be lost.

People can delay dementia by exercise to aid the brain's blood and oxygen supply. Other measures include sensible eating and mental stimulation, as from socialising, hobbies, card games, chess, Scrabble or other board games, learning new subjects, taking up new interests, doing crosswords, Sudoku or creative writing. If sports are beyond an older person's capacity, then bowls, putting, croquet or regular walking or gardening are good for mind and body, and brain-muscle co-ordination. Watching TV for hours does not help the brain or body, while reading stimulates the mind.

Alzheimer's is progressive, with increasing loss of memory, disorientation, getting lost, aggression, incontinence, loss of co-ordination, an inability to manage dressing and washing, and loss of coherence and even speech. As the symptoms worsen, institutional care often becomes necessary. It is devastating for a family if much-loved parents, through no fault of their own, become aggressive individuals who do not recognise them, are forgetful and a trial to all around. Treatment is palliative rather than curative.

Alzheimer's involves the loss of brain cells and volume. The brain develops characteristic plaques and tangles of starch and protein. The causes include head injury in some boxers (Muhammad Ali) and ball-game players. Alzheimer's is more apparent in families with good longevity as age is a major factor. Sufferers include former political leaders Harold Wilson, Margaret Thatcher and Ronald Reagan, novelist Iris Murdoch and actress Rita Hayworth.

Parkinson's disease is a progressive neurodegenerative disorder affecting movement, muscle control, balance and other functions, with tremors, rigidity and slowness in movement. It is debilitating, not fatal. There is no cure. Adult-Onset Parkinson's is the commonest form, with an average age of onset of 60, with an increasing incidence later. Young-Onset Parkinson's occurs between ages 21 to 40 and is frequent in Japan. Parkinson's greatly impairs the quality of life for patients, families and carers. According to the American Parkinson's Disease Association, there are approximately 1.5 million sufferers in America, 1 to 2% of the over 60s and 4% of the over 85s. About 50,000 new cases are diagnosed there annually. All races are affected.

Autism usually shows by 30 months. The children have communication difficulties, with half the sufferers hardly talking at all. They find it difficult to relate to others, may seem unaware of others, fail to make eye contact and are withdrawn, unwilling to participate in playing. The frequency has increased from about 3 per 10,000 in the 1950s to 21–50 (estimates differ) per 10,000. Treatments are often unsuccessful, with the child growing up with communication problems. **Asperger's syndrome** is a milder form of autism, with a frequency of about 1 or 2 per 10,000. For autism, 80% of sufferers are male; for Asperger's, 89% are male. Autistic children are often intelligent. In one family, the autistic grandfather and two autistic children had IQs above 150, but autism made them hard to live with because of their irrational behaviour.

Persistent bullying harms many children, who are often unwilling or too scared to report it to parents or teachers. It causes much distress but is not a common cause of mental illness. Physical or sexual abuse of children can but need not lead to psychiatric problems later.

In America, the commonest childhood psychiatric disorder is **Attention Deficit Hyperactivity Disorder (ADHD),** with 1 in 10 American children (1 in 20 in Britain), mainly boys, affected, usually by the age of seven. Affected children or adults cannot concentrate, are disorganised and give up easily. They are hyperactive, restless, disruptive, impulsive, disobedient, inattentive and bad at learning. They often later turn to drugs and drink. ADHD has a strong genetic component. Treatment with Ritalin is widespread in America and Europe, but controversial. ADHD's incidence has increased greatly in the last 20 years, perhaps from a lack of effective punishments for misbehaviour and a lack of imposed discipline by parents, schools and religious groups, resulting in a 'lack of moral fibre' and self-discipline.

Delirium is usually temporary with complete recovery, as when a fever-causing disease abates. Sufferers become confused and disoriented, often with hallucinations and trembling. Delirium may come from a high fever, starvation or extreme vitamin deficiency. Alcoholics can suffer delirium tremens, with shaking hands and other parts.

Although everyone gets depressed at times, it is usually temporary, but many suffer from sustained severe depression, at worst becoming suicidal. The depth of their depression seems to others disproportionate to its cause, and friends may try to help by telling the sufferer to "cheer up and snap out of it", which seldom works. In Western societies about one person in eight is depressed at any time, with 33% suffering significant depression some times. About 20% of teenagers have prolonged emotional difficulties, with anxiety, depression and low self-image. Women suffer more than men, old people more than the young; even children can suffer badly. About half the women who develop depression do so around the menopause. About 10% of new mothers suffer **postnatal depression**.

The symptoms of **severe depression** include extreme pessimism; looking miserable; loss of interest, motivation, self-esteem, concentration, sociability and appetite; bad temper; deep anxiety, often with little cause; feelings of guilt; headaches; pains; tiredness; emotional numbness; sleeping problems, and neglect of self and dependants. Sufferers may have delusions and paranoia. Feelings of utter worthlessness are common. About one in eight depressives commits suicide, especially men. A former colleague was a much admired, helpful, sociable extrovert. He became intensely depressed when his wife died of a brain tumour, and he committed suicide.

There are genetic factors in mental illnesses which often 'run in families'. People with a close relative who committed suicide have double the normal risk of killing themselves. According to the WHO, about 60,000 Russians commit suicide a year, with men six times more prone than women. Social and economic problems, depression and alcohol are the main factors there.

Depressed women and young people often attempt suicide, frequently unsuccessfully. One young lady in her late twenties became really depressed with feelings of worthlessness, that life was not worth living. Nothing seemed to cure her depression; one day she put her head in a gas oven and killed herself. Some intelligent, successful and attractive young women, whom others think have a bright future, attempt suicide, as in this account.

Personal Account. *I was bullied, anorexic, attempted suicide and had severe melancholic depression with acute anxiety*

I've always felt different, as though I didn't fit in. Not in the sense that I looked or seemed different from the outside, but that I was living on a different plane of existence. Now, at the grand old age of 22, it's nice to feel that I'm my own person, but when you're at school all you want is to be the same as everybody else.

I was always much encouraged as a child, by friends and family alike. My family is Scandinavian, so we've always felt tight-knit, to preserve our nationality. I was always taught that I could do anything, and I really believed that.

I had always been top of my class at school, but when I changed school at 8, I began to get bullied, and it was better to keep my head down and not make myself noticeable in any way, so I stopped working and my grades fell behind. I was underachieving in order to conform, and the older I got, the less 'cool' it was to be good at academic subjects.

The bullying continued at my first senior school and started to wear me down by the time I was 13. I started getting stomach aches and feeling sick for no apparent reason. I dreaded going to school and cried incessantly. I had every gastrointestinal investigation money could buy, yet they found no cause for my suffering. I changed schools to get away from the problem and start again. But my constant illness and unhappiness had begun to seriously affect my self-esteem, and starting a new school at a time when there was only one other new girl in the year was daunting. Things went from bad to worse, and by the end of my first year there I was constantly nauseous and upset — I had difficulty making friends as a result.

In an attempt to overcome nausea and bolster my self-esteem (having decided that I was too fat, even though I now concede that I was skinny) I pretty much stopped eating. By the end of that thoroughly miserable summer I was at my wits' end with nerves and misery, and was sent to a 'special doctor.' This psychiatrist with a speciality in 'nervous stomachs' took one look at me and admitted me to the psychiatric ward of a private hospital. I was drugged up to my eyeballs and the nurses threatened me with force-feeding and intravenous drips in an attempt to get me to eat. My stomach settled a little and I left after one week.

Getting through GCSEs was hell, but I managed 7 As, 1 B and 1 C, so my illness hadn't hampered me too much. My first year of A-levels was an improvement but I fell pregnant and had a miscarriage. I couldn't tell my parents for fear of upsetting them, and my boyfriend didn't want to know.

This all made my final year of school a living hell. I cried hysterically at least twice a day and the school nurse had to come and scrape me off the locker-room floor. It never occurred to anyone to talk to me about why I was behaving as I did. My parents assumed I was being over-emotional and having difficulty getting over the breakup with my first 'proper' boyfriend.

My result-driven school (a leading London independent girls' day school) wasn't interested in my education, or me, as I was destined to bugger-up my chances of going to Oxbridge, which is their only goal. I existed, rather than lived, through A-levels, achieving poor results. I took a gap year, and re-sat A-levels, achieving slightly better results. My health was not much improved. I took a second gap year to recuperate, but spent much of it ill, and a further 10 days in hospital. By some miracle I was accepted at Imperial College but felt very left-behind by my old school friends, and generally very lonely. I went downhill that year, and was in such a state by the end of it that I never took my exams, forcing me to retake the year.

This last year has allowed me slowly to get better. I've been seeing a psychiatrist and a psychologist once a week and have been put on a new antidepressant (I've been taking them constantly since my first stay in hospital). I'm improving daily, though I did go to pieces at exam time. I managed to pass all my exams (just) and am now looking forward to finally moving into 2nd year.

Sometimes I feel suicidal. *I cut my wrists several times during my first year at Imperial, but not seriously enough to do lasting damage. On the day of a resit exam in September last year, I ran away before the exam started and took 36 paracetamol tablets, 18 times the recommended dose. I realised the idiocy of what I had done immediately and took a taxi to the Accident and Emergency Department at Chelsea and Westminster Hospital. They made me drink lots of liquid charcoal and kept me* *for observation for 4 hours, after which they took blood tests to ensure that my liver wasn't damaged. They gave me a very stern talking to (time-wasting etc.). There were no lasting effects except for much egg-on-face and permanently lowered liver values.*

I don't suppose that I am 100% cured and still have my dark moments. But I am learning to be comfortable with myself, my abilities and failings, and to be less hard on myself: I have always been a perfectionist. I have found that having someone to talk to, who sees this affliction as the illness it is instead of as a character flaw, has made life immeasurably easier.

My father's side is riddled with perfectionists, and my grandmother still (at 84) has very low self-esteem. My mother's parents were both alcoholic, drinking to contain their misery. My maternal uncle has depressive tendencies and my maternal grandmother was probably an undiagnosed anorexic. My mother, due to her miserable family life, also suffered from nervous stomach aches and panic attacks (as I continue to), but she was lucky enough to grow out of them. My early childhood was very happy, and my parents very loving, if somewhat held-back as far as hugs and praise.

I believe I have a **strong genetic predisposition towards depression**, *exacerbated by my feeling of being an 'outsider' and bullied at school, of being foreign in a school full of English girls, and having a contemplative and analytical nature.*

Depression is a bigger problem in Scandinavia than in Britain. I know of more Scandinavian people with depressive disorders (including eating disorders), and it is generally accepted that people become very 'serious' after a long winter of almost 20 hours of darkness a day. But the situation for the sufferer in Sweden is easier as it is more openly discussed, and is seen for what it is — an illness and not a weakness. It is as common and mundane as colds and flu.

Swedes are very keen on light therapy to treat 'winter blues' — SAD, Seasonal Affective Disorder. It involves sitting in front of a very bright light (with the same wavelengths as sunlight) for about 15 minutes a day, trying to trick the body into producing more serotonin (a neurotransmitter associated with mood control). These readily available lights are very popular, most people eating their breakfast and reading the morning paper in front of them.

This account's writer was brought up in England with Swedish as her first language and the one she speaks at home. Her two secondary schools are high-pressure ones which score highly in academic league tables. **Prolonged depression** has many causes, with some genetic predisposition. Antidepressants such as Prozac are often successful. Psychotherapy to overcome negative thoughts, low self-esteem and repressed anger may be given. For severe depression, electro-convulsive therapy (applying electric shocks through the temples) often works and may be used if drugs do not. Severe depression tends to recur.

In **bipolar affective disorder**, commonly called **manic depression**, sufferers have periods of depression and periods of mania in which they are hyperactive, uninhibited, cheerful and outgoing, feeling that they can achieve great things quickly. They are often reckless during mania, and can lose friends, money and jobs. They may become angry if thwarted, and suffer delusions and paranoia. About 2% of the population have it, in both sexes and all racial types. Sufferers include Russell Brand, Frank Bruno, Lord Byron, Kurt Cobain, Edward Elgar, Carrie Fisher, Paul Gascoigne, Ernest Hemingway, Vivien Leigh, Florence Nightingale, John Ruskin, Frank Sinatra, Dusty Springfield, Vincent van Gough, Ruby Wax, Amy Winehouse, Virginia Woolf and Catherine Zeta-Jones.

Attacks of mania may be unpredictable, with no obvious cause. Psychological counselling is often ineffective but drugs such as lithium carbonate or anti-convulsant mood stabilisers and neuroleptics as calming agents usually work. Electro-convulsive treatment is a last-resort treatment.

These are extracts from a student's account of bipolar disorder in *Felix* 22/1/2001.

Bipolar disorder is damn serious and can take over every minute of your life, wreck your degree, friendships, relationships and even kill you. [In a manic phase] I'd just fling myself into work and rush through it like I was king of the world. …[In a depressive phase] Suddenly this wave would come over me, and take over all my thoughts. … It's like quicksand. You suddenly feel the world's going to end, listen to depressing music, stay in your room with the curtains drawn — you're just paralysed by this horrible black wave.

In his manic phase he started punching or flinging himself at solid objects, or people in clubs. It was only by cutting himself that he could calm down, and he 'spent an eternity washing blood out of his sheets'. Eventually he was treated with drugs and psychotherapy.

Narcolepsy is an uncommon dangerous condition in which people fall asleep easily during the day, even while sitting or standing, even if they are not lacking in sleep.

Obsessive-compulsive disorder (OCD) involves repetitive unnecessary actions, with rigid routines such as repeated washing of already spotless hands, ritually moving objects around, or checking that things are properly folded or closed. Sufferers are anxious and obsessive. Many are aware that their actions are irrational and unneeded, but cannot stop themselves. OCD can interfere with jobs and home life. Treatment is with psychotherapy or drugs, and may require a lot of support from friends and family.

Hermione Bailey had **multiple compulsive obsessions**. On reaching the door to go out, she would go back at least ten times to check that she had turned off the gas cooker and fire. She endlessly washed her hands and the kitchen floor in case she contaminated anything. She threw away food she had just cooked for her sons in case it was contaminated. She felt compelled to pick up broken glass. She thought she was going mad and became suicidal. After two years of drug treatment and cognitive behavioural therapy, she felt 90% better. Dr Samuel Johnson felt that he had to touch every lamppost that he passed, and would go back if he thought he had omitted one.

Tourette syndrome (TS) gives repetitive involuntary movements and vocalisations (tics), often with coprolalia (foul language). The average age of onset is between three and nine years. It affects males four times more often than females. About 200,000 Americans have the most severe form, while 1 in 100 exhibits milder symptoms. TS can last a lifetime but there is generally improvement in the late teens.

Paranoia makes sufferers pathologically suspicious, feeling that others are scheming against them, that no one can be trusted. They become angry quickly, are jealous, touchy, complaining, lacking in humour and tolerance.

Phobias involve extremely strong but misplaced fears of animals such as spiders, snakes or birds, of being in enclosed spaces (claustrophobia) or open spaces (agoraphobia), of heights (acrophobia), water (hydrophobia), or flying. Although many people dislike spiders or snakes, phobias cause strong irrational reactions. The sight of the hated thing causes reactions of avoidance, sweating, fast pulse, even breathing difficulties and extreme anxiety. Even images of the object or thinking about it can cause fear. Robert Mawdesley has a phobia of baked beans. Even pictures of them make him feel sick, causing him to shake uncontrollably. He avoids supermarket aisles that might stock them. One in five people has a phobia. Treatments may involve desensitisation, with gradual exposure in non-threatening situations to the feared object, perhaps starting with drawings or photos, or may involve medication or behavioural therapy. Classes are run for people with a phobia of flying.

Superstitions are culture-based, transmitted socially between generations and with little scientific basis. President Franklin D. Roosevelt suffered from severe triskaidekaphobia (fear of the number 13) and refused to travel on the 13th of the month. A silly superstition at primary school was that if you saw an ambulance, to avoid bad luck you had to hold your collar until you saw a dog. Superstitions about bad luck can involve breaking mirrors, spilling salt, walking under ladders, the number of magpies seen at one time, or which numbers bring bad luck. They vary greatly between people and cultures.

Schizophrenia is a psychosis with introversion, delusions and an inability to distinguish the real from the unreal. Its frequency is about 1%, affecting more men than women. Typical ages of onset are 17 to 24 in men, 25 to 30 in women. Afro-Caribbeans in Britain have twice the usual incidence, more than Afro-Caribbeans in the West Indies.

It is multifactorial, with genetic and environmental components. Someone with no affected relatives has a 1% chance of schizophrenia. For those with an affected relative, the risk goes up two-fold for first cousins, uncles and aunts, four-fold for nephews and nieces, five-fold for grandchildren and half-brothers and half-sisters, six-fold for parents, nine-fold for brothers and sisters, 13-fold for children, 17-fold for non-identical twins and 48-fold for identical twins, so if one identical twin develops schizophrenia, the other has a 48% risk of getting it.

Type I (positive) **schizophrenia** has a sudden onset, with strong psychotic symptoms such as hallucinations and delusions, profoundly illogical thoughts, odd moods and odd behaviour. One schizophrenic told me that he kept hearing very real voices in his head, and some schizophrenics who committed murder say that they did it because a compelling inner voice ordered them to. **Type II** (negative) **schizophrenia** is less obvious, with symptoms of inertia, lack of motivation, lack of focus on tasks in hand or routine activities like dressing, washing and going to work, and changed sleeping times.

Many schizophrenics are intelligent and keep good jobs if their condition is controlled by drugs and/or psychotherapy. Occasional relapses can be catastrophic for jobs and relationships. About one in three schizophrenics responds well to treatment but one in four hardly responds. About 8% of schizophrenics are violent, compared with 2% of the population. Antipsychotic drugs help many but not all schizophrenics by controlling some symptoms such as delusions, thought disturbances and hallucinations. Electro-convulsive therapy, with electric shocks to the temples, was used from the 1930s to help relieve depression, mania and schizophrenia. It is still used as mainstream therapy for patients who do not respond to drugs.

Depression and/or paranoia often affect schizophrenics. About 10% commit suicide, compared with 1% of the population. Male sufferers, on average, have about half the number of children fathered by normal men, and female sufferers have about four-fifths as many children as normal women. One might expect selection slowly to remove the genes from the population, but as the frequency remains similar, the removed genes must be replaced by new mutations.

A typical tragic case of a paranoid schizophrenic committing murder was described in the *Daily Telegraph* (11/10/2003). Paul Khan developed mental illness in his mid-20s, committing a series of crimes. In 1996, in Cardiff Central Library, he approached a stranger and slashed his neck with a razor, causing severe physical and mental scarring. The schizophrenic said that he had been directed by voices in his head and was given indefinite detention under the Mental Health Act. Paul spent the next four years in Welsh mental hospitals. In 2000, a mental health tribunal decided that he could be gradually released back into the community.

In 2003, he stopped taking his medication. In March, Brian Dodd, 72, was walking on Prestatyn Beach when Khan attacked and killed him with a kitchen knife, stabbing him ferociously 37 times. When arrested three days later, Khan said he had no memory of the motiveless murder. The victim's widow demanded an inquiry into why the killer had been freed from a mental hospital, given his record of violence.

Seasonal affective disorder (**SAD**), is where long, cold, dark, winter days make sufferers sad and depressed, but they recover as the number of daylight hours increases in spring. About 3% are affected in Britain and many more in Scandinavia where winters are longer and darker: see the above account by a Swedish woman. The symptoms include depression, lethargy, irritability, lack of effort and concentration, daytime sleepiness, and sometimes 'comfort eating' and weight gain. It is related to short daylight length, not cold. Treatment is by home-administered phototherapy, working through the eye and brain, not the skin.

Dyspraxia ('clumsy child syndrome') involves poor co-ordination. The children may be bright but are clumsy, may have difficulty writing, poor ball-game skills, poor balance and motor skills. According to the Dyspraxia Foundation, it affects up to 10% of the population and is severe in 2%. Males are four times more likely to be affected than are females. It affects perception, language, muscle-control and thought. Treatment is by co-ordination exercises. Dyspraxia can persist in teenagers or adults. One sufferer lamented that she had great problems with corkscrews, tin openers, laces, scissors, present-wrapping, sports such as tennis, opening cartons, threading needles, driving, etc.

Dyscalculia, an inability to do simple calculations, affects up to 6%. Curiously, some children with severe dyscalculia pass A-level maths, understanding abstract mathematics but being unable to do simple sums. **Dyslexia** was described in Chapter 6. **Visual stress** (Meares-Irlen syndrome) affects about 10% of dyslexics and some non-dyslexics. They have a hypersensitive visual cortex which generates too much electrical energy in response to particular light wavelengths. Wearing blue-tinted glasses solves the problem for some sufferers, including actor Johnny Depp. Others may need red or green glasses.

The **agnosias** involve impaired recognition abilities due to lesions in the cerebral hemispheres, not to disorders of sense organs or intellect. In **auditory agnosia** a lesion in the auditory cortex results in an inability to recognise sounds, words or music. **Finger agnosia** is the inability to name or recognise individual fingers, and **tactile agnosia** is the inability to recognise objects by touch. **Colour agnosia** is an inability to name or identify colours, while **visual agnosia** is an inability to recognise objects by sight.

Alana has good eyesight but suffers from **prosopagnosia** (difficulty in recognising familiar faces, in person or in photographs). This can be very embarrassing. She once chatted away to a stranger in a bar, mistaking him for her boyfriend. Hannah Ray could not even recognise her mother. An attack of encephalitis, brain inflammation, when she was eight, was responsible. **Face-blindness** has an incidence of 3% and can come from injury to the head, strokes or degenerative diseases, but usually occurs from brain lesions before birth.

One irritating thing which increases with age is **logamnesia** (Greek *logos*, word, *amnesia*, forgetting) a temporary, specific **word-blindness**. One knows a word, a name, yet cannot recall it when it is needed. One knew it before and will know it again without prompting. My mother in her 90s referred to 'thingamajig' or 'thingamabob' when she could not recall the words for a thing or person.

Self-harming is self-injury, often associated with low self-esteem. According to the National Self-Harm Network, Britain has one of the highest rates of self-harm in Europe. Research by the WHO in 2014 found that 20% of 15-year-olds in England had self-harmed, especially girls. It is often a coping strategy to provide release from emotional distress. It can enable an individual to regain feelings of control, but is often addictive and can increase in severity. Individuals go to great lengths to hide cuts, scars and bruises, wearing concealing clothing and avoiding activities which expose wounds.

Research by the Universities of Bath and Oxford found that one in ten schoolgirls harms herself each year, compared with one in forty boys. Cutting was the most prevalent (65%), followed by deliberate overdosing (31%). Coping with distress was the most frequently given reason, from being bullied, physically or sexually abused, or worrying about being gay. Self-harm may include cutting, burning, biting, substance abuse, head banging, scratching, pulling out hair, over-dosing and self-poisoning. While some self-harmers grow out of it, many require counselling or more severe psychological treatments. Some get blood poisoning from cuts and some commit suicide.

Six children in three families from the Qureshi birdari clan in northern Pakistan had many injuries because they were insensitive to pain, carrying a pain-killing mutation in a gene called *SCN9A*. One was a street performer who pushed knives into his arms and walked on burning coals. He died on his 14th birthday, jumping off a roof to impress friends.

The Gadd family in East Sussex have three children suffering from **congenital insensitivity to pain,** having two copies of the recessive gene, one from each non-suffering parent. Although the children have a sense of touch, the small nerves which register pain, heat or cold have not developed. Jonathan, 14, Jodie, 12, and Megan, 9, had many scars, had knocked out several teeth, bitten chunks out of their tongues, and two had broken several bones. Jonathan once leant on a radiator for too long and burnt the entire side of his face. The perception of pain is essential for human safety. All three children also have **anhidrosis**, an inability to sweat, so are in danger of overheating. (*Sunday Telegraph Magazine*, 7/8/2005)

Huntington's chorea has a very diverse age of onset, from the teens to the eighties, with an average of 37. It gives progressive nervous and mental deterioration, with chorea (involuntary leg and arm movements), and is eventually fatal. It affects about 1 in 20,000 people and is due to an autosomal dominant gene, so half the children are expected to be affected. It is caused by an unstable length mutation which can increase in severity between generations.

Migraines are severe, throbbing, disabling headaches which can dominate everything for hours or days. They affect three times as many women as men, with about 22% of women affected at some time. The headaches are more frequent during menstruation. There may be nausea, vomiting and extreme sensitivity to light and sound.

A study by Corrigal and Bhugra (2013) of ethnic variations in psychiatric admissions and detentions in London showed that young Blacks were six times more likely than young Whites to be admitted with psychosis to psychiatric hospitals but showed no increase in admission for non-psychotic conditions. Young Asians were more than twice as likely to be admitted as Whites for psychosis, but only one third as likely to be admitted for non-psychotic conditions. Other studies showed that in Britain, adult Blacks are more likely to be admitted to psychiatric hospitals than are Whites, with rates of compulsory detention being four times as great for Blacks and twice as great for Asians.

9.7. Synaesthesia

Synaesthesia (Greek *syn*, together; *aisthesis*, sensation) involves sensations of colours or tastes evoked by words or sounds. For James Wannerton, certain words evoke strong tastes, such as *motorbikes* suggesting wine gums; *Alan*, mucus; *cook*, wet nappies; *service*, a lovely toffee taste; *precise*, beautifully buttery. Others find that days of the week, numbers, letters or sounds are associated with colours. The brain region which deals with colour processing lies very close to the region dealing with numbers, so cross-wiring could account for some cases. Synaesthetes include artist David Hockney and musician Olivier Messiaen. Synaesthesia runs strongly in families and affects more women than men.

Meriem has strong synaesthesia, having had these associations since infancy: Monday, red; Tuesday, yellow; Wednesday, blue; Thursday, orange; Friday, silver-grey; Saturday, silver-white; Sunday, chocolate brown; number 1, black; 2, red; 3 and 8, which have similar shapes, yellow; 4, blue; 5, orange; 6, chocolate-brown; 7, black-brown; 9, black; 0, pearl white. The months have different colours.

A woman with Asperger's syndrome (she hated being hugged by her mother, disliking human contact) had synaesthesia. Tasting salt made her see yellow; petrol fumes evoked green, and love was a rich reddish purple.

9.8. Epilepsy

About 30,000 people a year are diagnosed with epilepsy in Britain, with about 600,000 sufferers. World-wide, the frequency is about 1%, some 70 million. More than two thirds can control seizures with medication but the rest cannot. Severe epilepsy affects employment, relationships, self-esteem and safety. If the cause is identifiable, say internal scarring after a blow to the head, a stroke or tumour, brain surgery may help. Usually the cause is unknown. Brain infections like meningitis can cause it. 9% of epileptics die young, compared with 0.7% of the population. Epileptics are four times more likely to commit suicide.

An epileptic seizure happens when there is a burst of intense brain electrical activity, causing a disruption to its function, and strong muscle contractions. In **tonic-clonic seizures**, there is contraction of the limbs followed by their extension and arching of the back. The sufferer goes stiff, loses consciousness and falls to the ground, making jerky movements, perhaps with a loss of bladder or bowel control. Tongue-biting is common. After a minute or two the jerking should stop and consciousness slowly returns. In **focal seizures**, sufferers may not be aware of their surroundings or what they are doing. They may pluck at their clothes, smack their lips, swallow repeatedly and wander around. Sufferers with **reflex epilepsy** have seizures triggered by particular stimuli such as flashing lights, shock or sudden noises.

9.9. Learning and memory

Learning is the impressing of items into the memory and their retention. Their recall and use help us remember them, but many items are swiftly forgotten. Memory processing during sleep seems important for what is retained and what is rejected, but how those enormous numbers of brain cells retain, represent and recreate particular facts or images is unknown. There is much diversity in people's ability to learn and to retain memories. I have a bad memory for names but a good one for faces. Some people have almost 'photographic memories', being able to learn huge amounts easily, perhaps reading them only once. Some people can remember card sequences in many packs of cards, many random numbers, etc.

Opera singers have to learn and appear to understand parts in Italian, English, French, German, Spanish, Russian, Czech, etc., with the correct words, notes, emphasis and actions. Concert pianists such as Rubenstein can often play many complete concertos from memory. Blind musicians remember many pieces just from hearing them, without ever seeing the music. Ballet dancers have to remember steps, positions and their relation to music and other dancers.

Sometimes memory readouts have to be extremely fast and coupled to physical abilities. Organists have to control ten fingers and two feet with several keyboard manuals, foot pedals and a battery of stops. In parts of Liszt's Paganini Etudes Number 6, the pianist has to play at a rate of 1,800 notes a minute, getting each note correct in sequence, loudness and duration, and with correct pedalling. It requires phenomenal mental and physical abilities.

In the elderly, short-term memory is often poor but some long-term memories are crystal clear. Childhood events may be remembered in detail when the previous day's activities are forgotten. **False memory** is where a person recalls an event that did not actually occur; it is often invoked in legal cases regarding childhood sexual abuse.

9.10. Creativity

People differ enormously in creativity, having original ideas, especially unexpected ones, and putting them to use. Such ideas can come uninvited into one's head, or from deliberate, intensive thought. Creative people such as writers, composers, artists and research scientists often have vivid imaginations. Sometimes they lose the ability to distinguish the real world from their imagined one and develop mental disorders such as schizophrenia. Vincent van Gogh's late paintings show a deranged perception of the world.

George Sand, a woman and Chopin's lover for nine years, gave this account of the Polish musician's creative processes:

> *His creation was spontaneous, miraculous. He found it without searching for it, without foreseeing it. It came to his piano suddenly, complete, sublime, or it sang in his head during a walk, and he would hasten to hear it again by tossing it off on his instrument. But then would begin the most heartbreaking labor I have ever witnessed. It was a series of efforts, indecision, and impatience to recapture certain details of the theme he had heard: what had come to him all of a piece, he now over-analyzed in his desire to write it down, and his regret at not finding it again "neat," as he said, would throw him into a kind of despair. He would shut himself up in his room for days at a time, weeping, pacing, breaking his pens, repeating and changing a single measure a hundred times, writing it and effacing it with equal frequency, and beginning again the next day with a meticulous and desperate perseverance. He would spend six weeks on one page, only to end up writing it just as he had traced it in his first outpouring.*

My best ideas for genetics research and short stories have come from my subconscious while I have been doing mundane things like washing shirts or having a bath. Some of Darwin's best ideas came while he was walking in his garden.

9.11. Happiness

Happiness is what people seek but unrealistic aims and desires are the source of much unhappiness. Many lottery winners find that riches do not bring the expected happiness and can lead to disaster through unwise behaviour. Being content with what one has and counting one's blessings sound trite, but they lead to more happiness than do unachievable ambitions.

Many families in less developed countries had a rich, enjoyable family and community life, then along came television. While bringing some happiness in the form of easy and cheap entertainment, it was often at the expense of traditional leisure activities. Even worse, TV — by showing people affluent Western lifestyles — made them unhappy with their standard of living and possessions. If they finally get cars, computers and washing machines, they may find elusive the expected happiness from material possessions.

There is a famous quotation, "Be careful what you wish for; you may receive it." What alcoholics, drug-takers and smokers most wish for — their next fix — harms them in the long term. At a Hindu wedding the priest asked the audience to chant a series of wishes for the young couple, who silently hoped that the wish for them to have eleven children would not happen. There is a traditional Indian wish for a woman: "May you be the mother of a hundred sons."

One's most urgent desire usually differs with age: a particular gadget for a child; a kiss or more with someone fancied as a teenager; then a good job, a love partner, a house, perhaps children, promotion, a higher salary and more status. With increasing age, good health, reliable friendships and a decent pension become more important.

Sadomasochists get happiness through inflicting and receiving pain. Hermits, anchorites and religious devotees get pleasure from self-denial, hunger, poverty and lack of possessions. In a Hindu procession, many men had stuck knives or splinters of wood in themselves and loved doing it 'for the Gods'. **Hedonists** want immediately pleasurable activities. Others seem happiest when grumbling and complaining. One can argue that the best kind of happiness comes from giving happiness to others.

Reference

Corrigall, R. and Bhugra, D., The role of ethnicity and diagnosis in rates of adolescent psychiatric admission and compulsory detention: a longitudinal case-note study. *J R Soc Med* (2013) **106**: 190–195.

Chapter 10

Sex, Attraction, Reproduction, Twins, Incest

10.1. Introduction

The biology of sex and reproduction was covered in Chapter 4, chromosome abnormalities affecting sex in Chapter 23, and relations between ethnicity, sex, lifetime numbers of sexual partners and sexually transmitted diseases in Chapter 11. Parental choice, religion and social custom determine whether male or female children are circumcised. In 1997 a survey in Egypt showed that 97% of married women had been circumcised (female genital mutilation, FGM), and that 80% of women there considered it good. A male Egyptian biologist said, "Those figures are realistic for rural communities but would be less in Cairo and Alexandria. The operation is done for the religion, Islam, to control the bad desires of women." The frequency of FGM in Somalia is 98%. In England, there are about 137,000 victims, mainly mutilated abroad. Some have the vagina partly sewn up. FGM increases the risk of serious complications during childbirth. About 140 million women worldwide have undergone FGM, usually without anaesthetic and by non-medically-trained practitioners.

Male circumcision has declined in Britain and America, although still common in South Korea and the Philippines. In Australia it has dropped from 90% to 15% over 40 years. About 15% of British males are circumcised; it is routine for Jews and Muslims. About one-third of males worldwide are circumcised. The NHS refuses to fund it, saying that it has no medical value. The WHO recommends that male circumcision, giving about 60% protection, should be considered for HIV prevention. The surgery covers the penis tip in scar tissue.

Personal choice (Chapter 5) influences sexual aspects of appearance. Menopausal or postmenopausal women may choose hormone replacement therapy. Many feel rejuvenated. Choice comes into the age at marriage and of having children. In 2014 the Office for National Statistics (ONS) reported that the average age of British mothers having their first child is 30, the highest in the world. It was 26 in 1975.

Much sex-equality legislation has laudable intentions, but ignoring the fact that the two sexes are different is damaging, especially to women. The British army introduced a 'gender free' policy in 1998, expecting female soldiers to follow the same rigorous training as males, but women's bodies are not the same as men's. An army doctor (Gemmel, 2002) wrote that less than 1.5% of army men were discharged with 'overuse injuries' such as tendonitis, while for women that figure was eight times higher. Women generally have shorter legs than men, causing problems when marching in step. Gemmel stated, 'It is clear that there are differences in muscle physiology, bone architecture and body composition that interact to place women at a substantial disadvantage when training or working to the same output as males. ... Health and safety guidance requires employers to make allowances for these gender differences, yet this has been overlooked in the interests of meeting equal opportunities legislation'.

In almost all societies, sexual intercourse is used for pleasure as well as reproduction. People worry terribly about the amount of pleasure they are receiving and giving during intercourse, with a big mutual orgasm desired. Most people imagine that others are getting more frequent and better sex than they are. For some women (and fewer men), sex is a duty — not a pleasure. The wife of Sir Edward Grey (Foreign Secretary, 1905–1916) returned from honeymoon and announced that she detested sex. Thereafter they lived as brother and sister.

A divorced mother, aged 51, with two children, had an unusual problem: up to 800 orgasms a day, even with no male present, and she had not dated for five years. **Persistent Sexual Arousal Syndrome** (PSAS) sounds terrific, but she found it extremely distracting. Until 43, she had normal orgasms with males, then PSAS started and increased. It made her feel angry, frustrated and dirty, and she never wants to be physically intimate again.

Britain has one of the highest rates of Caesarean deliveries, more than one in five births. While the operation saves many lives in obstetric emergencies, some women choose it because they are 'too posh to push', or fear the pain of a normal birth. A British obstetrician in Oman found that Indians there often wanted to bring in women for Caesarean deliveries at precise times, such as 3.24 a.m., auspicious times according to their horoscopes.

Morning sickness affects two thirds of pregnant women, with nausea and vomiting. It is worse in the first 12 weeks of pregnancy, and can occur at any time of the day. Although unpleasant, it usually indicates a healthy pregnancy.

Being heterosexual is the biological norm, keeping our species going, but there are variants, such as being homosexual. Bisexuals have mixed sexual preferences. Most people have some degree of autoeroticism, sexually stimulating themselves at least occasionally. Women may use mechanical vibrators. Even young children often find pleasure in rubbing their genitals. Transsexuals and eunuchs are covered later.

Same-sex civil partnerships and marriages are permitted in many countries, including Britain. The couples can adopt children or have children using opposite-sex volunteers. In 2013, Sir Elton John and his partner David Furnish were "overwhelmed with happiness" after becoming fathers to a second baby boy, born to a surrogate mother.

Which acts are regarded as criminal differ between cultures and religions. While nearly all regard murder as a crime, some cultures regard adultery as punishable by stoning to death; others regard it as trivial, almost normal, especially amongst French politicians. The Bible lays down in Exodus 31:15 that 'Whoever does any work on the Sabbath day, he shall surely be put to death', but law and public opinion do not accept Saturday or Sunday work as criminal, let alone deserving death.

For Muslims and traditional Jews, it is wrong to marry outside the religion. So-called 'honour killings' are not uncommon in Britain amongst immigrants, especially those from the Indian subcontinent. In 2003 a Pakistani bride, Sahjda Bibi, was stabbed to death on her wedding day by her cousin at her home in Birmingham. The family tradition was to marry a first cousin, but she chose a non-relative who was also a divorcee. Abdallah Yones was jailed for life after slitting the throat of his 16-year-old daughter because he disapproved of her boyfriend.

In some countries, especially in Africa and some Muslim countries, homosexual relations are a criminal offence, with male homosexual acts, especially anal intercourse (sodomy, buggery), being severely punished. In most Western countries, discrimination on the grounds of sexuality is illegal. Being women, lesbians cannot do penis/vagina penetration, but they commonly use dildos — substitutes for an erect penis.

In Britain in 2003, 1 to 4% of women and 3 to 10% of men were estimated to be exclusively homosexual. In one study, if one identical twin was homosexual, the other had a 50% chance of being homosexual, for males and females. Corresponding figures for non-identical same-sex twins were 22% concordance for males and 16% for females. Children adopted by homosexuals have an increased chance of becoming homosexual, so there are genetic and environmental factors.

10.2. Marital status, virginity, polygamy and polyandry

Single people may be single through choice, lack of opportunity, through not believing in marriage, or they may be cohabiting. A member of one couple, who had been living together for more than 20 years, asked another long-term cohabiting couple: "How long have you two **not** been married for?" Cynics like to joke that marriage is a sentence, not a word. While some single people envy the companionship that can go with marriage, many married people envy the freedom of single people. Some close cohabitations of more than two people have been amicable, such as Lord Nelson with Sir William and Emma Hamilton (her lover and her husband, respectively).

American research on couples who married between 2005 and 2012 found that 35% met through the internet. Most couples who first met 'in the flesh' did so at work (22%), through a friend (19%), at school (11%) or at social gatherings (10%).

Some people are forced to marry against their will. In Asia, the Middle East and Africa, very young children may be married without their understanding or consent. In Britain, the government is trying to stop forced marriages, which are often of young girls being sent to marry an older relative in Pakistan whom they have never met, then getting him a visa to live in Britain.

In some countries, mothers are keen to get daughters married off. One Sri Lankan friend was forced in her late twenties to marry a bus driver whom she did not like and certainly did not wish to marry. She soon divorced him for cruelty. Another friend there was told by her mother that she must marry before she was 39 as no one would want her after that. She was enjoying life and did not want to marry, but as an obedient daughter she married a relative of her parents' choice, and came to regret it.

There is a tendency for children of divorced couples to have an increased risk of divorce, as well as more learning problems, more depression and more difficulties in social adjustment.

Relationship figures were given for Britain in *UK 2002*. Out of 60 million people, the number of cohabiting couples was more than 1.5 million but those relationships often broke up quickly. There has been a striking increase in the number of people in their late 60s getting married, 90% of whom had been married before. The average age of those getting married rose eight years in a generation, with a typical groom of 32 and a bride of 30. The number of divorces was rising, with 155,000 in 2000. Teenage parenthood was rising, with 90% of births to teenagers being outside marriage. In America, achieving low teenage pregnancy rates has been helped by strong religious groups and movements such as the Silver Ring Thing, where girls make a commitment to chastity before marriage.

In 1938 Sarah Baring was a debutante and 'did the season'. She said that young ladies in her social circle then would dance the night away but were paragons of virtue: "Nobody told us anything about the facts of life. We were all ignorant, and if we had known, we'd have thought it disgusting. I and all my close friends would have considered ourselves defiled if we hadn't come to marriage as virgins. Even after you had become engaged, it made no difference. Virginity lasted right up until the wedding night." (from *The Daily Telegraph*, 16/2/2013) A psychologist and sex therapist said cynically in 2002, "Fifty years ago, for most people, marriage was where your sex life began. Today, for many people, it's where it ends."

In 2007, Unicef put Britain last out of 21 countries for children's well-being: more British children had sex by the age of 15 than elsewhere and had the highest rate of teenage pregnancies and sexual infections in Europe. Venereal diseases such as chlamydia, HIV, gonorrhoea and syphilis had dramatically increased in 12 years, with 10% of girls aged 16 to 19 having chlamydia. A report in 2003 from Canterbury University, New Zealand, found that girls who grew up without a father are more likely to become pregnant in their teens.

Polygamy is long-established in many societies, with a man having more than one wife at a time, as in Middle Eastern harems and in many Muslim communities. Amongst Christians, the Mormons in Utah were famous for polygamy, but the Mormons of Salt Lake City renounced polygamy in 1890. In Holdall and Colorado City in Utah, an eccentric, extremist, breakaway Mormon sect, the Fundamentalist Church of the Latter Day Saints, still practises polygamy and underage marriage (including forced marriages of 14- and 15-year-old girls), in spite of both being illegal. According to *The Sunday Telegraph* (19/10/2003), their leader, Warren Jeffs, had 12 wives. Sisters were often required to marry the same man. According to one woman who left the sect, "This place is like a mixture of the Taliban and the Mafia. It runs on intimidation, ignorance of the outside world and fear."

Some men have a series of different wives — serial monogamy. A Malaysian, Kamarudin Mohammed, 72, had 52 failed marriages, and ended up by remarrying his first wife when she was 74. As a Muslim man, he obtained divorces just by saying "I divorce you", three times.

Polyandry, a woman having more than one husband, is rare. Men dislike sharing women and want to be sure who the father of their wife's children is. Writer Manel Ratnatunga (2010) stated that 'Polyandry was common in Sri Lanka. Many brothers sharing one wife meant money remained in the family, reduction of population, and when the males go to work or war, there is always one brother at home to look after the family.' She retells a folk tale, *How eleven Brothers found a Wife*, but that one wife's feelings are not mentioned.

A survey for British Satellite Broadcasting (BSB)'s TV series, *Sex, Lies and Love*, found that one in eight British women would pay for sexual intercourse. They would pay £200 for 'passionate sex and fulfilment of all their fantasies', £100 for 'something to remember ... a bloody good time', but, strangely, only £60 for 'absolute ecstasy'.

10.3. Attractiveness and factors affecting the choice of mate; marriage and divorce

It is said that "opposites attract", but marriages usually show attraction of similar types ('positive assortative mating'). Tall men tend to marry tall women and small men tend to marry small women. Other recorded similarities between mates are in weight, interests, hair colour, deafness, education levels, social status and parental income.

Positive assortative mating occurs for intelligence. The closeness of IQ scores within 51 American married couples is really striking in Outhit's data (1933), with very bright usually marrying very bright and less bright marrying less bright. Of 18 men with IQs between 120 and 140, 15 married women in that range.

Positive assortative mating can arise from conscious choice, subconscious choice, increased meeting opportunities due to common interests, similar jobs, type of education or membership of a subgroup — undergraduates tend to meet undergraduates. Immigrants or ethnic minorities often marry within their group, as do members of religious groups, castes or sects.

Positive assortative mating for genetic traits increases the frequency of **recessive characters** such as red hair. Another usually recessive trait with strong positive assortative mating is deafness, where deaf individuals often form social groups, with special schools, training and easier communication within the group than with people with normal hearing. The normal alleles at different loci are complementary, so two deaf parents usually have hearing children.

Assortative mating for one character can have secondary effects on others. Positive assortative mating for one Negroid characteristic, such as very dark skin, could cause apparent positive assortative mating for associated characters such as hair type, nose shape and blood groups.

Examples of opposites attracting include Mills Darden, 72 stone 12 lb (1,020 lb, 463 kg), and his wife Mary, 7 stone (98 lb, 44 kg), so he was ten times heavier. She had three children by him before dying in 1837. A case of dissimilarity in height was Bao Xishun, a giant at 7 ft 7 in (231 cm), who married Xia Shujuan, 5 ft 5 in (165 cm), in Inner Mongolia in 2007.

Factors important in some cultures are: female virginity, social status/ caste, domestic abilities, size of dowry, and horoscope. An attractive 40-year-old Sri Lankan woman had for many years been unable to find a husband because astrologers gave her a bad horoscope based on her time and place of birth.

In human diversity seminars with students, we discussed factors influencing one's choice of long-term partner. They included age, sex, weight, body shape/figure, height, attractiveness, facial features, wealth/ future earning capacity, fitness, sporting ability, mental stability, intelligence, sense of humour, lifestyle, temperament/personality, niceness, nationality, ethnicity, religion, personal hygiene, health, where they want to live, upbringing, sexually transmitted diseases, originality/creativity, kindness to animals, diet, criminal record, drugs, ability to hold alcoholic drink, disabilities, desire for exercise, desire for children, sexiness/libido, common interests or hobbies, common language, taste in music and dancing.

All these undergraduates wanted someone reasonably intelligent. They sometimes distinguished between factors in initial attraction, such as looks, dress and behaviour, and ones of longer-term importance, such as lifestyle and religion. An American proverb draws a distinction between short- and long-term activities: *Kissin' don't last. Cookin' do.*

None of the women would marry a man much shorter than themselves, and most men were against marrying a taller woman. When they were asked to rank 16 characteristics in order of importance in their preferences in a long-term partner, the results differed between the sexes, as shown in Table 10.1.

The men were more interested than the women in being amused and excited. Both sexes rated kindness and understanding highly; physical attractiveness was fairly important. While intelligence was highly rated, especially by women, university-level education was unimportant, as were race, religion, good heredity and good housekeeping/cooking. The rankings would probably change dramatically after several years, with earning capacity and good housekeeping moving up the order. Disagreements about having children could have severe consequences.

Table 10.1.   Preferences in the choice of a long-term partner of the opposite sex.

Characteristic	Ranking by women	Ranking by men
Adaptability	10th	10th
Amusing company	6	**2**
Creativity	13	9
Exciting personality	4	**1st**
Faithfulness	**3**	4
Good earning capacity	9	12
Good health	7	7
Good heredity/genes	11	13
Good housekeeper/cook	12	14
Intelligence	**2**	6
Kindness and understanding	**1st**	**3**
Matching desire for number of children	8	8
Physical attractiveness	5	5
Race	14	16
Religion	16	15
University-level education or equivalent	15	11

I asked these 20–22-year-olds what age-range partner they would consider if they were to marry within the next two years. 40% of the women would not marry men younger than themselves, while 35% would marry men up to one year younger. About 8% would marry men two years younger, and 8% men three years younger. Only 2% would marry men four years younger. Three out of 40 would only marry men older than themselves, specifying minimum age gaps of two, four and seven years. All the men would marry younger women, from one to four years younger, with two and three years being the most frequent.

All the students would consider older partners. The maximum age gap for older men considered for marriage by the women ranged from one to 15 years, with four to eight years listed most often. For men, the maximum age gap considered ranged from two years to a massive 16 years, with two to three years being commonest.

Data from the ONS in 2003 showed that more women were marrying men younger than themselves. For first marriages, the figure was 20%, with more than half involving a gap of five or more years. Celebrity women doing that include Emma Thompson, 44, marrying Greg Wise, 37, and Madonna, 45, marrying Guy Ritchie, 35. Joan Collins' fifth husband is 32 years younger than she is, and Vivienne Westwood's partner is 25 younger than she is. Men much older than their wives (or ex-wives) include Michael Douglas with Catherine Zeta Jones (gap 25 years), Ronnie and Sally Wood (31 years) and Rupert Murdoch with Wendi Deng (38 years).

The Daily Telegraph (13/2/2007) ran an online survey of attractiveness. Men and women put sense of humour as most important, followed by intelligence and then looks, then giving emotional support, then being good in bed. The bottom three factors were religious/spiritual belief, financial resources and interest in politics. The 1% of women who had slept with more than 51 partners rated their own looks very highly. The average number of sexual partners so far was 11 for heterosexual men and eight for heterosexual women.

In some cultures, arranged marriages are normal. In some Pakistani families in Britain, this causes great friction between traditional parents who often choose a bridegroom from Pakistan who is unknown to their daughter, who may have to go to Pakistan to marry him whether she wants to or not, and the daughter who wants to make her own choice as do her British friends. Millions of couples have never seen each other before the arranged marriage ceremony (Chapter 2.1, a Hindu marriage), while sometimes the couple have lived together for years.

Most marriages in the West are love marriages. Things which encourage romance include proximity and common interests. Our church youth club was known as 'the marriage bureau' as so many marrying couples became friends there. A strange common interest led to the marriage of Lorna Tolchard and John Wing who went to University College Hospital in 1949 to train as doctors. They were allocated the same corpse for dissection.

In Britain in the last 20 years, divorce rates among those over 60 have increased by 75%. Divorce does not put people off marriage: a third of all British weddings are now second marriages for bride, groom or both. According to a *Reader's Digest* poll, four million married people in Britain would not choose the same partner again.

People will go to great lengths to make themselves seem more attractive. See Chapter 5 for clothes, make-up and cosmetic surgery, including

Chinese girls having leg lengthening operations. A pretty 19-year-old said, "Sometimes this is painful, but I am too short. Now in China, taller is more beautiful." In contrast, a tall Hong Kong student (172 cm, 5 ft 8 in) told me that she was too tall for boyfriends there and would like to be 160 cm (5 ft 3 in).

An American study showed that females were rated more attractive when moving with a seductive hip sway, a wiggle in their walk, like Marilyn Monroe. A survey in Australia by the University of Queensland showed that the most important thing in finding a partner was that they should love and care for them. Wealth, looks and brains were rated lowest (*Marriage & Family Review* 2006, **40**: 5–23).

In speed dating, Essex University found that for every inch (2.54 cm) taller a man was than his rivals, the number of women wanting to meet him again increased by 5%. A man's being fat made little difference, but a woman being overweight resulted in her being selected by 70% fewer men.

In 2014 it was announced that the Japanese government was funding matchmaking events in a desperate attempt to boost a birth rate which has halved over 60 years. Japan has one of the fastest ageing populations, shrinking by a record 244,000 in 2013. Births in 2013 were only 1.03 million, the lowest since records began in 1899.

The 2001 census showed that the number of single parents in Britain was four times higher amongst Blacks than amongst Whites. The Government encouraged black couples to attend relationship counselling in an attempt to reduce the number of single mothers and the high rate of teenage pregnancy among Blacks.

Law Smith *et al.* (2006) found that the more fertile a woman was, the more attractive she was to men. Higher levels of sex hormone oestrogen restrict growth of bones, reduce the size of nose and chin, increase the size and thickness of lips, and increase fat deposits in the cheeks, hips and buttocks, making women look more feminine as well as boosting reproductive health. In males, testosterone causes the jaw and eyebrow ridges to become more prominent.

Dr Linda Boothroyd of Durham University found that women who had had good childhood relationships with their father tended to select partners resembling their father, while those with poor paternal relationships did not.

I did an experiment on attractiveness with undergraduates, getting them to assess each other anonymously for having an **attractive personality** (excluding looks), **being pretty** (females) or **handsome** (males) (excluding personality), and **being beautiful** (females) or **ultra good-looking** (males) (excluding personality). There was no correlation between attractiveness of looks and of personality. There was good agreement between males and females that one female had an attractive personality, that another was pretty and beautiful, and that one male was handsome and ultra good-looking. However, one female was put as beautiful by females but not by males; one male was highly rated for looks by males but not by females, and one female had half the males putting her as having an attractive personality but none of the females did. A Chinese female student told me that it was difficult for a really good-looking girl to be popular with other girls. She said of one attractive man-loving Hong Kong girl, "The girls don't like her because the blokes all like her too much."

Although Dorothy Parker (1893–1967) wrote 'Men seldom make passes/At girls who wear glasses', it is not true, or the vast numbers of young Orientals who wear glasses would have a difficult love life. Other Dorothy Parker comments include, 'Brevity is the soul of lingerie', and, 'I require only three things of a man. He must be handsome, ruthless and stupid'.

Chapter 5.4 described some of the things people choose to do to **make themselves more attractive to the opposite sex**. One thing omitted was the extensive use of scent, aftershave, deodorants, etc. In 1770 an English law was passed to prevent women using artifices such as heady perfumes to 'seduce and betray' men into marriage. Its list (Thompson, 1927) makes interesting reading today, but fortunately for most women the law was impracticable to enforce. It included:

That all women, of whatever age, rank or profession or degree, whether virgins, maids, or widows, that shall from and after this Act impose upon, seduce and betray into matrimony any of His Majesty's subjects by the use of scents, paints, cosmetic washes, artificial teeth, false hair, iron stays, hoops, high-heeled shoes, or bolstered hips, shall incur the penalty against witchcraft, and the marriage ... shall be null and void.

(i) From Plates 2–20, choose the two most attractive females and males.
(ii) Compare your choices with those of your family and friends.
(iii) Try to give reasons for your four choices.

10.4. The first period, menopause, hysterectomies, miscarriages, hormone replacement therapy, endometriosis

The age at which girls have their first period varies even within families and has decreased. In America it averages 12, varying from 8 to 15. It starts about two years after breasts begin developing. In America, black girls tend to develop a year earlier than white girls. About half the former start to develop breasts or pubic hair by the age of eight, compared with 15% of white girls. In most Western communities, a girl's first period is often commented upon among female relatives: "She's a big girl now", but is not marked by ceremonies. In some communities there are elaborate ceremonies, as in this one described by an attractive 13-year-old.

Personal account. *A village girl's coming of age in Sri Lanka*

I don't go to school these days because I attained age on the 16th October. My festival is on the 29th October. I can't go to school until then. My sister helps with everything. When a girl attains age [has her first period] she is put into a room and isn't allowed to face any males. Only females are allowed to see her and talk to her. If she wants the toilet, she is covered with a sheet from head to toe and led there so that any male cannot see her face.

Her mother and an elderly lady go to an astrologer who asks the mother when she got to know about her daughter's period. He adds, divides, subtracts and multiplies numbers and tells the exact time she attained age, what she was wearing and if she was alone or with others. He forecasts her future and specifies an auspicious day and time to bathe her, and what colour frock she should wear.

Sometimes a girl will have to stay in the room for a week till the auspicious day. Usually the auspicious time is early morning at 5.45 a.m. or 6.00 a.m. It differs. On that day a bath is filled with water. A new small clay pot is brought and in it they put sandalwood powder and jasmine flowers, then the first pot of sweet-smelling water is poured at the auspicious time on her hands. All this is done in an enclosure where nobody can see her. She is bathed by a dhobi [washer] woman or by her own mother, as decided by the soothsayer.

After that she is covered with a white cloth and led to the main door of the house. On the threshold she splits a coconut in two with a large knife. They say that this action foresees how many sons and daughters she will have, then she is led to the table full of sweetmeats and milk rice.

Here she is at last free of the sheet covering her and is allowed to gaze at the fully laden table. She lights the traditional oil lamp and is clothed in the new dress of the colour specified by the soothsayer. Presents are given by parents and other relations. Gold earrings, a gold chain and bangles are put on her by her mother or a relation.

The girl worships her elders [prostrates herself before them] and gets their blessings. She can now meet and talk to male family members. Until she is allowed to look at the specially arranged table, lights the oil lamp and is appropriately dressed, she isn't allowed to see any males. All the male family members stay aside until these traditional customs are done. After that they wish her well and give her presents of money, dress material, etc. A lunch or dinner is given to invited guests like neighbours and relations.

H. G. Melani Dinusha

In many communities, including some Hindu ones, women are deemed unclean during their periods, with many restrictions on what they do, where they go and who sees them. A week before the first day of a woman's period, she may suffer from premenstrual syndrome (PMS, premenstrual tension). It affects about 70% of women of child-bearing age, with one in 20 severely affected physically and mentally. Symptoms can include weight gain, weepiness, itching, headaches, irrational behaviour, and even aggression and violence. Hormonal treatments may be given, or diuretics to reduce water retention.

The **menopause** occurs between 45 and 55 in most women, with an average of 51. As women age, more and more ovarian follicles degenerate. When few are left, oestrogen production by the ovaries diminishes and eventually ceases. Women differ enormously in menopausal effects. Some have no symptoms except irregular periods and then no periods. Others have hot flushes, night sweats, big mood swings, emotional effects, anxiety, headaches, etc. After the menopause, a lack of oestrogen leads to vaginal dryness, gradual shrinking of the sex organs, diminished libido and some loss of bone calcium. In Britain, the average age for the menopause is 51 and women suffering from symptoms such as hot flushes are usually offered HRT for two to five years.

Some women have early menopauses: one in a 100 before 40; one in 1,000 before 30, and one in 10,000 before 20. Liz Fraser went through the menopause at 16, with mood swings, hot flushes, aching joints, shouting at people for no reason and frequently bursting into tears. She married, was put on the Pill to replace her sex hormones, and had a testosterone implant to increase her sex drive.

The frequency of miscarriages rises with a woman's age: 14% of pregnancies for women under 35; 25% by 40; 50% by 47. Many miscarriages between weeks eight to 12 of pregnancy are of babies with chromosomal defects, the frequency of which increases in older women, especially after 40.

Women who have very heavy periods or other womb problems often have the womb removed, usually with the ovaries to avoid ovarian cancer. Such hysterectomies cause an early menopause from removal of the ovaries and the women are usually given sex hormones and can continue to be sexually active. As one put it: "The surgeon took away the cradle but left the playpen!" In Britain, about 80,000 women a year have hysterectomies, at an average age of 45. A Dutch study found that the operation usually improved their sex lives.

Hormone replacement therapy (HRT) involves giving postmenopausal women sex hormones, especially oestrogen. It can increase libido and retard bone demineralisation but carries health risks as well as health benefits. It has large numbers of satisfied recipients.

Endometriosis affects about two million women in Britain and 70 million worldwide. Endometrial cells lining the womb can migrate and become attached to organs in the pelvic cavity. Every month, they break down and bleed as if still in the womb. The clots accumulate and can cause acute pain, internal spasms, bowel and bladder difficulties and infertility. Laparoscopic surgery and hormone treatments can help but are not always satisfactory.

10.5. Natural and assisted reproduction, abortions, infertility, contraception

Natural reproduction involves a male's sperm being ejaculated into a female's vagina at a fertile period in her menstrual cycle. The classic mating position, the 'missionary position', has the female on her back, with the male on top, but there is a big diversity of positions. Some are described in the ancient Sanskrit *Kama Sutra* and in books such as *The Joy of Sex: A Gourmet Guide to Love Making* by Alexander Comfort (Beazley, 1972). Fertilisation and development are described in Chapter 21.

Technology has enabled the survival of even more premature babies, although very premature ones often suffer physical and mental defects. Amilla Taylor was the world's youngest surviving baby, born at 21 weeks 6 days after *in vitro* fertilisation (IVF), weighing 10 oz (284 g), 9.5 in (24 cm) tall. She was breathing unaided at her Caesarean delivery. After four months in an incubator in Miami, she had grown to 26 in (66 cm) tall and weighed 4.5 lb (2 kg), but still needed oxygen supplied.

A study commissioned by www.evriwoman.co.uk found that 40% of pregnancies in Britain were accidental, in older as well as younger women. The majority of mothers-to-be wanted to keep the baby but one fifth did not. Some accidental pregnancies came from unprotected sex but half were due to contraception problems — forgetting to take the Pill, split condoms, etc.

There have been campaigns to reduce the number of unwanted teen-age pregnancies, as children start having sexual intercourse at earlier ages. A report from the Family Education Trust in 2004 found that teenage pregnancies had risen fastest in UK areas where the Government had specifically targeted resources to reduce them. Explicit sex education leaflets and issuing free condoms to underage children had encouraged them to have sex. One in 10 babies in England is born to a teenage mother.

The Department of Health reported in 2006 that more than 18,000 girls under 18 had abortions the previous year, with more than 1,300 having their second abortion and 90 having their third. In China, under the one-child policy, there have been at least 350 million forced abortions and sterilisation of 200 million men and women.

Artificial insemination by donor (AID) involves injection of sperm by syringe into the vagina and cervix at the right phase of the menstrual cycle. In Britain from 1991 to 2003, 55,936 women were treated by AID, resulting in 13,512 babies. A Danish sperm donor fathered about 100 babies before he was found to be carrying a gene for neurofibromatosis 1, and ten of the babies developed this disease which causes nerve tumours and other defects (Chapter 11.11). A spokesman for the Danish clinic said that it tests for a range of genetic disorders but could not test for all with 100% certainty. According to the Human Fertilisation and Embryology Authority (HEFA), 47,800 women in the UK underwent fertility treatment in 2012, a rising trend. In the UK, there are 1.9 million single-parent families, 29% of all families. The youngest single parents in Britain had a daughter in 2014 when the mother was only 12¼ and the father was 13.

When interviewed in *The Sunday Times* (6/4/2014), a single nurse from Leeds, Naomi Watson, said that at 35 she found she had low fertility and decided to pay for IVF. She chose a sperm donor from a catalogue, checking the men's details. She had twin boys.

Through IVF of donor eggs, some women have given birth to their grandchildren. A woman from Essex, Lata Nagla, had no womb but produced eggs. Her mother, aged 43, who already had four children, volunteered to carry her daughter's child. After the daughter's two rounds of IVF in India with her husband, Aakash Nagla, five embryos were transferred to her mother's womb. Healthy twins were born by Caesarean section and were registered as the grandparents' children but were adopted by the genetic parents, Lata and Aakash.

About 1 in 5,000 women in the UK have Mayer-Rokitansky-Küster-Hauser (MRKH) syndrome and are born without a womb. Many have functioning ovaries and hormonal cycles, but not periods. By 2014, nine women in Sweden had received womb transplants from altruistic live donors, and one woman had successfully had her own baby, from her own fertilised egg.

Male infertility causes a couple's infertility in one third of cases. The quality and number of sperm produced have decreased over 40 years. Possible reasons include pesticides, female hormones in the water supply from women taking them, and higher scrotal temperatures from more heated homes and tighter male underwear. Low fertility occurs in 9% of British men. Men are usually unaware of their low fertility.

Heat impairs sperm production. Long exposure to hot showers or baths can have that effect. Men who drive for long periods often have low sperm counts as sitting raises scrotal temperatures. French research showed that professional drivers have lower sperm counts, more sperm abnormalities, and that their partners take longer to conceive. After two hours of driving, mean scrotal temperatures rose from 34.2° to 36.2°C.

Erectile dysfunction (ED), failure to get and sustain an erection, affects many men, especially older men and diabetics. It is acutely embarrassing. Its causes can be hormonal, psychological or physical, including high blood pressure, high cholesterol, smoking and narrowing of the penis arteries. In 1998, Viagra was launched to combat ED. Newer drugs include Cialis and Levitra.

Some couples have a big decline in the frequency of sexual intercourse after the arrival of their first baby, with stress, tiredness and children often given as reasons. According to the ONS in 2006, one in eight women claimed to have had no sex in the past year. After 35, the 'no sex' figure is about 10% for men and women, rising to 12% for women between ages 40 and 44, and 10% for men. The lack of eroticism for couples is not bad if mutually agreeable, and comes as relief to many who are satisfied with companionship.

For couples unable to have children, adoption is an option. The number of adoptions in England and Wales in 2011 was only 4,734. Adoption rules are very restrictive, greatly reducing the numbers of adoptions. In Australia in 2013, only 339 children were adopted, 129 of them from overseas, especially Asia.

The frequency of **induced births** has been rising in England, reaching 22% in 2012. The Royal College of Obstetricians and Gynaecologists recommends that women with uncomplicated pregnancies should be offered induction between weeks 41 and 42 'to avoid the risks of prolonged pregnancy', which can harm mother and baby.

Many people want **sex without reproduction**, with contraception available. There is the Pill for women and attempts at male hormonal contraceptives. There are spermicides and barrier methods such as condoms for males. For females there are condom-like devices to fit inside the vagina, or diaphragms, or the Dutch cap which fits over the cervix. The coil, worn inside the womb for long periods, is cheap, preventing fertilised eggs from implanting. Permanent methods are cutting the man's vas deferens or tying the woman's Fallopian tubes. The Roman Catholic Church is generally against contraception but permits the use of the 'safe period' (rhythm method), restricting intercourse to the non-fertile part of the menstrual cycle. It has been called 'Vatican roulette' as it is unreliable. Withdrawal of the penis before ejaculation is liable to failure as some sperm may have already been shed.

There are sexual activities which do not risk conception, including kissing, cuddling and 'heavy petting', such as stimulation of the male or female genitals by hand or mouth. Anal intercourse can transmit venereal diseases such as HIV.

10.6. Twins and other multiple births; Siamese twins

Twins, triplets, quadruplets and higher multiples occur naturally. With fertility treatments and implantation of several embryos after *in vitro* fertilisation, multiple births are common. Multiple Births Canada gave these figures for multiple births with no fertility treatment. Twins, 1 pair in 90 births. Triplets, 1 set in 8,100. Quadruplets, 1 set in 729,000. Quintuplets, 1 set in 65,610,000. A figure for sextuplets, from a different source, is 1 set in five billion births, with only six authentic records before fertility treatments began. The number of multiple births in Canada is rising, with 16% of them coming from fertility treatments, including 60% of triplets, 90% of quadruplets and 99% of quintuplets. In 1895, Hellin suggested that twins occurred once in 89 births, triplets once in 89^2 births (1 in 7,921), quadruplets once in 89^3 births (1 in 704,969), quintuplets once in 89^4, (1 in 62,742,241). Sextuplets would be expected at a frequency of 1 in 5.6 billion. These predictions agree well with those Canadian data.

The frequency of twins differs between populations. A Nigerian Yoruba population had 4.5% of all births being of twins, while this was lower than 0.8% in several South American populations and among Chinese and Japanese.

As multiple births involve higher mortalities than single births, the frequency of multiple births at conception is higher than at birth, with many spontaneous abortions. American research suggests that twin embryos are present in about 15% of conceptions. In most cases, however, one twin is lost early in pregnancy without this being noticed.

Julia Cook (*The Daily Telegraph*, 28/4/2014) had a routine ultrasound scan at 12 weeks of pregnancy. It showed a healthy daughter and also a twin, one third smaller, curled up in its own sack, having died about four weeks earlier. The sonographer assured Julia that normally the dead twin did not miscarry and was slowly absorbed into the body. The daughter was born healthy. The dead twin rarely causes problems for the live twin unless they share a placenta.

Twins can be **identical**, from the division of a single fertilised egg, so they carry identical genes and are of the same sex, or **non-identical**, from two separate fertilised eggs, with different sperm, with only half their genes in common. Non-identical twins can be same or opposite sexed; apart from their time of birth, they are no more alike than siblings. Identical twins occur at similar frequencies all over the world, at about 3.5 per 1,000 births.

Identical twin formation from one egg can occur at different stages up to two weeks after fertilisation. If the daughter cells separate after the first division of the egg, identical twin embryos could implant far apart or close together. If separation occurs at a later stage, there may be a single implantation, with two embryos developing. Two thirds of identical twins share a placenta and one third have their own, as do non-identical twins.

One complication which sometimes arises from sharing a placenta is **twin-to-twin transfusion**, caused by the development of bloods vessels linking them, so blood can transfuse between them. One twin may lose blood to the other, fail to thrive and even die.

Non-identical twins come from the mother maturing two or more eggs in one menstrual cycle, with fertilisation. Non-identical twins have variable frequencies, from a low rate of 2 to 7 per 1,000 births in many

Oriental communities, to 45 per 1,000 in Nigeria. In interracial crosses in Hawaii, the frequency of non-identical twins was correlated with the mother's race, not the father's, because multiple egg production is the determinant. Multiple births from fertility treatments are mainly non-identical, from implantation of separate fertilised eggs.

Apart from fertility treatments, the mother's age is the main factor in the frequency of non-identical twins, which rises rapidly as women increasingly delay pregnancy, say for career reasons. For adolescent mothers, the chances of a multiple birth are 6 per 1,000 pregnancies, rising to a peak of 16 per 1,000 for ages 35 to 39, going down to 13 per 1,000 at 40 to 44, and 8 per 1,000 in women of 45 and over.

The Dionne girl quintuplets, born in Canada in 1934, were the first recorded case of all five quintuplets surviving birth and adolescence. By blood and other tests, they were shown to be identical, from one fertilised egg. Most cases of quintuplets are combinations of identical and non-identical babies. **X-inactivation** usually occurs before separation of identical twins so that identical female twins who are heterozygous for X-linked characters such as haemophilia, Duchenne muscular dystrophy or colour blindness may differ in those characters. Two of the Dionne identical quintuplets were colour blind and three were not.

Triplets can come from one, two or three separate eggs. In 2003 in Turkey, Fatma Saygi was pregnant with triplets for the sixth time, with no hormone treatments. An Italian woman, Madalene Granata, bore 15 sets of triplets between 1839 and 1886. The Walton girl sextuplets were born in England in 1983, the world's first all-female surviving sextuplets, after their parents' 13th attempt at fertility treatment.

Twins of opposite sex are non-identical. If twins are of same sex and differ in many characteristics, they are probably non-identical. DNA testing is the best way of finding out, although blood groups and finger prints have been used. About 95% of the variation in the number of fingerprint ridges is genetic, with 5% being environmental or chance, so identical twins do not have identical finger prints but they are much more alike than for non-identical twins.

Non-identical twins may have different fathers if the mother mated with different men in a short space of time. In Austria, after its take-over by Nazi Germany, it was advantageous not to have Jewish origins. A non-Jewish mother of non-identical twins had a Jewish husband but claimed that her 25-year-old twins were conceived during an affair with a non-Jewish man, giving the twins no Jewish ancestry. All five were tested for blood groups ABO and MN. The husband was B, M; the mother was O, M; the ex-lover was A, MN; the twin brother was B, M, and the twin sister was A, MN, consistent with the husband being the boy's father and the ex-lover being the girl's father, providing her with the A and N alleles which her parents lacked. In a different case, one father was white and the other was black, giving very different non-identical twins.

Identical twins may show physical diversity. One may get more nutrients in the womb or be better placed for space. One may suffer birth trauma when the other does not. After birth, twins may get different diseases or live in different environments. With a pair of Asian boy identical twins at the age of 14, the top of the shorter one's head only reached the bottom of his twin's nose. There was a case of two white boys where one identical twin had a cleft lip and palate and the other had neither.

There are cases of extraordinary similarities between identical twins reared apart, including this from the Minnesota Study of Twins Reared Apart. Jim Springer and Jim Lewis are identical twins who were separated at four weeks and did not meet again until they were 39. Both were christened Jim by their adoptive parents, both worked as part-time deputy sheriffs, had holidays in Florida, drove Chevrolet cars, had a dog called Tony, and had married and divorced a woman named Betty (different women). One named his son James Allan and the other named his James Alan. Both twins were good at maths but not spelling, and liked carpentry and mechanical drawing. Both chewed their finger nails down to the quick, had almost identical smoking and drinking habits and sleep patterns. Both had haemorrhoids and gained 10 pounds in weight at the same time.

There have been trials where having a twin allowed a criminal to get off. One of the infamous Kray identical twins was alleged to have acted alone in a crime. The defending barrister persuaded the jury that it could not be proved which twin committed the offence, so the defendant was acquitted. French police were hunting a serial rapist in 2013 and had DNA evidence which led to a pair of identical twins. The normal analysis of 400 base pairs was not enough to decide which twin was guilty. Similarly, in

England, identical twins Aftab and Mohammed Ashgar could not be separated on DNA evidence in a 2013 rape trial.

Apart from new mutations, identical twins have identical DNA sequences, but DNA is modified by methylation of cytosine bases and by histone acetylation. These epigenetic changes can affect gene action. Fraga *et al.*, 2005 found that identical twins have similar epigenetic changes in their first years but progressively diverge. There were different gene expression profiles between 50-year-old identical twins, with four times as many differences as between three-year-olds. The earlier the twins had separated, and the greater the difference in lifestyle, the greater were these epigenetic DNA changes. This could be used forensically.

Multiple births put more strain on mothers and babies than do single births, with some effects on fathers. As a mother has limited resources, the more babies per pregnancy, the lower the average birth weight of each baby and the higher the risk of defects and the greater need for neonatal intensive care. Almost half of all twins are low birth weight and/or premature, rising to more than 90% for triplets, quads and quins. Children in multiple births are five times more likely than those in single births to have birth defects and/or disabilities. Twins, triplets, etc., have to compete for food and space within the womb, and for food, care and attention after birth. The IQ of twins is on average five points lower than for non-twin siblings. Identical twins suffer more than non-identical twins from a higher neonatal death rate and from a higher frequency of congenital malformations, possibly relating to having more crowding than for non-identical twins. Most twins are healthy and catch up to normal weights as they grow.

About 50,000 children a year are born worldwide from **IVF**; they are sometimes called '**test-tube babies**', although fertilisation is usually in a Petri dish. Louise Brown, the first test-tube baby, born in 1978, was still healthy at 36 in 2014, when there were about six million IVF babies. A study of Swedish children (Healy and Saunders, 2002) found that 2% of babies there were born after IVF, with 45% of those by intra-cytoplasmic sperm injection.

Maricia Tescu in Romania has two wombs and became pregnant in both. One womb produced a healthy son in December 2004, and the second produced a son nine weeks later, in 2005. One woman in 50,000 is born with a complete **double uterus**, and 5% of women have some degree of womb division. Hannah Kersey in Bristol has two wombs. Identical twin girls were born from one womb and one girl from the other womb on the same day in 2006.

Siamese (conjoined) twins arise from incomplete splitting of one embryo. Mary and Elizabeth Chulkhurst were born in 1100 in Biddenden, Kent, joined at the shoulder and hip, legend says. They lived for 34 years, then Mary was ill and died. Elizabeth was advised to be separated from her twin's body but refused, saying, "As we came together, so we will go together." She died a few hours later.

Siamese twins are named after Chang and Eng, born in 1811 in Siam (Thailand). They were joined only by a tissue bridge 10 cm (3.9 in) wide from the lower end of the breast bone to the navel. It contained liver tissue connecting the two livers, and separation surgery would probably have been unsuccessful although it would succeed today. They married a pair of sisters, and being joined did not inhibit their love lives as Chang had 10 children and Eng had 12. Eng was unaffected when Chang had a stroke at 61, but when Chang died of bronchitis two years later, Eng died two hours later.

The success of surgery to separate Siamese twins depends on what parts are joined or held in common. In 2001, staff at Birmingham City Hospital made a successful separation of twins joined at the spine. The girls had emergency surgery to separate a joined piece of gut, then were allowed to grow stronger for three months before the main operation. A silicone balloon was placed under the skin 10 weeks before the operation to stretch the skin to cover the separation wounds on both twins. In the successful 16-hour operation, the surgeons separated their two spinal canals and their lower spinal cords which were joined in the lower back. The mother noticed that her identical twins had different personalities, with Emma more dominant, vocal and demanding than Sanchia. In Sudan in 1986, boys Hassan and Hussein Salee were born joined from chest to hip. A year later they were successfully separated in London in an operation lasting 16 hours, leaving them each with one leg. They had their livers separated and a mutual limb removed.

Our knowledge of human genetics has benefited enormously from the **study of twins**. See heritability, Chapter 1.3. By comparing the differences between identical and non-identical (of the same sex, to avoid sex differences biasing the results) twins reared together and reared apart, one can study the extent to which variation for particular characteristics results from genetic and from environmental variation. Many countries keep twin registers of twins who are willing to participate in such studies. There are twin registers in Britain (at the Twin Research and Genetic Epidemiology Unit, St Thomas' Hospital, London), Sri Lanka and Australia. The Information Leaflet from the St Thomas' Unit states that: 'osteoporosis is 75% genetic; disc degeneration causing back pain is 60% genetic; genes control body shape; timing of the menopause is genetic; musical ability is 80% genetic; 'wacky' humour is due to upbringing; short-sightedness is 85% genetic; blood clotting properties are mainly genetic; moles and freckles are strongly genetic'.

10.7. The sex ratio

The normal sex ratio of one male to one female is fine for finding partners. In India, decades of sex determination before birth, a strong preference for sons, and the abortion of female foetuses, plus selective infanticide, have left a huge imbalance between the sexes, leaving many males unable to find brides. In the Fatehgarh Sahib district, there were 754 girls per 1,000 boys. In some villages dominated by Jat Sikh farmers, there are only 550 to 600 girls per 1,000 boys. The sex tests before birth costing only £8 are illegal but widely used.

In India, there is a history of female infanticide by poisoning, drowning, starvation or abandonment. The Indian Government estimated in 2007 that about ten million girls had been killed by their parents in the last 20 years. Girls are considered a liability because of expensive dowries required at weddings. Dowries cause enormous trouble in many cultures, but would not be a problem to Mr and Mrs Schwandt of Michigan. In 2013 the wife gave birth to her 12th successive son, but has no daughters.

Because of different amounts of DNA in X-bearing and Y-bearing sperm, DNA staining, measurement of individual sperm's DNA and extremely rapid cell-sorting machinery, one can have selective enrichment of sperm for female- or male-producing sperm for IVF. In San Diego in 2013, a private clinic offered sperm with an 82% chance of a boy being conceived, or sperm with a 93% chance of a girl.

At conception there are about 130 male embryos per 100 female embryos. Male embryos and foetuses suffer more spontaneous abortions than do female ones, with a sex ratio at birth of about 105 males per 100 females. There is diversity in sex ratios, with the number of males per 100 females being 115 in Korea, 105 in America and 100 in the West Indies, at birth. As men tend to die younger than do women, the sex ratio changes with age. Figures for America in 1977 for people aged 65 or more were 68 men per 100 women, so different from 105 males per 100 females at birth.

In Britain, rural areas tend to have the most available single males per single female, while cities tend to have more single females. In the 2001 Census, females in their twenties outnumbered men in their twenties by 81,300 in England and Wales, while women in their thirties outnumbered men by 132,633.

Wars kill far more men than women and can seriously affect the sex ratio. The deaths of ten million military men in World War 1 (1914–1918) left many unattached females who would otherwise have married.

Many parents want a balanced sex ratio amongst their children. Suppose a couple has five children. The chances of five boys is $(\frac{1}{2})^5$, 1 in 32, so is the chance of five girls, so the chance of all five children being of the same sex is 1 in 16. Mandy and Bill Cribben had seven girls, a 1 in 128 chance if at random, but there could have been a recessive sex-linked lethal gene on one of the mother's X chromosomes, which would kill half her sons, with the daughters being unaffected carriers or normal. Mandy conceived again and, to her and Bill's relief, had a boy. Bill commented that with seven girls (aged 18 months to 18) it was already bad enough getting into the bathroom without another girl coming along.

10.8. Hormone imbalance; men with breasts and women with beards; sex tests and sports

In development, the pituitary gland stimulates testicles to produce male sex hormones and ovaries to produce female ones. The adrenal glands in women normally produce small amounts of male sex hormones. With age, there may be over-secretion of male hormones by the female adrenal cortex, giving a deeper voice, heavier facial hair and other masculine signs.

Sex hormone imbalances cause oddities such as **gynaecomastia**, where men have over-development of one or both breasts, which may have a watery secretion, or women may develop a beard. **Hereditary early baldness** usually only affects men, as it needs sufficient testosterone. Women can show hair thinning if they have raised levels of male hormones.

When contraceptive pills containing female sex hormones were first manufactured, men in those factories sometimes absorbed the hormones through the skin or from dust in the lungs. They began to lose sex-drive and to develop breasts. People undergoing sex-change operations take opposite-sex hormones, which influence behaviour as well as physique.

In sports, men have a physical advantage over women. Sometimes the sex of certain competitors has been doubted, especially rather masculine 'females'. In the 1936 Olympic Games, two American women, Stella Walsh and Helen Stephens, were sprint rivals and had masculinised muscle patterns and faces. Stella was nicknamed 'Stella the fella' and always changed by herself. Stella accused Helen of being male but a physical examination showed that Helen had female external genitalia. Stella had ambiguous genitalia and abnormal sex chromosomes. Irina and Tamara Press were Russian sisters who won many gold medals in track and field events. Their sex was often doubted and they stopped competing when being examined naked was introduced in many competitions. It was joked that they were suspected of shop-lifting jock-straps.

The **intersex condition**, with sexually ambiguous external genitalia, has a frequency of 1 in 5,556. There are also chromosomal conditions (Chapter 23) and testicular feminisation (see below) which can cause confusion about sex. At Olympic Games from 1960 until 1996, methods of testing 'female' competitors have included body examination, looking under the microscope for Barr bodies (indicating two X chromosomes), and the polymerase chain reaction on DNA to test for the *SRY* gene which is needed for male sexual development. Compulsory gender verification was abandoned for the Olympics in 1999 as no method was entirely satisfactory (Ritchie *et al.*, 2008).

One female athlete who was banned was Polish sprinter Ewa Kłobukowska who won gold medals from 1964 to 1966 but failed a gender test for the European Cup in Kiev in 1967. She was a genetic mosaic of XX/

XXY, and had previously had testicles removed and was being treated with oestrogens. If she had been tested one year later at the Mexico Olympics she would have been eligible as she had Barr bodies. 'Female' athletes without a Barr body (inactive X-chromosome) were suspended from competition by 1968 in Mexico City. That test would have allowed Klinefelter syndrome males to compete as females (Chapter 23).

10.9. Sex-change operations and hormone treatments

Some people are unhappy with their biological sex, e.g., feeling that they have 'the mind of a woman in the body of a man'. About 1 in 12,000 people is thought to be strongly **transsexual**, adopting the dress and mannerisms of the opposite sex. In Britain, there are about 300 gender reassignment operations a year, two thirds male-to-female, one third female-to-male. Patients range from 18 to 75. About 70% of the female-to-male **transgenders** have sexual relationships with women, and of the male-to-female ones, about 50% have sexual relations with men, 30% are lesbian, 10% are bisexual, and 10% abstain from sex relations.

Male-to-female operations involve the removal of the penis and testicles, construction of a vagina and female genitalia, preserving the nervous supply for the construction of a 'pleasure centre', and perhaps cosmetic facial surgery and operations to reduce the Adam's apple and raise the voice pitch. Female hormone treatments give breast development and reduce facial hair. **Female-to-male operations** involve removing breasts and the womb. Male hormones give a deeper voice and facial hair. The operation may involve **phalloplasty**, construction of a penis from forearm tissue.

George Jamieson was three when he felt that he was a girl born into a boy's body. At 14 he joined the Merchant Navy to prove his masculinity. He hated living a lie and twice attempted suicide. In a mental hospital he was given male sex hormones and electric shock treatment. In Paris he saw a man who had had the operation to become a woman; he saved £3,000 to go to Casablanca for the nine-hour operation when he was 25. He became April Ashley and two months later lost her/his virginity to an American man. Back in London, she became *Vogue's* most popular underwear model until a friend exposed her as a former man. She married an Old Etonian but the marriage was eventually annulled on the grounds that she had been born a man. Even in her late 70s, she looked a very glamorous lady.

Lisa-Lee Dark was brought up as a boy but after the age of 14 was diagnosed as having intersex features. She was given male hormones to make her more masculine. At 20, she was found to have **congenital adrenal hyperplasia (CAH)** which results in excessive production of male sex hormones in early foetal life. It can lead to fusion of the female genitalia and an enlarged clitoris, although the ovaries and womb are intact. She is XX, chromosomally female, and is now treated as a female and feels that she is one. In boys, CAH can cause precocious puberty, with pubic hair, an enlarged penis and rapid physical growth at the age of three or four. Girls have unwanted hair growth, irregular periods and fast growth. CAH affects about one birth in 4,500, with varying degrees of masculinisation.

For the 0.6% of men born with a very small penis (shorter than 7 cm, 3 inches, long when erect, instead of about 12.5 cm, 5 inches), an operation is performed to enlarge it using a flap of tissue, including skin, from a forearm. A urethra is provided and an inflatable penile prosthesis to allow erection for intercourse. The operation was performed at University College London in 2005 on nine men, including three hermaphrodites and two men with hormone deficiencies.

10.10. Castration — eunuchs, castrati and eugenics — and 'the third sex'

The removal of boys' testicles gives a higher adult voice. There is some breast development, sterility and a tendency to grow fat. **Castrati** were valued in Italy for their strong, high voices for church music and operas. The operation was usually done at age six to eight. In Italy, the Catholic Church permitted the castration of boys for singing in church choirs, and only banned it in 1878. Italian castrati were extremely popular in 18th century operas. Their voices had extraordinary power, range, purity and breath control. The last known castrato singer, Alessandro Moreschi, sang in the Sistine Chapel choir until 1913.

Eunuchs were valued as 'safe' attendants on women in harems in the Middle East. Castration has been used to stop men from breeding, for example slaves from rival tribes. 'Chemical castration' with drugs is sometimes used to restrain the sex drive of convicted rapists.

In 2014, the Indian Supreme Court recognised the country's two million *hijras* as a **third sex**, listed as a 'backward caste' entitled to reserved places in universities and government jobs. It includes transsexuals, men with deformed genitals and castrated males. Many *hijras* live in groups controlled by a *hijra* guru, living by aggressive begging. They are feared and can be disruptive at weddings and parties, demanding much money to give their blessing and go away.

The **eugenics movement** believed in improving the human race by eliminating unfit individuals. In 1910, playwright and socialist George Bernard Shaw recommended using lethal chambers to kill off many people because it wasted other people's time looking after them. Charles Davenport, director of the Carnegie Institution of Washington Station for Experimental Evolution at Cold Spring Harbor (which existed from 1904–1921), said that defectives should be allow to die when ill. Others wanted the feeble minded to be stopped from breeding. In a Supreme Court case, the State of Virginia wanted to sterilise a black girl, Carrie Buck, as she was the daughter of a feeble-minded mother and the 'potential parent of socially inadequate offspring'. Virginia won, and as a result, nearly 30,000 American men and women were castrated or sterilised up to 1940, especially in California.

10.11. Hermaphrodites and testicular femininism

In Greek mythology, Hermaphroditos, son of god Hermes and goddess Aphrodite, grew together with nymph Salmacis into one person, with male and female sex organs. So-called **'true' hermaphrodites** (incidence 1 in 1,000) have both testicular and ovarian tissues but do not have full genitalia of both sexes. They are usually mosaics, with mixtures of XX and XY cells. In **pseudohermaphrodites** (incidence 1 in 1,000), only one type of sex-gland tissue, male or female, is present, but one or more features of the opposite sex develops. Males are affected eight times more often than females, usually developing breasts. Female pseudohermaphrodites have excess male hormones, tending to have normal oviducts and womb, but the vagina is small and external genitalia have male tendencies.

Intersex bodies have a frequency of about 1 in 100,000 and are usually diagnosed early in life. A remarkable case was reported in *The Sunday Telegraph* (8/2/2015) of a man, Rob, who had normal sexual relations as a male, but at age 37 an MRI scan showed that he had a functioning womb, ovaries and cervix, but no vagina, in addition to external male genitals. He was booked in for a hysterectomy. About 120 babies a year are born in Britain with this Persistent Müllerian Duct Syndrome, with external male genitals and internal female organs. **Testicular feminisation** (androgen insensitivity syndrome) is a fascinating case of male **pseudohermaphroditism**, affecting 1 in 62,000 XY individuals. Sufferers are chromosomally male, XY, but the X chromosome has a recessive gene giving insensitivity to male hormones (androgens). They look feminine, with good breast development and external female genitalia. They usually have unseen small internal testicles and a shallow vagina. Being unaware of their condition, they often marry as females and have normal sexual relations with men, but lack menstruation and cannot have children.

10.12. Nature is sexist — sex differences in disease susceptibility

Females live longer than males, on average (Chapter 22). The following figures are very approximate and differ between societies. In Western communities, men are six times more prone than women to alcoholism, three times more likely to get skin cancer, 25 times more likely to get hardening of the heart arteries, seven times more likely to get duodenal ulcers, twice as likely to get leukaemia, and one and a half times as likely to commit suicide as are women. Females are 10 times more likely than males to get gall bladder cancer, much more likely to get anaemia (menstruation is one reason), twice as likely to get influenza, six times as likely to get migraine, and six times as likely to get whooping cough.

Possible reasons include contact with infected individuals, with mothers catching diseases from children, stress patterns in different occupations, and different social expectations, such as some men living up to a hard-drinking, go-getting, stressed, aggressive stereotype. Some men claim that women's behaviour drives them to alcoholism.

Some diseases are sex-specific, such as cervical cancer in females and prostate cancer in males. The incidence of diseases caused by sex-linked recessive genes is very different in the two sexes, with males suffering much more frequently than females (Chapter 8.3).

10.13. Polycystic ovary syndrome and ovarian tumours

About 7% of women suffer from **polycystic ovary syndrome** (PCOS), yet some sufferers do not know they have it. Even with one and a half million sufferers in Britain, it is not widely known. It is a leading cause of infertility. The most frequent symptoms are irregular or absent periods in 75% of sufferers, reduced or no fertility in 50%, acne, oily skin, excess weight, excess hair on the face, chest and abdomen, decreased breast size, enlarged clitoris, increased muscle size, a low-pitched voice and male pattern baldness, with individuals suffering to different extents. The main effects come from raised testosterone levels. PCOS can have serious psychological effects, making sufferers feel 'fat, hairy and infertile'. It has genetic and environmental components.

Polycystic ovaries have small fluid-filled cysts around their outside from failed egg follicles. The ovaries are usually twice the normal volume. Treatments are directed at particular symptoms, e.g., male-hormone-blocking drugs, and the contraceptive pill can make periods regular. Sufferers are more likely to develop type 2 diabetes and are often given an insulin-sensitising drug to help with weight control and ovulation.

Ovarian tumours may be benign or cancerous, often resulting in the overproduction of testosterone, with similar symptoms to PCOS. Malignant tumours are surgically removed. **Ovarian cancer** is hereditary in 5 to 10% of cases. Mutations in breast cancer gene *BRCA1* are associated with about 40% of hereditary ovarian cancers. Female carriers of *BRCA1* mutations have a 23% risk of ovarian cancer by 50, and 63% by 70. About 1 in 350 people have such mutations, usually autosomal dominants.

Survival for five or more years after diagnosis is just 29% in Britain, 33% in Europe. More than 14,000 British women are diagnosed with gynaecological cancers each year. Almost half are ovarian cancers, with a poor outlook because most are discovered too late.

The frequency of ovarian cancer is higher in affluent white women than in others. The risk goes down if a woman has children and is substantially reduced in women who take or have taken the contraceptive pill. Ovarian cancer has a life-time risk in Europe and America of 1 in 48 in one study, 1 in 75 in another, compared with a lifetime risk of 1 in 11 for breast cancer.

10.14. The effects of human inbreeding; incest; first-cousin marriages

Most cultures and religions ban intercourse between close relatives, e.g., the Bible, Leviticus 18:6, 'None of you shall approach to any that is near kin to him, to uncover their nakedness', or, more bluntly, 'No one is to approach any close relative to have sexual relations'. Incest leads to inbreeding and the **expression of harmful recessives**. There are about three recessive lethal genes per person, in the heterozygous, unexpressed, condition, but the true number must be much higher as those giving death long before birth are largely undetected.

In Britain there are bans on marriage of an adopter and the person adopted. There were bans on marriage between in-laws but that was overturned by the European Court in 2005 for those who had separated from their original spouses by divorce or the spouse's death. The case was brought by a divorced man of 59 who wanted to marry his divorced daughter-in-law, who was 20 years younger. Step-children can only marry a step-parent if they are grown up and have never lived together.

In Japan after World War 2, studies were made on survivors of the atomic bombs, giving data on the effects of inbreeding as well as of radiation. Where parents were first cousins, the frequency of malformations in the children rose by (not to) 48% compared to children of unrelated parents; stillbirths rose 25% and infant deaths rose 35%. The inbreeding in first-cousin matings was clearly deleterious.

First degree relationships, with parent-child and full-sib (siblings are brothers and/or sisters) matings, are defined as incestuous, with half their genes in common. Incest is usually banned but many communities tolerate or encourage first-cousin marriages, especially in India, Pakistan and the Middle East. The term **consanguineous**, for matings between close relatives, extends as far as second cousins.

There are confidential incest registers, recording the consequences of mating between close relatives. Data from 31 children from father/daughter and brother/sister matings showed that only 13 (42%) were normal and 18 were handicapped, with six dying as young children. Deleterious recessives, e.g., cystic fibrosis, PKU and galactosaemia, will be expressed more often in the offspring of consanguineous marriages than with unrelated parents.

The frequency of marriages between relatives differs greatly between communities, depending on religion, customs, laws and population size. In 1950s, in Baltimore, USA, the frequency of first-cousin marriages was 0.05% while in rural India, in Andhra Pradesh, it was 33%, with 9% of maternal uncle/niece marriages, but a much lower incidence of paternal uncle/niece marriages, as if it didn't matter if the relationship was on the female parent's side. In a very small population or a small religious group, even random mating may involve relatives such as first or second cousins. In Hopi Indians, with much marriage between close relatives, albinism (recessive) has a frequency of 1 in 121 compared to that in Whites, 1 in 20,000. Bittles (2002) gave an account of that frequency in India. It differs greatly between religions, castes, social levels and regions, being common for Muslims and South Indian Hindus (especially uncle-niece and first-cousin unions between a man and his mother's brother's daughter) but not North Indian Hindus. In those Southern Hindus, consanguineous marriages average 20 to 45%, and when they migrate to Britain or America, they often retain that.

Figures for consanguineous marriages given in www.consang.net (2005) include Srikakulum, India, 72%; Nyertiti, Sudan, 71%; Nubia, Egypt, 65%; gypsies in Boston, 62%; Riyadh, Saudi Arabia, 31 to 55%. While there are clear genetic disadvantages to such marriages, benefits can include retention of property and goods within the family; the woman's position is strengthened by kinship ties as a deterrent to ill-treatment by the husband or his family; dowry demands may be lessened for relatives; there are fewer problems in finding a mate of a suitable religion, culture, language and social group; better knowledge of the spouse's family background.

In Britain, **first-cousin marriages** are most common amongst Muslims, with the deleterious genetic consequences being partly offset by earlier arranged marriages, earlier first pregnancy and more children per family than in other groups. Here Pakistanis account for 30% of births with recessive genetic disorders but only 3.4% of births. About 55% of Pakistanis in Britain are married to first cousins, and more than 75% in Bradford.

Offspring of first-cousin marriages on average have slight reductions in height, girth and aptitudes (IQ down about 4 points), compared to non-inbred children, and suffer 4% more pre-reproductive deaths. The frequencies of sufferers from genetic disorders who have first-cousin parents are 15% with albinism, 10% with PKU, 26% with xeroderma pigmentosum (Chapter 13) and 54% with microcephaly (very small head). In Iran, 46% of children with hearing loss were offspring of first cousins.

Inbreeding is useful in exposing harmful recessive genes to selection, so reducing their frequency. Continued inbreeding should eventually adapt a group to inbreeding. The pharaohs of ancient Egypt practised brother-sister marriages for many generations and yet were generally successful. One cannot be sure whether the stated royal parents were the actual father and mother.

An example of a rare recessive gene made homozygous and being expressed after inbreeding is that of the blue-skinned people of Troublesome Creek, near Kentucky. Martin Fugate married Elizabeth Smith and both were unaffected carriers of a rare recessive gene for being unable to convert methemoglobin to haemoglobin. There was much mating of relatives. After many years some Fugate descendants had blue skins as they were homozygous for the recessive gene, accumulating blue methemoglobin.

A nasty case of incest and murder is that of Marcus Wesson, 57, in California. He had nine children by six mothers, two of whom were his daughters and one was his niece. He sexually abused his daughters, and at his trial in 2004 he was accused of murdering all nine of his children. He had told them he was God.

10.15. Enforced sex

Men rape women and other men; less frequently, women rape women or, rarely, women force men to have sex with them. A frequent legal issue concerns 'date rape', where a woman claims that her boyfriend raped her and he claims that intercourse was consensual, after affectionate behaviour.

Throughout the centuries, military men have frequently committed rape, including the Viking invaders of Britain. Human diversity includes some very different attitudes to sexual offences within and between cultures, for example, in whether a man's forcing himself on his unwilling wife is a crime or his right. 'Marital rape' is permitted in most Muslim communities. Under Islamic law, rape can only be proven if the rapist confesses or if there are four male witnesses, which really puts raped women at a huge disadvantage. In the Church of England marriage ceremony, 'to have and to hold' is open to different interpretations.

At the lesser, but still wrong, end of offences, Italian men have been renowned for many years for groping and pinching the bottoms of foreign women tourists, regarding it as harmless fun. At a much worse end, in India in 2012, a 23-year-old student in South Delhi was gang-raped on a bus by seven men and eventually died of the appalling injuries inflicted on her. This received widespread international coverage and was widely condemned. It led to public protests in many cities in India against state and central governments for failing to provide adequate security for women, and for not taking seriously the enormous number of sexual offences against women in India. There were much more widespread crimes of rape in times of war, as in these examples from World War 2, which show how soldiers of some nationalities behaved when they felt that they would never face prosecution.

In her book *War in Val D'Orcia* (Penguin, 1956), Iris Origo wrote about the troubles in Tuscany in World War 2 from the invading Germans and the liberating Allies:

> *In the lower part of the property, where the French coloured troops of the Fifth Army have passed, the Goums [Moroccans] have completed what the Germans have begun. They regard loot and rape as the just reward for battle, and have indulged freely in both. Not only girls and young women, but even an old woman of eighty has been raped.*

More than 7,000 Italians — including women, children and men — were raped by Moroccan troops. In Esperia, 700 women out of 2,500 inhabitants were raped, some dying as a result.

In World War 2 and before it, the Japanese enslaved 300,000 to 400,000 foreign women to provide sex for Japanese troops. They treated these 'comfort women' appallingly. The victims included about 200,000 Chinese plus Koreans, Indonesians, Filipinas and some Dutch. Here are two Chinese examples. Huang Youliang was abducted at 14 by Japanese soldiers and taken to a military camp where she was raped for four years. Yang Abu was repeatedly raped by Japanese soldiers and went through several pregnancies. The brutal tortures left her sterile.

References

Bittles, A. H., Endogamy, consanguinity and community genetics. *J Genet* (2002) **81**: 91–98.

Fraga, M. F., Epigenetic differences arise during the lifetime of monozygotic twins. *PNAS* (2005) **102**: 10604–10609.

Gemmell, I. M. M., Injuries among female army recruits: a conflict of legislation. *J R Soc Med* (2002) **95**: 23–27.

Healy, D. L. and Saunders, K., Follow-up of children born after in-vitro fertilisation. *Lancet* (2002) **359**: 459–460.

Law Smith, M. J. *et al.*, Facial appearance is a cue to oestrogen levels in women. *Proc R Soc B* (2006) **273**: 135–140.

Ritchie, R. *et al.*, Intersex and the Olympic Games. *J R Soc Med* (2008) **101**: 395–399.

Thompson, C. J. S., *The Mystery and Lure of Perfume.* (1927) John Lane, London.

Recommended readings

Origo, I., *War in Val D'Orcia.* (1956) Penguin, London.

Ratnatunga, M., *Tales From Sri Lanka: Folk and History,* Rev. ed. (2010) Vijitha Yapa Publications, Colombo.

Chapter 11

Diseases, Disorders, Immunity, Cancer

11.1 Diversity in disease

Humans differ greatly in respect of disease: in susceptibility, diseases suffered, inoculations received and in locally frequent diseases. Children are in close proximity in schools and families, so childhood diseases spread easily. Racial features such as skin colour or nose length influence the chances of skin cancer or respiratory diseases. Even within one family, different body odours affect the likelihood of mosquito bites. Differences in public sanitation (sewage disposal, water treatment) and personal hygiene (washing hands, washing and cooking food) affect health. In 1854, outbreaks of cholera in London's Soho district were traced by physician John Snow to contamination from a single public water pump in Broad Street. Diversity within diseases includes an alarming rise in antibiotic-resistant strains.

Inherited disorders caused by defects in genes and chromosomes in the nucleus are mainly covered in the relevant chapters, such as those on limbs, digestion, skin and blood. Ones caused by the very small fraction of DNA in mitochondria are dealt with in Chapter 11.13.

While some diseases are deadly, many involve temporary discomfort. As a child I had whooping cough, mumps, measles, German measles, tonsillitis, chicken pox, flu and many colds, with no lasting damage, but other sufferers may have serious complications.

German measles, rubella, is caused by a virus transmitted in inhaled droplets from the nose or throat. It is usually unimportant in young children, causing mild fever, swollen lymph nodes and an itchy rash of pink-red spots. In older children and adults, there may be headaches, appetite loss, eye inflammation, a runny nose and joint pains. However, in pregnant women it can cause **congenital rubella syndrome**, with devastating consequences for the foetus. Children infected before birth may suffer growth retardation, mental retardation, deafness, malformations of the heart and eyes, etc.

Diseases are incredibly diverse. Their cause can be minute animals called protozoa (malaria, toxoplasmosis), nematodes (round worms such as hookworm), cestodes (flat worms such as tapeworms), larger animals (mites giving scabies; lice), and microbes. Insects can carry malaria, sleeping sickness and bubonic plague; dogs can transmit rabies; cats can transmit toxoplasmosis, while sheep and water-snails transmit liver fluke.

Microbial agents include viruses (flu, HIV, chickenpox, rabies), bacteria (pneumonia, tuberculosis, syphilis) and fungi (athlete's foot). Viruses can cause **cancer**, e.g., papillomaviruses causing cervical cancer. **Antibiotics** which work against bacteria are powerless against viruses (colds, influenza). Some diseases do not involve infectious organisms but are caused by asbestos, lead, arsenic or UV light.

Populations may tolerate diseases which have been present for hundreds of years but have little resistance to different diseases brought in by conquerors or immigrants. Aztecs, Incas and other inhabitants of Central and South America suffered terribly from smallpox and measles brought in by the Spanish conquistadores in the sixteenth century, with a decline of 93% from the pre-contact populations.

Medical staff may transmit diseases from their hands, clothes or breath. A gynaecologist infected patients with hepatitis C virus. Some hospital staff are HIV positive. In olden days, medical and ancillary staff often did not wash their hands between patients, and the frequently fatal bacterial puerperal fever was spread by midwives' hands and equipment to new mothers.

The approximate fatality rates from some diseases are: swine flu, 0.02%; measles, 0.1%; Spanish flu (1918 strain, 75 million infected, 1918–1920), 2.5%; diphtheria, 5%; smallpox, 30%; bird flu, 60%; ebola, 70%; AIDS, 100% (40 million infected).

It is not possible to cover all diseases here, so a representative sample is given. Others are considered in chapters on skin, brain, lungs, digestion, etc.

11.2. The body's protective mechanisms; the immune system

Skin is our first line of defence against harmful organisms. We have active systems to deal with some attacks, such as vomiting up bad food, sneezing to get rid of nasal irritants, coughing up bacteria-laden phlegm, tear-production to flush out the eyes, or scratching to remove skin irritants and ticks. Tears, saliva and mother's milk contain the enzyme lysozyme which kills many bacteria. Lung airways have billions of tiny cilia (hairs) which move mucus, pollen, dust and micro-organisms out from the lungs to the throat, where they are swallowed. The stomach's hydrochloric acid destroys nearly all of them.

An **antibody** is a protein able to recognise and adhere to part of another protein, normally one not made by the body, such as part of an invading micro-organism. We can make more than a million different antibodies, but mistakes can produce antibodies which attack our own proteins as in **auto-immune diseases** such as multiple sclerosis and rheumatoid arthritis.

A major source of antibodies is the **B-cell lymphocyte**, a white blood cell made in lymph tissues. While we are in the womb, genes which specify antibodies in B-cells undergo recombination, so that in the baby each of millions of early B-cells makes a different antibody, specific for a different protein fragment. In the foetus, B-cells at that stage die if they are specific to any of the baby's own proteins, so only ones responding to foreign proteins are left.

Antibodies have a cavity which fits the shape of their target protein, its antigen. If an invading micro-organism has the target antigen on its surface, the antigen and antibody bind together, stimulating the B-cell to divide repeatedly, building up millions of activated B-cells of that type. The B-cells eventually turn into **plasma cells**, releasing antibodies. They bind to the invader's antigens, 'labelling' them so that other parts of the immune system, especially large white blood cells, can recognise and engulf them. Some activated B-cells continue dividing indefinitely as '**memory cells**', the basis for long-term immunity to foreign proteins which have been present before, e.g., from disease organisms.

One may be vaccinated against diseases such as polio, mumps, whooping cough and tuberculosis. This is done by injecting killed organisms, or mutated, weakened, live organisms, or protein parts of them, to stimulate memory B-cell production so the body is ready to fight that organism even though we have not before had its disease. Immunisation gives us the right memory cells and pre-existing antibodies for that disease, ready to multiply hugely if needed. We each have an enormous diversity of antibodies.

People are diverse in their reactions to vaccines in general and to specific vaccines. The elderly do not respond as well as younger people as their immune systems have reduced functioning. Those with egg allergies have to avoid vaccines produced by culture in eggs. Any vaccine can cause side effects but these are usually minor, such as a sore arm, a headache (especially from ones which include phenol) or a low-grade fever, but these disappear after a few days.

Here, from the US Centers for Disease Control and Prevention website, is a list of diverse symptoms which some children suffer from the **diptheria, tetanus and acellular pertussis (whooping cough) vaccine**. Mild problems (common): fever, swelling and soreness at the injection site. They occur more often after the 4th and 5th doses than after earlier doses. Tiredness affects 1 child in 10, and vomiting 1 child in 50. Moderate problems (uncommon): seizure, about 1 child in 14,000; non-stop crying for 3 hours or more, up to 1 child in 1,000; high fever (105°F, 40.6°C, or higher), 1 child in 16,000. Severe problems (very rare): serious allergic reaction, less than 1 in a million doses. Extremely rare problems include long-term seizures, coma, or lowered consciousness, or permanent brain damage.

An international campaign to eliminate polio (infantile paralysis) through vaccination was largely successful. The more than 350,000 cases a year worldwide in 1988 were reduced to 5,000 in 1999, making many regions polio-free. However, in May 2014, the WHO declared polio's renewed spread a world health emergency, with infections in Nigeria, Afghanistan, Pakistan, Somalia, Kenya, Ethiopia and Syria. Polio is an acute infectious viral disease, spread through faecal contamination of food and water. In the 1940s and 50s in England, there were many affected children with partly paralysed legs, wearing iron callipers for support. Inflammation of the spinal cord and brain could necessitate the use of an iron lung for breathing. Surprisingly, 90% of polio infections cause no symptoms while others cause degrees of paralysis.

For some diseases, but not others, one attack can give immunity for many years, but those causing the common cold and malaria mutate or recombine, with sufficient changes to evade our defences. One can catch colds many times.

As well as antibodies, we have a **cell-mediated immune system** involving the major histocompatibility complex (MHC), but details are beyond the scope of this book. The MHC is vital in transplants as different tissue types between donor and recipient lead to transplant rejection. Corneal transplants are usually successful without tissue matching because the cornea has no blood or lymph vessels to carry cells of the immune system. Metallic implants such as artificial hips, knees and elbows, and plastics such as Teflon for heart valves and Terylene for blood vessels, are not recognised by the protein-based immune system, so are not rejected.

The MHC has an enormous amount of genetic variation in its human leukocyte (white blood cell) antigen (HLA) genes. Bateman and Howell (1999) suggested that different genes in this region account for many ethnic and individual differences in the incidence of cancer and other diseases. Here are examples of HLA alleles and the increased risk of diseases: allele A3, risk of haemochromatosis increased 6-fold; B27, ankylosing spondylitis, 150-fold or more; DQ6, narcolepsy, 38-fold; DQ2, coeliac disease, 250-fold; DR2, multiple sclerosis, 12-fold; DQ8, type 1 diabetes, 14-fold. There are strong links between HLA genes and cancers.

A destroyed immune system (as in AIDS) renders one liable to many infections, some lethal. Milly Smith was born with no functioning immune system. Until she was six, she had to live in a sterile 'bubble'. Her sister Kacey did not have that and was a perfect match. The bone marrow transplant was a success, so in 2014 Milly was able to go out and play like a normal girl. The genetics of some diseases is very complicated: in inflammatory bowel diseases such as ulcerative colitis and Crohn's disease, more than 150 genes influencing them had been discovered by 2001, and the number increases.

Seasonal environmental effects influence mortality, with 30,000 more deaths in winter in Britain than in summer, with the over 55s most affected.

However, the well-prepared people of Yakutsk in eastern Siberia, the coldest city in the world, wear massive fur clothing with many layers and showed no increase in mortality as the temperature fell to minus 48°C. Deaths from heart disease generally occur one or two days after cold exposure, deaths from strokes after five days and deaths from respiratory problems about 12 days later.

Autoimmune diseases occur when the immune system fails to recognise some cells as being 'self' and attacks them as if they were invading bacteria. Healthy tissues are destroyed mistakenly. **Lupus erythematosus** is one such condition, affecting about 1 in 1,100, especially females, Asians and Afro-Caribbeans. Lupus can cause general weakness and scaly red plaques on the skin especially where it is exposed to sun, with reddening of the cheeks and nose. Autoimmune disorders may affect blood vessels, connective tissues, glands such as the thyroid or pancreas, joints, muscles, red blood cells and skin. Disorders with autoimmune components include Addison's disease, coeliac disease, multiple sclerosis, rheumatoid arthritis, thyroid problems and type 1 diabetes.

Autoimmune diseases are less common in less developed countries. The '**hygiene hypothesis**' is that high standards of hygiene reduce young people's exposure to bacteria, leaving the immune system with not enough to do, so it is more likely to attack their body. Countries where populations have heavy infestations with round worms have fewer cases of autoimmune diseases. Thanabalasuriar and Kubes (2014) suggested that the first two weeks of life in newborns form a 'developmental window' in which exposure to bacteria was required. Over-clean conditions in childhood increased the likelihood of asthma and other allergy diseases.

11.3. Diseases caused by protozoa

Toxoplasmosis is caused by a single-celled parasite, *Toxoplasma gondii*, which infects cats, dogs, sheep and humans. The symptoms are usually mild so it often goes unnoticed, but 10 to 15% of infected people get high temperatures, sore throats and muscle aches. Severe toxoplasmosis can be debilitating, causing prolonged tiredness, sore throats, loose bowels, swollen lymph nodes and headaches; it can cause inflammation of the brain and affect the heart, liver, ears and eyes. It can be lethal.

Personal Account. *I have had toxoplasmosis since before birth, and it could kill me at any time*

I was born light, 4 lb 11¾ oz (2.15 kg). When I was five, an optician saw scarring which caused a large blind spot in my left eye. He diagnosed toxoplasmosis. It affects the placenta and stops the child from receiving nutrients. My body had been forced to feed on itself for the last weeks of pregnancy.

Living with toxoplasmosis didn't have much effect on my early life. However, there is the possibility it could have a huge effect on my later life. I may have cysts in both eyes and in my brain. They could burst at any time, causing blindness or death, but haven't yet. I count my blessings frequently.

Andrew Savage

About 30% of Britons acquire toxoplasmosis at some time. **Severe congenital toxoplasmosis**, from maternally infected babies, affects about 3 in 100,000 and can cause brain damage, blindness or death. Cats pass the parasite into the soil in their faeces, where it can get onto people's hands or food. The disease can come from eating raw meat such as lamb, a frequent cause in France. Up to a third of the world's population carries *Toxoplasma*, with about 350,000 infections a year in Britain. Sufferers with weakened immune systems, as with AIDS, may become seriously ill.

Malaria has killed more people than all wars and plagues combined, with up to 1.2 million people a year dying from it, especially African children. Although once common in London and marshy areas of Europe, it is now largely confined to the tropics and subtropics where conditions suit mosquito larvae. The five species of *Plasmodium* differ in virulence. See Chapter 17 for the legacy of harmful genes causing inherited blood disorders such as thalassaemia, selected for partial malaria resistance.

Malaria is transmitted through bites of infected female *Anopheles* mosquitoes. The parasites multiply in the human liver and then infect red blood cells. Symptoms include fever, headache and vomiting, usually appearing 10 to 15 days after the bite. If untreated, malaria can quickly become life-threatening by disrupting the blood supply.

11.4. Diseases caused by parasitic worms (helminths)

These include round worms (nematodes), tapeworms (cestodes) and flukes (trematodes). At least one and a half billion people suffer from intestinal worm infections. They are endemic in warm countries, as in sub-Saharan Africa. Children are often too ill to go to school. Symptoms include malnutrition, anaemia, diarrhoea, lethargy and gut inflammation. Eggs excreted in faeces contaminate the soil. Infections follow from contaminated water or food, or finger-soil contact. Roundworm eggs were discovered in the remains of King Richard III of England. Anti-helminth drugs are cheap and effective but do not prevent reinfection.

The **ascaris roundworm** infects about a billion people. It is cylindrical, pale, and 15 to 35 cm (6 to 14 in) long. Light infections cause little trouble but heavy infestations can block the intestines and retard growth. **Liver flukes** are common in livestock, including sheep, but can attack humans where they cause liver damage. About 700 million people suffer from **hookworm**, which lives in the small intestine. It is white to pink, with males 0.5 to 1 mm long; females are larger. Infection can be by the faecal-oral route, or larvae can penetrate the soles, so going shoeless and/or working in flooded paddy fields promotes infection. Bad infections cause anaemia because hookworms are vicious blood-suckers. They are a leading cause of maternal and child morbidity in the tropics and subtropics, causing mental and growth retardation, and low birth weight in babies of infected mothers.

Pinworm mainly affects children and institutionalised adults where infection frequencies can reach 50%. Mature gravid female worms (8 to 13 mm, 0.3 to 0.5 inches long) move from the small intestine to the colon and at night lay eggs around the anus, causing itching. The eggs are so small that they can be breathed in as well as swallowed.

Tapeworms (*Taenia*) are long, flat, segmented worms which live in the intestines of animals, including man. Infections arise from eggs via the faecal-oral route, or from larvae in raw or undercooked meat, especially pork or beef. Symptoms are usually minor but larvae can move from the gut to the muscles, and in severe cases damage the liver, eyes, heart and brain. Because tapeworms can lead to weight loss, some women deliberately infect themselves in a slimming attempt. The only sign of infection may be worm segments in the faeces; in bad cases, the gut may become blocked. The **pork tapeworm** is white, hermaphrodite, attached to the intestine wall by suckers around its head, and is usually two to three metres (6.5 to 10 feet) long but can be up to eight metres (26 feet) long.

11.5. Diseases caused by viruses

Rabies targets the brain and nervous system, and is normally fatal unless treated very early. It kills about 25,000 people a year in Africa and is frequent in India. It is passed to humans from infected animals, such as dogs and squirrels, in their saliva which gets into bites or scratches. There is a vaccine for humans and animals. Rabies ('mad dog disease', 'hydrophobia' — fear of water) is much rarer in Europe than in Africa and the six months quarantine period for bringing dogs into the UK has been lifted for vaccinated animals.

Warts are caused by viruses, particularly papilloma virus, on any part of the body, often toes or fingers. Warts usually disappear spontaneously over months or years. There are chemical and laser treatments, and freezing.

Measles is highly infectious but was almost eradicated in Britain after the introduction of the MMR (measles, mumps, rubella) vaccine in 1988. Because of rumours that this could cause autism, many parents chose not to have their children immunised. Measles predictably made a comeback, with more than 2,000 cases in 2012 in England and Wales. One in five affected children develops complications such as ear infections, vomiting, pneumonia or meningitis, with 10% needing hospital treatment. Measles kills about a million children a year worldwide. The virus grows in the lungs and throat. There is a high fever lasting four to seven days, often with a runny nose, a cough, red, watery eyes and small white spots inside the cheeks. After several days, a rash erupts on the face and neck, spreading to the hands and feet. After five to six days, it fades.

An outbreak of measles in the USA in 2015 was traced to infection from Disneyland, with 142 of the 173 cases up to March 6 being tied to the Disney outbreak that began in December 2014. It was probably a direct consequence of the growing **anti-vaccination movement** in the United States. The rapid spread of the disease indicated that most of those infected were not vaccinated or were incompletely vaccinated. The national average for the Measles, Mumps and Rubella vaccination in the USA is about 92%, but pockets of under-vaccinated children are breeding grounds for outbreaks. Because measles is so contagious, a very high level of vaccination is needed for population protection. "If the anti-vaccination movement in the U.S. continues to grow, the likelihood of outbreaks will increase — as will their scale and scope," the researchers said. (reported in the 16/3/2015 online edition of the journal *JAMA Pediatrics*)

Measles is a leading cause of death among young children although a safe, cost-effective vaccine is available. In 2012, there were 122,000 measles deaths in the world. More than one billion children in high-risk countries have been vaccinated against it.

Mumps is very infectious and was common in schoolchildren in Britain and America before the introduction in 1988 of the MMR (mumps, measles and rubella) vaccine. There were about 2,500 cases in England and Wales in 2012. The virus gives painful swellings under the ears (the parotid glands). There is usually a headache, joint pains and a high temperature. Although not usually serious, it can lead to meningitis on reaching the brain. Infection usually gives life-long immunity. The virus is spread through infected droplets of saliva that can be inhaled or picked up from surfaces and transferred to the mouth or nose. There is no cure but it should pass within one or two weeks.

Cold sores and genital herpes come from the herpes simplex virus. With cold sores, small fluid-filled blisters erupt, particularly on the lips, but can affect the cheeks and inside the nose. Blisters turn darker and scabby, usually healing in 10 days. Infections recur, with a variety of triggering factors. The virus is often passed by skin contact from mother to baby or between children.

In **chicken pox**, varicella zoster virus gives a rash of itchy small blisters which burst and form scabs, often leaving small white scars. It is highly contagious but not usually dangerous. The virus can persist in nerves long after the disease has gone, giving rise to **shingles** in middle age or later. Shingles causes burning sensations, rashes, and pustules along the nerve paths, giving crusting. It can be extremely painful and is contagious (Chapter 13.7). **Smallpox** causes a flu-like fever and a pus-filled rash. It killed Queen Mary II of England (1662–1694), Tsar Peter II of Russia, King Louis XV of France and 300 million people in the 20th century.

In 1762, smallpox was the leading cause of death among children in London, killing one in four. Inoculation against it started in 1721, placing fluid from a smallpox pustule in a small puncture on the upper arm. About one in 50 of the inoculated people died, compared with one in six uninoculated people who caught smallpox. By 1765, improved techniques meant that only one in 500 treated people died.

Edward Jenner (1749–1823) pioneered the smallpox vaccine, the world's first. His work is said to have 'saved more lives than that of any other human'. He was apprenticed to a country surgeon near Bristol, England, and heard a dairymaid say, "I shall never have smallpox for I have had cowpox." In 1796 he found a dairymaid with fresh **cowpox** lesions on her hands and arms. Using their pus, he inoculated an 8-year-old boy who developed a mild fever. Jenner later inoculated him more than once with matter from a fresh **smallpox** lesion. No disease developed. The cowpox inoculation gave complete protection against smallpox. Jenner called the process *vaccination* (Latin for cow, *vacca*). A worldwide vaccination campaign began in 1966 and the WHO declared that smallpox had been eradicated by 1980.

11.6. Diseases caused by bacteria

Bacterial skin diseases are covered in Chapter 13.7. The bubonic plague last occurred in Britain in 1666, with the Great Fire of London destroying the squalid, overcrowded, rat- and flea-infested housing which promoted epidemics. The plague is caused by *Yersinia pestis*, carried on fleas on small rodents, especially rats, mice, rabbits and squirrels. It kills about two thirds of infected humans within four days if untreated. It causes swollen lymph nodes (**buboes**) especially in the armpit and groin. The same bacterium causes septicemic plague and pneumonic plague, which makes patients highly infectious by going into the lungs where it is spread by coughing. *Yersinia* was responsible for the Black Death that swept through Europe in the 14th century and killed 25 million people, 30 to 60% of the population. Plague reached America when trade ships from Asia brought infected fleas and rats to San Francisco in 1899. There have been more than 376 cases in America since 1970.

Leprosy is caused by *Mycobacterium leprae*, transmitted through droplets from the nose and mouth of patients with severe disease but is not highly infectious. The symptoms include granulomas (nodules) of the nerves, respiratory tract, skin, and eyes. The face can look appallingly diseased. Leprosy can cause nerve damage, leading to muscle weakness, atrophy and permanent disabilities. Nerve damage can result in an inability to feel pain and thus loss of fingers and toes from repeated injuries. Leprosy can be treated effectively with a 6 to12-month course of multidrug therapy.

The WHO has a programme to eliminate this scourge which remains a problem in parts of Africa, Indonesia, Brazil, India and Nepal. There are still more than half a million people infected a year, particularly in India where leprosy carries an enormous stigma, especially for women. Female sufferers there are often disowned by their husbands, thrown out of their homes, ostracised by the community and cut off from their children. They lose their jobs and get deeply depressed, often having to live by begging. In Britain in the Middle Ages, lepers were usually confined to isolated leper colonies.

Tuberculosis (TB) is caused by a related bacterium, *Mycobacterium tuberculosis*, caught by inhaling droplets from the coughs or sneezes of an infected person. It mainly affects the lungs but can invade any part of the body. It causes a persistent cough, weight loss, night sweats, fever and tiredness. Prolonged exposure to someone with TB is the usual cause, as in crowded workplaces or dwellings. Often the immune system clears the disease, but sometimes bacteria remain dormant in the body, with **latent TB**, which can become active, particularly if the immune system becomes weakened. One third of the world's population is infected with latent TB; about 10% of those cases will later get active TB. TB can be cured with antibiotics, usually taken for six months. Drug-resistant forms need longer. The Bacillus Calmette-Guérin (BCG) vaccine can provide effective protection against TB in 80% of people. It was given routinely in schools when I was young.

TB, then known as **consumption**, was a major health problem in Europe, featuring in plays and operas, with the lady dying in the last act. Today it is much less common, yet recently it has increased, particularly among members of ethnic minority communities from countries where TB is common. In 2011 in Britain, 8,963 cases of TB were reported, of which more than 6,000 were in people born abroad. In America in 2004,

TB frequencies in Whites were 27 per 100,000 but were 20 times higher among Asians, and eight times higher in Blacks and Hispanics; 95% of affected Asians were foreign-born.

In Harrow, near London, where 25% of the population are Gujarati Indian immigrants, its incidence is high, more than 800 cases per 100,000 a year. Most are strict vegetarians, getting very little vitamin D precursor; what there is would be converted to vitamin D in Gujerat's sunny climate but not in Harrow. Half of the tuberculosis patients in Harrow had undetectable levels of vitamin D, giving a ten-fold increase in the risk of developing tuberculosis which had been latent since they left India.

Streptococcus pneumoniae causes **pneumonia, blood poisoning** (septicaemia) and **meningitis** (inflammation around the brain). Any of these can be lethal. The bacteria are spread by coughing, sneezing or close contact. They may persist in the nose and throat with no effects, or invade the lungs giving pneumonia, or the blood, giving septicaemia. People over 80 are particularly vulnerable, especially if they have a weakened immune system, e.g., after cancer treatments or transplants. **Pneumonia** in the lungs causes high fever, coughing, shaking, chills, breathlessness, chest pains and confusion. It has been called 'the old man's friend' as it can kill old people quickly and less painfully than some other causes of death. **Pneumococcal meningitis** causes a severe headache, a stiff neck, high fever, confusion, and sensitivity to light.

Various strains of the intestinal bacterium *Escherichia coli* are pathogenic, although most are harmless and normally present. Babies have their intestines colonised by harmless strains very quickly. '**Travellers' diarrhoea**' can be caused by *E. coli* strains or other bacterial genera. Although **scarlet fever** was a threat in the 1940s, it receded in Britain until a sudden upsurge in 2014. By May 2014 there had been 8,300 cases, four times the usual number.

11.7. Diseases caused by fungi

Fungal diseases particularly affect skin (Chapter 13.7) where humidity builds up, such as the groin and between the toes. In Britain, about 15% have fungal foot infections, especially '**athlete's foot**' which is easily picked up in changing facilities and hotels. About 40% of those affected also have nail infections. So-called '**ringworm**' of the body and scalp is caused by fungi, giving red rings as they grow outwards from infection sites.

11.8. Problems from arthropods (crustaceans, insects, spiders, scorpions and mites)

Bites and stings by scorpions, spiders and insects such as wasps are not infectious although they can be lethal. Insects like mosquitos and fleas are important **vectors** for diseases. Painful bites can become infected, and some people have serious allergies to bites and stings from particular species. Scabies and lice are covered in Chapter 13.7.

11.9. Venereal diseases

Sexually transmitted diseases are embarrassing, infectious and can be lethal. **HIV (human immunodeficiency virus)** can cause the **acquired immunodeficiency syndrome (AIDS)**, leading to a progressive failure of the immune system, allowing life-threatening opportunistic infections and cancers to flourish. Infection with HIV occurs through contact with an infected person's blood, semen, vaginal fluid or breast milk. Some anti-retroviral drugs are effective in prolonging the life of sufferers.

It is difficult for the body to fight HIV because this virus has a **high degree of genetic diversity**. It can replicate so fast that 10^{10} (ten billion) virus particles can be produced in one day, with a high mutation rate and recombination. This gives many variants of HIV in a single patient. At least 100,000 people are infected in Britain, with an estimated 25% being unaware that they have HIV. One infectious person can infect many sexual partners. In parts of Africa, AIDS deaths have caused marked reductions in population size. AIDS was first observed in 1981 in America among drug injectors and gay men who showed symptoms of a rare kind of pneumonia which occurs in people with compromised immune systems. It was then found that an unusually high number of gay men developed a rare skin cancer, Kaposi's sarcoma. Infected mothers can transmit HIV to their babies, before or after birth. In Britain in 2005, 7,450 new cases were diagnosed, including 5,050 heterosexuals.

Chantelle was diagnosed with HIV at the age of 13 and takes 23 tablets a day. She was born with it and both her parents died from it. She joined a London-based charity, Body & Soul, with 400 teenager members, 70% of them black like herself. Chantelle said that she would have gone mad without the support of the charity which reassured her that HIV was something you lived with rather than died from.

In Kenya, about 11% of adults are infected with HIV. In the Luo tribe where polygamy is practised, one in four women is HIV positive, compared with one in 20 men. In that tribe, a widow customarily becomes the wife of her dead husband's brother, so if the husband died of AIDS, she often passes HIV on to her new husband and he may infect his other wives.

Gonorrhoea, 'the clap', is caused by *Neisseria gonorrhoeae*. It is mainly transmitted through mating, although mothers can pass it to their babies. This bacterium attacks mucus membranes in the urethra, cervix and Fallopian tubes. It can grow in the mouth, throat and anus. It is very common in America, with about 700,000 new cases a year, especially among sexually active teenagers. There were 29,300 cases reported in Britain in 2015.

Gonorrhoea is called 'the drip' because of a pale greenish-yellow discharge from the penis or vagina. Both sexes may experience pain on urination, and oral sex can result in burning sensations in the throat. Symptoms usually appear two to 14 days after infection but may be so slight in females that they do not realise that they have it. Prompt treatment with antibiotics can cure it and prevent complications such as pelvic inflammatory disease.

Syphilis, 'the pox', is caused by *Treponema pallidum*. It usually starts with an **ulcerating chancre** on the male or female genitals and can progress through the lymphatic system to other parts, giving serious symptoms and even death. This bacterium is highly contagious during first and second stage infections, entering the body after sexual contact with an infected person, or by close contact with a sore or rash caused by syphilis. It is easily curable if detected then but the early stage often produces few or no symptoms, so many who are infected are unaware of it. It can go undetected for months. Even when no symptoms are present, carriers are contagious. In 2015, there were 3,249 reported cases in Britain.

Within weeks or months of exposure, a sore may develop at the site of infection on the penis, in the vagina, mouth or anus. This sore eventually heals spontaneously but the bacterium is still present. Second stage symptoms include a highly contagious rash that can appear in the mouth, on the sexual organs or anus. Symptoms such as fever, headache, loss of appetite and weight, sore throat, and muscle and joint pain are common. Syphilis can damage other parts including the liver, kidneys and eyes. If left untreated, it can progress to the third stage, damaging other parts such as the nervous system. There can be memory loss, balance and walking problems, loss of sensation, poor vision and poor mental function. Antibiotics can be effective in the early stages.

Mental illness from late-stage syphilis was once common. Historical figures suspected of having had syphilis include Charles VIII of France, Hernán Cortés of Spain, Benito Mussolini, Ivan the Terrible, Napoleon Bonaparte, Cesare Borgia, Al Capone, Frederick Delius, Paul Gauguin, Édouard Manet, Franz Schubert, Robert Schumann, Leo Tolstoy and Oscar Wilde.

Condoms offer some protection against unwanted pregnancies and venereal diseases. There is an illustration of one from Egypt from about 3,000 years ago when linen ones were used against venereal diseases. The syphilis epidemic that spread across Europe in the 1500s gave rise to a published account of the condom. Gabrielle Fallopius described a sheath of linen which he claimed to have invented to protect men against syphilis. Only later was its usefulness for contraception recognised.

In the late 1500s, the linen sheaths were sometimes soaked in a chemical solution and then allowed to dry — **spermicides** on condoms. In the 18th century, the Italian womaniser Casanova used linen condoms. Rubber ones were mass-produced after Charles Goodyear patented rubber vulcanisation in 1844. The female condom has been available in Europe since 1991. According to the United Nations Population Fund, an estimated 10.4 billion male condoms were used worldwide in 2005, around 4.4 billion for family planning and 6 billion for HIV prevention.

A survey in *The Lancet* (Fenton *et al.*, 2005) found the following ethnic and sex differences in the frequencies of having, or having had, sexually transmitted diseases in Britain. Men: White, 11%; Black Caribbean, 20%; Black African, 16%; Indian, 3.4%; Pakistani, 3.2%. Women: White, 12%; Black Caribbean, 23%; Black African, 14%; Indian, 8%; Pakistani, 4%. There were also large differences in the median number of lifetime sexual partners. Men: White, 6; Black Caribbean, 9; Black African, 9; Indian, 2; Pakistani, 1. Women: White, 5; Black Caribbean, 4; Black African, 3; Indian, 1; Pakistani, 1.

11.10. Non-infectious disorders

Hay fever, caused by immune reactions to pollens, can make life hell for sufferers who may get a streaming nose, sneezing fits, breathing difficulties and swollen, itchy eyes. Those with it are usually aware of what season, weather and pollens are worst for them. The allergies can be to tree pollen, released during spring, grass pollen, released during late spring and early summer, or to weed pollen, released any time from early spring to late autumn. About one person in five is affected. Chemicals called **phthalates,** found in clingfilm, plastic bottles and nail varnish, are suspected of being partly responsible for the increase in childhood allergies, including eczema.

Metals such as copper, lead and arsenic cause disorders, including poisoning. Nickel can cause **allergic skin reactions** (Chapter 13.6). Arsenic causes poisoning and death, depending on the dose, with nausea, vomiting, stomach pains, skin discolouration, neurological disorders, cancer and organ failure. Arsenic is common around old copper-smelting plants and where it has been used in agricultural sprays. Levels in water in parts of Taiwan, Argentina and Bangladesh are hazardous. In countries like Bangladesh, labourers in the heat may drink up to 20 litres of water a day, with as much as 0.5 mg of arsenic per litre, risking cumulative poisoning.

In Bangladesh surface water is often polluted, so Unicef led a campaign to switch to groundwater 56 years ago, with 10 million rural shallow tube wells constructed. Unfortunately about a million of them draw water from arsenic-rich sediments, resulting in skin, bladder and lung problems, slow poisoning and many cancers. Brake fern (*Pteris vittata*) is effective at removing arsenic from drinking water, so bioremediation is possible.

Attempts have been made throughout history to treat diseases. The dosage is crucial. Arsenic is very toxic, used in many murders, but has been used in medicines. In China it has been used for more than 5,000 years against syphilis, parasites and leukaemia. Other forms have been used in Britain against infectious diseases, intestinal parasites, fevers and headaches. Its brightly coloured compounds have been used in paints (Titian and Michelangelo used toxic arsenic sulphide in orange and yellow paints), fabrics, cosmetics, glass, soaps and wallpapers.

11.11. Cancer

Overall frequencies of cancer per 100,000 for some countries are: Denmark, 326; Australia, 314; Belgium, 306; France and USA, 300; Germany, 282; UK, 267; Spain, 241; Portugal and Poland, 223. American cancer frequencies per 100,000 are: lungs, 62; mouth, 11; larynx, 2 in women, 8 in men, with more in Blacks than in Whites. In Britain, South Asians have higher rates of coronary artery disease and diabetes, but lower rates of cancer, 68% lower in males, 48% lower in females, than population averages.

The highest frequency of **skin cancer** is in Australia, with 280,000 new cases and 12,000 deaths a year. Australia is sunny, with a tradition of out-door life and exposing flesh. The high levels of skin cancer there have led to campaigns such as 'slip, slop, slap': slip on a long-sleeved shirt; slop on sunscreen, and slap on a hat to shade the face and neck.

In Britain, skin cancer affects 5 per 1,000, increasing with age from cumulative sun exposure and poorer immune responses. Of those cancers, 1 in 50 is a **malignant melanoma**, where cancer starts in melanocytes in the dermis, forming dark, irregular tumours which have a very high rate of spreading (**metastasis**) to other body parts where they are often fatal. There may be inflammation, oozing or bleeding, and a mild itch. The thicker the tumour, the poorer the survival. The most frequent cause is mutational UV damage from the sun. Predisposing factors are having more than 100 moles, a skin type which sunburns easily, and long sun exposure, even several decades before the tumour becomes apparent.

As people live longer, **age-related cancers** become more common. Figures from Cancer Research UK for cancers per 100,000 people in rela-tion to age in years were: 15 cases, ages 0 to 20; 72, 20 to 40; 401, 40 to 60; 1,536, 60 to 80; 2,555, over 80 years. There about 6,000 new cases of malig-nant melanoma a year in Britain, and about 1,700 deaths. Malignant skin cancers in British men have increased threefold over 20 years to 12 cases per 100,000. In 2015, Cancer Research UK stated that British pensioners were seven times more likely to suffer malignant melanoma than they were 40 years ago because of the rise in package holidays to sunny destinations.

Of non-melanoma skin cancers, 75% are basal cell cancers and 20% are squamous cell cancers.

Basal cell carcinoma (BCC) is the commonest skin cancer, with about 40,000 cases a year in Britain, especially in the over-50s, affecting males more than females and occurring mainly on the face. About 2% of the population are affected, particularly those with fair skins. The tumours can be disfiguring but are usually not lethal because they do not spread far. They arise from the epidermis or hair follicles, growing slowly into adjacent tissues. Because of this gnawing at surrounding areas they are called 'rodent ulcers', with a tendency to ulcerate and bleed in the centre. Lesions have a rolled edge. Long-term sun exposure is the main cause. Tumours may be red, pink, brown or flesh-coloured patches, or ulcer-like growths, and are treated by surgery, lasers, radiation or abrasion.

When I was 71, the rodent ulcer on my left temple (Plate 16B) was treated with surgery, but radiotherapy was offered as an alternative. Sufferers include Ronald Reagan and Elizabeth Taylor. Princess Michael of Kent had three unrelated BCC tumours, with treatment by surgery to her face and leg, and laser to her arm.

Squamous cell carcinoma (SCC) also has a sun-exposure origin, usually forming a lump which may bleed, become painful or crusty. It has a small risk of spreading to other parts through the blood. About 10,000 cases a year occur in Britain. SCCs are more aggressive and dangerous than BCCs, capable of **metastasis** to lymph nodes and beyond. They are most frequent on the head and neck, with ulcerating tumours taking only a few months to develop. They arise from squamous (scale-like) epidermal cells and often have hard, raised edges. They are dangerous when they penetrate the skin's basement membrane.

Researchers at the University of Reykjavik found that the risk of skin cancer was ten times higher than normal in aircraft pilots, and fifteen times higher in those flying across several time zones. This was attributed to disturbances of the body's biological rhythms, although exposure to increased radiation at high altitudes is another possibility.

Irradiation with different wave lengths of ultraviolet light, UVA and/or UVB, is used to treat skin conditions such as vitiligo, psoriasis and eczema. Like natural sunlight, such treatments can have adverse effects such as ageing the skin and slightly increasing the risk of skin cancer. Skiing and other snow sports carry risks of sunburn because snow reflects about 85% of UV, and UV intensity from the sun increases by 5% for every 1000 ft (300 m) of altitude.

Internal tumours, such as prostate or ovarian cancer, most kinds of breast cancer, or cancers of the liver, gut or kidneys, are caused by viruses or spontaneous mutation of several to many genes, including tumour-suppressors. One very serious inherited disease is **xeroderma pigmentosum**, an inability to repair DNA, leading to multiple malignant melanomas and early death.

Causes of cancer include UV light, radon gas, medical X-rays, smoking, alcohol, polycyclic aromatic amines and heterocyclic amines found in charred food, obesity (especially for hard-to-diagnose kidney cancer, which kills more than 1,300 women a year in Britain), chance mutations in tumour-related genes, chromosome aberrations, viral infections (including sexually transmitted ones), certain pollutants and industrial chemicals, and various food ingredients, natural and man-made. Scientists at Johns Hopkins University School of Medicine found in 2015 that two thirds of cancers (especially brain and bone cancers) were due to **random chance mutations**, not diet or lifestyle. Tumours and cell cultures from them often show several complex chromosome rearrangements, with duplications, deletions, insertions and translocations (illustrated in Plate 1E). **Particular cancers** are strongly associated with certain genes (alleles), especially **breast cancer** when it runs in families. Some healthy individuals with cancer rife in their families opt for preventative surgery to remove susceptible organs (Chapter 5.3). In 2000, Ruth Michael, a 38-year-old healthy mother of three from Norfolk, England, chose to have both breasts, her womb and ovaries removed. Her mother had had one breast removed because of cancer when she was 40 and the other at 50, but died of cancer when 57. Ruth's sister died of breast cancer at 44. Ruth was advised that she had an 85% chance of developing breast and/or ovarian cancer, because she carried faulty copies of genes *BRCA1* and *BRCA2*.

BRCA1 mutations are responsible for 10 to 40% of hereditary breast cancers and *BRCA2* for 10 to 35%. Cancer-causing mutations of those two genes are autosomal dominants. For most cancers, it takes an accumulation of three or more mutations in a cell before the cancer becomes overt.

About 5% of women in Britain diagnosed with cancer have a mutant *BRCA1* gene. Possession of such faulty genes does not mean that a woman will get breast cancer, but gives an increased predisposition. Breast removal is not too outwardly disfiguring as there are good artificial breasts. Women with single or double mastectomies are often embarrassed about their condition, feeling less complete as women and less attractive to men. Breast cancer is strongly age-related. The risks are 1 in 15,000 up to the age of 25, 1 in 200 by age 50, 1 in 23 by 60, 1 in 15 by 70, and 1 in 11 by 80.

Men can get breast cancer. In Britain, there are about 370 cases a year and about 120 deaths, as men are slow to seek diagnosis and treatment. Those affected are usually between 60 and 70, with obesity and higher than usual levels of oestrogen as factors.

Stomach cancer kills about 6,000 people a year in Britain. Emma O'Connor's family had a history of stomach cancer. Her father had it and DNA testing showed that he carried a mutation in the E-cadherin gene which can lead to stomach cancer. It is dominant and Emma had a chance of 50% of inheriting it. When she was found to carry it, she had her stomach removed preventatively, at the age of 27.

Lung cancer is the leading cause of cancer deaths in America, accounting for 28% of cancer deaths, killing about 157,000 Americans a year. The male death rate from lung cancer is 83% greater than that of females and 13% higher in Blacks than in Whites, with lower rates among Hispanics, Native Americans and Asians. 342,500 Americans had lung cancer in 2000. According to the American Lung Association, the main lung cancer factors in America are smoking (87%), radon gas (10%), air pollution (1 to 2%), and occupational exposure to carcinogens (9 to 15%). Those estimates add up to more than 100% because of interactions between factors. In Britain each year, lung cancer kills about 36,000 people, with 41,000 newly diagnosed, of whom 5% never smoked. The most common form of lung cancer kills 80% of sufferers within one year, with only 5% surviving for five years. Worldwide, about one million people a year die from lung cancer, and more than three million have it.

Malignant mesothelioma occurs mainly in males whose jobs exposed them to certain types of asbestos, but is rare, about 1 in 100,000. It kills about 3,500 people a year in Britain. The common white asbestos is very much less dangerous than blue asbestos.

Prostate cancer is common, affecting 40,000 men a year in Britain, with 10,000 deaths. One in six men gets prostate gland cancer, mainly after 65, but only 3% die from it. In the early stages, men may have no symptoms. Later, symptoms can include frequent urination, especially at night; difficulty starting or stopping urination; weak or interrupted urinary stream; pain during urination or ejaculation; blood in urine or semen. Spreading cancer can cause deep pain in the lower back, hips, or upper thighs.

Most cases involve a slow-growing tumour, with no treatment needed but with periodic checks made. Serious tumours need surgery or radiotherapy, which can leave erectile dysfunction or incontinence. A blood test for high levels of **prostate-specific antigen** is useful but not conclusive, and palpation through the rectum can identify larger tumours. Biopsies are usually required. Having an affected father or brother more than doubles the risk. Prostate cancer occurs 60% more often in African-American men than in white American men. Japanese and African males in their native countries have a low incidence but it increases sharply when they emigrate to America.

Ovarian cancer starts in an ovary or Fallopian tube, affecting about 20,000 women a year in America. Women have a roughly 1 in 54 lifetime chance of suffering from it. The risk is increased two to three times for first-degree relations of sufferers. Having genes which stimulate breast cancer can promote ovarian cancer.

There are few symptoms initially and the cancer can spread lethally to neighbouring parts before detection. The symptoms may include bloating, pain in the abdomen or pelvis and urinary problems, but these are not unique to ovarian cancer. A lump in or on an ovary may be palpable, or detected by ultrasound, but most lumps are not cancerous. A blood test for high levels of **cancer antigen 125** can indicate ovarian cancer but is not specific. Biopsies are the surest test. Treatments involving removing one or both ovaries and the Fallopian tubes can be successful if the cancer is found early. Chemotherapy is usually given too. Survival rates are low because of late detection.

Cancer survival rates are usually measured by the percentage of sufferers still living five years after diagnosis. American figures are: 15% for lung cancer, 62% for colon cancer, 87% for breast cancer and 97% for prostate cancer, with lower rates for Blacks than for Whites for all these figures. Survival depends on how far the cancer has spread before diagnosis, treatment quality, the will to survive, diet, whether the patient smokes, etc.

There are viruses which selectively attack cancer cells and spare normal cells. Donnelly *et al.* (2013) reported that stage III trials were in progress with such oncolytic viruses, including using a modified herpes virus to attack malignant melanomas.

Environmental factors include **carcinogens in the diet**, such as naturally occurring chemicals, synthetic ones in food additives and pesticides, and compounds produced in cooking, such as heterocyclic amines. Colon cancer incidence rises with red meat consumption, being low in Japan, China, Nigeria and Finland, high in Canada, America and New Zealand. Food carcinogens include fungal toxins (aflatoxins and ochratoxins) from *Penicillium* and *Aspergillus* which can contaminate maize, peanuts, some fruits and decaying vegetation. Some processed foods and drinks, including bacon and salami, are treated with nitrites which can give rise to N-nitroso compounds which can cause gastric cancers. Some food additives can cause cancer but levels are usually very low. Using no preservatives in foods and drinks can lead to food poisoning and very short shelf lives. The first environmental factors proven to be carcinogenic were soot and coal tar, causing scrotal cancer in chimney sweeps, as found by Sir Percivall Pott in 1775.

Cancer treatments are improving. Devices such as the CyberKnife™ can target radiation therapy with extreme accuracy, greatly reducing damage to surrounding tissues. Biopsies of tumours can identify their different properties, such as dependence for growth on particular hormones, and lead to appropriate drug treatments.

The 10-year survival rates for most cancers have increased markedly in Britain to well over 50% in the last 40 years, except for brain cancer which only rose from 6 to 13%, compared with rises from 69 to 98% for testicular cancer and from 40 to 84% for breast cancer. About 16,000 people a year are diagnosed with brain cancer in Britain but it receives only 1.3% of cancer research funding.

Neurofibromatosis (types 1 and 2) is caused by an autosomal dominant gene with very variable effects. The prevalence is 1 in 3,500, with half the cases being new mutations. It causes light brown ('café au lait') spots from early childhood. About 65% of sufferers have none of the complications which include learning difficulties (30%), mental handicap (3%), and tumours of the skin and nervous system, including ones on the optic chiasma. The nerve tumours (neurofibromas) may be benign or cause serious damage by compressing nerves and other tissues. Treatments of cancerous tumours include surgery, radiation or chemotherapy. Optic nerve tumours need treating when vision is threatened. Average life expectancy is about 65 years.

11.12. Genetics and personalised medicine

It is cheap **to test for known genes** (alleles) or proteins associated with particular forms of certain diseases. The breast cancer drug Herceptin is very effective in one quarter of sufferers. That quarter has a gene making a particular protein for which a test is available. Abacavir is an excellent treatment for HIV but causes fevers and painful rashes in 5% of sufferers. Genetic tests enable likely side-effect sufferers to be identified. The big hope is that analysis of tumours' genetic make-up will enable the most effective drugs to be used.

Gene chips containing many thousands of DNA sequences are in frequent clinical use for identifying particular forms of genes in patients. Analysis of messenger RNA and proteins can be used to study disease gene expression. Some diseases and disorders are mainly controlled by detectable single genes of major effect, as in cystic fibrosis and the thalassaemias. Unfortunately many diseases are influenced by many genes of small effect, interacting with environmental influences, making personalised medicine very difficult. There is diversity in types of breast cancer tumour, such as oestrogen receptor positive (ER+), lymph node negative (LN-) or human epidermal growth factor receptor 2 negative (HER2-). These are identified by tests for the genetic make-up of tumours removed during surgery. This information, plus details of tumour size and grade, enables predictions of the likelihood of the tumour spreading to other parts. A British trial in 2011 found that 46% of women with ER+ cancer, the most common form of early breast cancer, could avoid needless gruelling chemotherapy.

Many older people are advised to take low-dose aspirin to lessen the risk of bowel cancer (as well as to thin the blood). Studies by Nan *et al.* (2015) at Massachusetts General Hospital (*JAMA* 2015, **313**: 1133–1142) confirmed that for most people, taking aspirin and non-steroid anti-inflammatory drugs reduces that risk. For about one in nine people, their genetic make-up means that such drugs have no effect on bowel cancer. However, for 1 in 25 people, with a particular genetic make-up, such drugs actually increase the chance of getting bowel cancer. In Britain, more than 40,000 individuals develop bowel cancer each year, with about 16,000 deaths. These findings show that personalised genetic testing can indicate which treatments can be helpful or harmful for particular individuals.

11.13. Inherited disorders from mitochondrial DNA defects

While most inherited disorders are caused by defects in genes and chromosomes in the nucleus of the cell, a minority are caused by mutations in the DNA of the mitochondria, the very small organelles which produce a cell's energy. About 1 in 6,500 babies is born with one of many possible inherited serious mitochondrial diseases, some of them lethal. Symptoms include muscle wastage (types of muscular dystrophy but not sex-linked), heart or liver disease, sight or hearing problems, poor growth or poor muscle co-ordination. For example, Lily Merritt had a series of seizures, convulsions and heart attacks, dying when eight months old with a hugely enlarged heart. Her mitochondria failed to produce enough energy (reported in *The Daily Telegraph*, 3/2/2015). Mitochondria are inherited exclusively from the mother, going down the maternal line, with defects affecting male and female children. The mothers may or may not show some symptoms, depending on their proportion of mutant mitochondria.

In February 2015, the UK was the first country to approve a technique of **mitochondrial donation** to correct such disorders. An egg is taken from a woman with healthy mitochondria; its nuclear DNA is removed and replaced by the nuclear chromosomes from a fertilised egg from a woman with defective mitochondria, or just with the latter woman's DNA, with subsequent fertilisation by sperm in the lab. Less than 0.001% of the resulting baby's DNA will be from the egg donor, and all the baby's characteristics from nuclear DNA will be unaffected. There is much diversity of view about advances in reproductive technology, with opposition from the major churches and other groups and individuals to mitochondrial donation. The media fuss about 'designer babies' and 'three-parent babies' is unwarranted; this technique does not improve any normal characteristics but promises to minimise the production of seriously ill babies, including about 125 a year in the UK with mitochondrial disorders.

References

Bateman, A. C. and Howell, W. M., Human leukocyte antigens and cancer: is it in our genes? *J Pathol* (1999) **188**: 231–236.

Donnelly, O., Live viruses to treat cancer. *J R Soc Med* (2013) **106**: 310–314.

Fenton, K. A. *et al.*, Ethnic variations in sexual behaviour in Great Britain and risk of sexually transmitted infections: a probability survey. *Lancet* (2005) **365**: 1246–1255.

Nan, H. *et al.*, Association of Aspirin and NSAID Use With Risk of Colorectal Cancer According to Genetic Variants. *JAMA* (2015) **313**: 1133–1142.

Thanabalasuriar, A. and Kubes, P., Neonates, antibiotics and the microbiome. *Nat Med* (2014) **20**: 469–470.

Recommended reading

Bolouri, H., *Personal Genomics and Personalized Medicine*. (2009) World Scientific, Singapore.

Lamb, B. C., *The Applied Genetics of Humans, Animals, Plants and Fungi*, 2nd ed. (2007) Imperial College Press, London.

Chapter 12

Eating, Drinking, Diet, Digestion, Liver, Cystic Fibrosis, Diabetes, Allergies, Food Intolerances, Anorexia

12.1. Introduction

Food and digestion affect our health, mood and waistlines in ways which depend on our genes, age, sex and exercise. Medical problems of the alimentary tract such as piles and diverticulitis are more frequent in Western societies than in less affluent groups whose diet contains more fibre. In many regions of Africa and Asia, the drink problem is getting enough safe water. In Pakistan, about 40,000 children under the age of five die a year from diseases linked to unsafe drinking water.

Food additives can be controversial, as in the case of adding fluoride to water to reduce tooth decay, but few people object to breakfast cereals having added vitamins and iron. According to the American Thyroid Association, iodine deficiency was common in the Great Lakes, Appalachian, and Northwestern U.S. regions and in most of Canada before the 1920s. Its treatment by the introduction of iodised salt has virtually eliminated the 'goitre belt' in those areas. However, many other parts of the world do not have enough iodine in people's diets, so iodine deficiency is still an important public health problem globally. Approximately 40% of the world's population is at risk of iodine deficiency, with goitre (an enlarged thyroid gland) being common.

In Canada, the addition of folic acid to flour, cornmeal and pasta reduced by 78% the incidence of neural tube defects in babies, e.g., spina bifida. Norwegian research showed that 400 micrograms a day of folic acid for pregnant women reduced the incidence of cleft lip and palate in babies by 40%.

Share some **asparagus** with friends, getting them to report back later. Some people quickly detect strong, coarse, asparagus-like smells in the urine, but others do not. Everyone produces this pong but only 10% can smell it (Lison *et al.*, 1980).

12.2. The digestive tract

This tube runs from the mouth to the anus, about 30 feet (9 m) long when relaxed, but usually half that length because it contracts. It includes the mouth, throat, oesophagus (about 10 inches, 25 cm, long), stomach, small intestine (duodenum, jejunum and ileum, about 20 feet, 6 m, long in total, 1 inch, 2.5 cm diameter), large intestine (about 5 feet, 1.5 m long; 2.5 inches, 6.5 cm diameter), which consists of the caecum, appendix, colon and rectum (about 8 inches, 20 cm, long), ending in the anus.

The inside of the tube can be considered as **outside the body**, with only digested products crossing the gut wall entering the body proper. **Conditions inside the gut** are often too extreme to be tolerated within body tissues but are necessary to break down food. Most body fluids have a pH (a measure of acidity, going from 0, extremely acid, to 7, neutral, to 14, highly alkaline) of 6.8 to 8.0, but secretion of hydrochloric acid into the stomach gives that a pH of 2, very acid, and capable of breaking down body tissues, food or bacteria.

Enzymes in saliva start breaking down carbohydrates such as starches during chewing. A study from the University of California, Santa Cruz, found that human populations traditionally eating a high-starch diet had an increase in copy number for the *AMY1* gene, giving an increase in starch-digesting enzyme amylase. The stomach absorbs alcohol and some fat-soluble items including aspirin. The stomach's gastric juice is rich in hydrochloric acid and in pepsin enzyme which starts digesting proteins. Carbohydrate breakdown continues. In the duodenum, food receives pancreatic enzymes to break down proteins, carbohydrates and fats, while alkalis are added to neutralise most of the stomach acidity, to get the right pH for later enzymes. Bile secretion from the liver aids fat digestion through emulsifying large fat droplets into many very small ones, with a large surface area for fat-digesting enzymes to attack. In the small intestine, nearly all remaining carbohydrates, proteins and fats are digested and their useful products (sugars, amino acids, glycerides and fatty acids) are absorbed. Its inner surface is covered with folds, villi (small finger-like projections) and microvilli, together giving it an area about 600 times greater than if the surface were smooth, with an area equal to that of a tennis court. If the inner surface of the small intestine were smooth, it would need to be 2¼ miles (3.6 km) long instead of 20 feet (6 m)!

After the small intestine, no major food-product absorption occurs. The large intestine absorbs water and salts, and receives indigestible food residues (cellulose, lignin and other fibrous matter), and moulds them into faeces to be stored in the lower end of the colon and in the rectum before being voided from the anus. About 30% of the dry weight of faeces is bacteria, with undigested cellulose fibres adding bulk.

We have about 100 trillion bacteria in our intestines, weighing about three pounds (1.4 kg), with more than 1,000 species, with *Escherichia coli* (non-pathogenic strains) prominent. The colon has about 10^{12} bacteria per gram. They aid digestion as some can break down compounds our enzymes cannot. Bacteria make B vitamins and provide butyryl coenzyme A, needed by cells lining the colon. Babies acquire bacteria from the mother's vaginal and anal areas on birth, and from her breasts. Taking antibiotics can change our bacterial flora adversely because some bacteria help fight pathogenic bacteria. Fluctuations in gut bacterial populations influence the composition, smell and hardness of faeces. Gut bacteria can be encouraged by prebiotic foods such as inulin. Bananas, artichokes and leeks are high in such nutrients. So-called probiotic (live) bacteria in fermented foods such as yoghurt are claimed to do one good.

The body **recycles water** very efficiently. This is essential because the body has only about 4¾ pints (2.7 litres) of blood plasma which provides the water for digestive secretions. Average amounts taken in daily are 2.2 pints (1.25 litres) of food (about 80% of solid food is water) and 2.2 pints (1.25 litres) of fluid drunk. Typical daily secretions are 2.6 pints (1.5 litres) of **saliva**, 3½ pints (2 litres) of **gastric juice**, 2.6 pints (1.5 litres) of **pancreatic juice**, 0.9 pints (½ litre) of **bile** and 2.6 pints (1.5 litres) of **intestinal juice**, a total of nearly 17 pints (9.5 litres) into the guts. The small intestine daily absorbs about 16 pints (9 litres) of water, leaving 0.9 pints (½ litre) of digestive residues to enter the **colon**, which absorbs another 0.6 pints (350 ml), leaving about ¼ pint (150 ml) in **faeces**.

The **appendix**, near the junction of the large and small intestines, is about 3½ inches (9 cm) long, a worm-shaped pouch with no known function. If it gets blocked, infected by bacteria and inflamed, **appendicitis** results, with severe pain and fever. Antibiotics can be used, but usually the inflamed appendix is surgically removed so that it cannot burst and infect the abdomen, causing peritonitis.

12.3. An adult problem with milk — lactose intolerance

Babies digest milk using the enzyme **lactase** which splits the milk sugar **lactose** into simpler sugars, glucose and galactose, which are absorbed through the gut wall, but the gene is usually switched off around the age of four. In most white Europeans and Americans, a gene for adult lactase is by then switched on, but **many adults are lactose-intolerant** in other populations, especially among Africans (100% in Zambia), Chinese and Thais. Charles Darwin was probably a sufferer. Lactose intolerance generally acts as an autosomal recessive, with the gene on chromosome 2. Lactose tolerance arose independently in Europe and Africa.

Drinking milk can cause sufferers to vomit or have diarrhoea, with embarrassingly large amounts of gas (carbon dioxide, methane and hydrogen) in the lower intestines, usually half an hour to two hours after consuming it. One Egyptian had not drunk milk since childhood. After one drink of it as an adult, he was ill for four days with gas, vomiting and diarrhoea. Cheese and yoghurt are usually safe as bacteria used to make them have pre-digested the lactose. Milk was probably included in adults' diets about 10,000 years ago, when cattle, sheep, goats and reindeer were domesticated.

Lactose is the principal energy source for the first few months of life, being 7% of milk. In young babies, lactase is secreted in large quantities, reducing to about 10% of that level when they are weaned off milk onto solid food. The continued drinking of fresh milk causes continued production of lactase in people who are not lactose intolerant. Even pastoral societies often consume little fresh milk, which spoils very rapidly unless chilled, pasteurised, or fermented, e.g., to form cheese.

McCracken in 1971 published data on the frequency of lactose intolerance in adults: Czechs, 0%; Danes, 2%; Arabs, 14%; Greenland Eskimos, 72%; American Negroes, 74%; Australian Aborigines, 85%; Chinese, 85%; Bantu, 89%; US Asians, 97%; Thai, 99%; American Indians (Chami, Colombia), 100%. Populations with low frequencies of lactose-intolerance have a history of dairy herding. An Asian dermatologist with lactose intolerance eats hard cheeses but soft cheeses and milk caused internal gas problems until he tried Lactaid®, which contains a fast-acting lactase-like bacterial enzyme. He chews the tablet thoroughly with the first mouthful of a lactose-containing food, which avoids symptoms.

Other causes of intestinal gases — hydrogen, carbon dioxide, methane and hydrogen sulphide (bad-egg smell) — are foods promoting gut bacterial fermentations, including beans, lentils, cabbage, sprouts, onions, apricots and cucumbers.

Food and drink (garlic, onions, curry, strong cheese, milk, beer, whisky, etc.) can affect one's smell to others and hence social acceptability in different cultures. Some Chinese claim that Westerners smell unpleasantly of sour milk.

12.4. Baby problems with milk

Babies with lactose intolerance are put on low-lactose milk, with glucose supplements to avoid severe malnutrition as 40% of milk calories are in lactose. Even babies who can digest lactose in to glucose and galactose may then have problems digesting **galactose** through lack of the relevant enzyme. This causes **galactosaemia**, an autosomal recessive condition affecting 1 in 60,000 European babies. Heterozygotes have about half the normal amount of enzyme and are usually unaffected. Sufferers show distress within weeks of birth with weight loss, vomiting, jaundice, and galactose in the urine. If untreated, they have stunted growth, enlarged liver, mental retardation and cataracts. It can be detected in the first week of life by testing for galactose levels in blood from a heel prick, and affected babies can be put on milk substitutes.

Because galactosaemia is rare, it would be expensive to screen for separately. In many countries, a number of inherited diseases are tested for together about day three after a baby's birth. Drops of blood from a heel prick are tested for galactosaemia, PKU (about 1 in 5,000; see below), maple sugar urine disease (1 in 250,000) and homocystinuria (1 in 100,000); these diseases can be controlled to some extent by dietary changes.

12.5. A failure to break down a dietary amino acid — phenylketonuria (PKU)

About one baby in 5,000 cannot change into tyrosine the amino acid **phenylalanine**, which is present in milk and almost all foods. Phenylalanine accumulates in the blood, leading to brain damage, often with seizures and eczema. Sufferers are homozygous for a recessive autosomal gene.

PKU is detected by testing for high levels of phenylalanine in a small blood sample as described in 12.4. Untreated babies become severely mentally retarded. They have lighter than usual skin colour as tyrosine is needed for melanin pigments. Sufferers can be treated by immediately switching them to a diet low in phenylalanine for about 15 years while the brain develops, giving almost normal intelligence. They receive a tyrosine supplement. The frequency of PKU is about 1 in 10,000 in most white Europeans, but 1 in 4,000 in the Irish and 1 in 3,000 in Turks. It is rare in Afro-Caribbeans, Indians and Jews.

12.6. Missing digestive enzymes, cystic fibrosis; Johanson-Blizzard syndrome

Cystic fibrosis is a serious inherited disease affecting digestion, nutrition, lungs, liver, salt excretion and male fertility. There are big racial differences in frequency. In white Caucasians, 1 in 2,500 suffers, with 1 in 25 being a symptomless heterozygous carrier, *Cc*, of the autosomal recessive gene, *c*, on chromosome 7. Sufferers (*cc*) have a high frequency in Afrikaners, 1 in 622, but only 1 in 17,000 American Blacks and 1 in 250,000 Orientals. About 75% of European and American sufferers carry the same mutation, a three base-pair deletion at position 508 in the *CFTR* gene, although more than 1,000 different mutations are known.

Sufferers used to die before 10, with a life expectancy of five years but at least 70 % now survive to adolescence, with an average survival of 30 years or more if they avoid lung infections. Some have their lives prolonged by heart and lung transplants. Life expectancy is only 16 years for sufferers attacked by a bacterium, *Burkholderia cepacia*, which usually attacks onions but is an opportunistic human parasite.

Death was usually from malnutrition because sufferers lack three vital pancreatic digestive enzymes: amylase (starch digestion), trypsin (protein digestion) and lipase (fat digestion). They are now given pig digestive enzymes in acid-resistant capsules which break down in the right part of the gut, so they can digest food if they take capsules with each meal.

Cystic fibrosis causes excessive excretion of salt in sweat, so salt tablets are given. The disease was known for hundreds of years from salt over-excretion. Old Northern European folklore has the saying: "Woe to that child which when kissed on the forehead tastes salty. He is bewitched and soon must die."

The gene also causes thick mucus in the lungs, which can lead to infections such as pneumonia. Very vigorous physiotherapy is given to clear the lungs, and antibiotics are used when necessary. Males are sterile through lack of a vas deferens connecting the testicles to the penis, but females are usually fertile and some have children.

The gene was cloned in 1989 and many attempts have been made at gene therapy, e.g., by blowing liposomes (fat vesicles) containing non-mutant copies of the gene into the lungs. Unfortunately, cells lining the lungs are shed continually, so treatments are needed every few weeks. Gene therapy does not yet work well, and treating the lungs does not cure the salt or digestive problems.

There are DNA tests for cystic fibrosis mutant genes, used for prenatal diagnosis or after birth. Unfortunately they sometimes fail to detect a mutant gene because there are so many possible mutations, and each DNA probe is specific for one mutation. A probe for the commonest mutation, affecting 75% of Western sufferers, fails to detect the other 25%.

Personal account. *I have cystic fibrosis, affecting my everyday life and life expectancy*

I suffer from frequent chest infections and cannot digest food without enzyme capsules. I go through a daily routine that involves taking a large assortment of tablets (antibiotics and dietary supplements) and spending time inhaling nebulised drugs and completing physiotherapy.

I try not to let my condition prevent me from achieving ambitions but I need to be organised and plan carefully. Sometimes just getting out of the door can be a trial; morning medication means I can't skip breakfast even if I'm running late. I need to get up very early to do nebulisers and other treatments. Turning up at a friend's house for a weekend often causes raised eyebrows as I come with more bags than most people take for a two-week holiday — not clothes, but medication and devices to keep me healthy.

Approximately three times a year I need to go on intravenous antibiotics (IVs), so I need to be more regimented, with doses at 6 a.m., 2 p.m. and 10 p.m. I have learnt to self-administer IVs, so I no longer need hospital stays. During the two-week courses I have been able to go to school, university and work. They make me tired and rather sick. I have a 'porta-cath' fitted under the skin, which when I need IVs is accessed by a needle. It means that I do not need peripheral cannulas and is much less painful.

Although my disability is not visible it is something which I need to explain to people after becoming acquainted with them. Taking such a large amount of medication each time I eat raises questions as do frequent bouts of coughing. I prefer that people get to know me as a person first rather than find out about my illness straight away as this can make them interact with me differently (usually out of concern but this is still irritating).

Emily, age 22. She preferred not to write about the reduced life expectancy.

Alice Martineau was a pop singer who died, aged 30, in 2003 from cystic fibrosis after no donor was found for a triple transplant of heart, lung and liver. She needed oxygen night and day, could only digest two thirds of her food and had a tube in her stomach. Each day there were three rounds of chest physiotherapy, 40 pills to take, and a 14-day course of intravenous antibiotics every month.

Emily Hoyle was just weeks away from death from cystic fibrosis in 2012. It had been diagnosed when she was three months old and had caused many crises from lung infections. Luckily, a pair of compatible lungs became available then from an 18-year-old boy and she received them in a six-hour transplantation operation. It transformed her life. Now she is fully fit and training to climb the 6,000 m volcano Cotopaxi in Ecuador, along with 12 other heart or lung transplant recipients.

The **Johanson-Blizzard syndrome** is a very rare autosomal recessive condition, with a prevalence of about 1 in 250,000.

Personal account. *A genetic condition affecting digestion, hearing and the thyroid gland: Johanson-Blizzard Syndrome — my son nearly died in infancy*

Our sixteen-year-old son, Simon, was born in Bermuda, two weeks overdue, weighing 8 lb. From the outset, Simon failed to thrive. He was unable to feed and lost weight rapidly, surviving from regular blood transfusions. The Bermuda hospital was unable to diagnose or control his condition, and we returned to Britain for treatment. Simon was two weeks old.

Our local GP and hospitals had Simon in their care for some weeks, with regular blood and physical checks, including lumbar puncture, in an effort to discover the cause of his trauma, to no avail. He lost half his birth-weight and became increasingly weak, surviving only on blood transfusions. Simon was referred at 10 weeks to the Great Ormond Street Hospital for Sick Children.

The tests there were unsuccessful and the prognosis was bleak. The breakthrough came when Simon was visited by Professor Bereitzer who a few weeks earlier had read Johanson and Blizzard's account of a genetic condition tending to occur in families which interbreed, especially in Iran and Iraq. Prof. Bereitzer immediately prescribed treatment by pancreatic enzyme supplements, and Simon's health improved rapidly. His pancreas did not exist other than some fatty tissue, but the insulin site was intact. In addition to pancreatic medication, he takes thyroxin for hypothyroidism, along with vitamin supplements. He was diagnosed with the J-B syndrome.

He has no prospect of growing adult teeth, is deaf, and has extremely limited smell and taste. His mental function is good but Simon is small since food absorption is poor. He took longer than average to crawl/stand/ walk and is going through puberty now.

Alan Brooks

12.7. Inflammatory bowel diseases: coeliac disease (gluten intolerance), ulcerative colitis, diverticulosis and Crohn's disease

Coeliac disease affects 1 in 2,000 white Europeans, causing poor absorption of food. The sufferer's small intestine is sensitive to alpha-gliadin, one of the proteins in **gluten**, which occurs in wheat and some other cereal grains. It damages the intestinal villi, impairing the uptake of all nutrients. Treatment is by elimination of gluten from the diet. There is a multifactorial genetic component, with brothers and sisters and offspring of sufferers having a 1 in 33 chance of suffering.

Ulcerative colitis affects 1 in 1,500 people and is autoimmune. It is multifactorial with a 15-fold increase in risk for the children, brothers or sisters of sufferers. It causes ulceration of the colon and rectum, giving rectal bleeding, abscesses, abdominal pains and diarrhoea. The bleeding frequently causes anaemia. If the rectum is affected, there can be loss of bowel control. Simon Rogers was super-fit but was struck with ulcerative colitis at the age of 17. It caused him abdominal pain, bloody diarrhoea, fevers, weight loss, and an embarrassing need to find a lavatory urgently. He had his colon removed when he was 19 and is super-fit again, not needing a colostomy bag.

Diverticulosis involves diverticula — small bulging pouches that can form in the lining of lower parts of the large intestine and are common, especially after age 40. They seldom cause problems but inflammation can cause severe abdominal pain, fever, nausea and a change in bowel habits. If antibiotics do not work, surgery may be needed.

Crohn's disease usually requires surgery to remove inflamed regions of the gut, and several successive operations may be needed as new areas become affected. Patchy deep ulcers may cause fistulas, with narrowing and thickening of the bowel. Symptoms include fever, cramping abdominal pains, unpredictable, persistent diarrhoea, and weight loss. Many sufferers have much of the gut removed, with a stoma in the abdomen wall and colostomy bags to collect faecal wastes. It affects 1 in 5,000. It is multifactorial, with an increased risk to relatives.

12.8. Piles (haemorrhoids)

Piles are enlarged and inflamed veins in the lower rectum or anus which can cause extreme irritation and embarrassing itching. It runs in families and is multifactorial. Its frequency rises with age.

Piles are often caused by straining during bowel movements (or lack of them), by prolonged sitting, or by diets low in fibre or with insufficient liquids, causing constipation and straining. Pregnancy can cause piles. They are usually an irritant rather than serious, although prolonged blood loss causes anaemia. Treatment is by injections to shrink distended veins, by suppositories to reduce swelling and itching, or increasing dietary fibre and fluids, stool-bulking agents, or tying off piles to remove their blood supply. Pricking them is not recommended.

> **Personal Account. *I have piles***
>
> *Having piles runs in my family. Mother used to prick father's piles. In my thirties, I felt something soft and small sticking out of my anus. The doctor said it was piles, with an enlarged vein protruding. A consultant injected it and some others which he could see developing in my back passage. He recommended a high-fibre diet, with five dessert spoonfuls of bran every breakfast.*
>
> *Over the next 35 years, I have had many more piles, some protruding and some being an enlarged vein round the outside of the anus. I get intermittent bleeding from inside the rectum, especially when a faecal pellet is too large and too dry for easy passage. It is alarming when rectal bleeding*

leaves the lavatory bowl bright red. The enlarged external veins around the anus usually cure themselves within a week, shrivelling away.

The consultant ordered X-rays and internal examinations to check for cancer. The barium enema was no fun, with a heavy barium salt introduced through the anus into the large intestine. X-rays were taken with me in strange positions, strapped to a frame which could be rotated to put me upside-down. The sigmoid colonoscopy, where an endoscope tube with a light, lens and viewing port was pushed impossibly far up my guts from the rear end, was distinctly uncomfortable but revealed nothing nasty.

This was written by a man in his sixties, anonymously in case his friends laugh unkindly. Piles are not funny.

12.9. Alcohol, enzyme deficiencies, the flushing reaction

Alcohol affects sociability, mood, appearance, health and intelligence. It provides calories, giving energy; non-distilled alcoholic drinks such as beer and wine are good sources of vitamins and minerals. Excessive alcohol slows down thought, reduces co-ordination and can cause nausea and vomiting.

Some good mental and physical effects of moderate alcohol consumption (two glasses of wine or a pint (0.57 litres) of beer a day) are reduced anxiety and tension, better sociability, thinning of the blood, reduced fatty clogging of the arteries, and a reduced risk of diabetes, dementia, kidney stones, gall stones and heart attacks. Queen Elizabeth the Queen Mother (1900–2002) liked her gin and Dubonnet. In her later years she pointed out that the doctors who had advised her to cut down on alcohol were themselves long dead.

The '**French paradox**' is that the French generally have a fatty diet but do not have a bad record of heart disease. A high intake of red wine and garlic may help to protect them. The polyphenol tannins and red pigments in red wines act as antioxidants.

Alcohol affects women more than men because, on average, men have larger bodies, proportionately more water in their bodies and more detoxifying alcohol dehydrogenase enzyme in their liver. There is a huge diversity between individuals in alcohol tolerance. A short, slim lady who can drink a lot of wine with no adverse effects explained that she has 'hollow legs'. Other friends feel tipsy after one glass of wine. Figures from Columbia University Medical Center, New York, revealed that 30% of Americans have some form of alcohol-use disorder at some time, including 18% with alcohol abuse and 13% with alcohol dependence.

Professor Corder (*The Red Wine Diet*, 2007) gave a talk in 2009 to the Royal Society of Medicine on health aspects of wine and chocolate. In France, the highest levels of heart disease are those where the main drinks are beer or white wine, as in Alsace, and those with the lowest levels (and best longevity) are red wine areas such as the Midi and South West France. For people who have had a heart attack or stroke, alcohol decreases the chance of another attack. Blood vessel linings were healthiest with red wines and red wine extracts, with red grape juice much less effective. White wines and rosés had no benefit. The good effects were due to flavonoids, especially procyanidins, extracted by alcohol from grape pips during long fermentations. He recommended drinking half a bottle a day of red wine (11 to 12% alcohol good, 15% not so good).

He said that people on Kuna Island drank about five cups a day of crude cocoa and were very long lived. A woman in France ate one kilogram a week of dark chocolate. When she was 119, a doctor told her to cut back: she did, and quickly died, aged 120. Professor Corder recommended eating dark chocolate for its health-giving properties.

Excess alcohol is poisonous. In 2003 there was a vodka-drinking competition in Volgodonsk, Russia. The winner downed several pints of vodka, then died. The other five contestants needed hospital treatment.

Gin is sometimes called 'mothers' ruin'. In 1688, King William of Orange banned the import of French brandy and it was considered patriotic to drink Dutch gin instead. Between 1688 and 1730, the consumption of gin reached 327 pints (186 litres) a head a year, leading to drunkenness, child neglect and poverty, with 75% of children dying before the age of five.

Scientists at Cancer Research UK and Oxford University (Munafò *et al.*, 2003) showed that genetic factors strongly influenced personality traits linked to unhealthy lifestyles. One version of the serotonin transporter gene, *5HTT-LPR*, was associated with anxious personalities, likely to find social interactions stressful, seeking refuge in alcohol, drugs or smoking.

Although some alcohol is excreted in urine and a little is breathed out, more than 80% is oxidised by the enzyme **alcohol dehydrogenase,** mainly in the liver. The **acetaldehyde** produced is further oxidised to acetate by two **aldehyde dehydrogenase enzymes.** Strangely, the liver alcohol dehydrogenase of 90% of Orientals is about 100 times more active than that of most Caucasians, but about half of Orientals lack active aldehyde dehydrogenase-2 in their mitochondria. Even small amounts of alcohol cause intense red flushing of the face (Plate 3D) and chest in affected individuals, from the acetaldehyde. Virtually all Caucasians have an active form of this enzyme, although some 5 to 10% have the flushing reaction. American Indians are Mongoloids, having about 60% of flushers.

Jennifer Cummins found the following frequencies of flushing reaction in Imperial College students, in response to small amounts of alcohol: Blacks, 0%; Orientals, 47%; other Asians (Indians, Pakistanis, Sri Lankans), 0.5%; Caucasians, 32% for females, 7% for males. The difference between Caucasian Asians and Mongoloid Asians was dramatic. Leo Xenakis described his strong flushing reactions even to very small amounts of alcohol: "I quickly develop red rashes on my face but spreading all over my body. I have nausea, sometimes vomiting, dizziness but no loss of movement. It is like almost immediate extreme drunkenness but without enjoyment. My Japanese mother reacts similarly but my Greek father is unaffected. The problems come from acetaldehyde in the blood, from insufficient aldehyde dehydrogenase-2."

With five students of Chinese ethnicity, Xi Yu Phoon and I ran an experiment on the effects of drinking 25 ml of neat whisky (40% alcohol) every 20 minutes, for four drinks. Two students proved to be non-flushers, two showed some flushing and one showed strong flushing, light to mid-purple on her head and neck. Cheek temperatures rose from 37° to 39°. After four drinks, symptoms included feeling tipsy, tired, peaceful, relaxed, getting a headache, feeling very hot, light-headed, calm but deliriously happy, feeling drunk and giddy, feeling OK, could fall asleep, blurred vision, can see but the images don't come together. One female wrote: 'Second drink, much laughs had by all. Feeling hot. Third drink, pretty happy with the world. Hot, red, good mood. Mental arithmetic impaired. Fourth drink. Felt the love in the air. Everyone was a buddy. All was good in the world. Possible slurred speech'.

Alcoholics often have unsteady hands and feet and may look unkempt, with red faces and bad skin, reduced intelligence and responsiveness.

Prolonged excessive alcohol consumption causes many problems, especially liver cirrhosis. Alcoholism tends to run in families. Danish studies showed that being reared by alcoholic parents was of less significance than being the biological child of an alcoholic parent. In mice, one can do selection over several generations to get strains in which nearly all individuals prefer alcohol (10% in water), and other strains which are teetotal.

Many sportsmen, actors and politicians have had problems with alcoholism. Football star George Best had a history of excessive drinking. He was rushed to hospital in 2000 after 10 days of drinking with little food. He spent six weeks there and was told that his liver was so badly damaged that even one more alcoholic drink might kill him. It didn't. After further binges and hospital admissions, he had Antabuse pellets implanted into his stomach. They are sometimes given to alcoholics to make them feel extremely ill if they drink alcohol, but having them implanted in the stomach is extreme. In 2002, he had a successful liver transplant on the NHS, but that was controversial as his need was self-inflicted. He died in 2005, aged 59. This unrepentant drinker often said, "I spent a lot of money on booze, birds and fast cars — the rest I just squandered."

Women who drink large amounts of alcohol whilst pregnant can stunt the baby's physical and mental development. In severe cases, the child is born with strange facial features, part of the **foetal alcohol syndrome.** Some people cannot enjoy red wines because they get headaches and hangovers from tyramine which constricts the brain's blood vessels. Writer Jung Chang loved champagne but became allergic to alcohol.

Alcohol is slowly absorbed through the stomach wall, but is absorbed faster in the small intestine which has a much greater surface area. Emptying of the stomach into the duodenum is delayed by high-fat foods, delaying the absorption and effects of alcohol.

12.10. The liver; jaundice, hepatitis and cirrhosis

The liver lies opposite the stomach and is the largest, most important organ of metabolism, weighing about 3 pounds (1.4 kg). It is about 8 to 9 inches wide, 6 to 7 inches high and 4 to 5 inches thick (20 to 22, 15 to 18, 10 to 13 cm). It is unusual in that its cells can divide so well that a liver lobe, if destroyed, can be naturally regrown.

The by-products of digestion, when absorbed through the gut walls, are not ready for general circulation but must be processed by the liver. Blood from the intestines goes into the hepatic portal vein to the liver. As well as processing the products of digestion, the liver detoxifies and breaks down blood-borne body wastes, alcohol, hormones and drugs. The liver's own leukocytes (like white blood cells) remove bacteria and worn-out red blood cells. The liver excretes bilirubin from haemoglobin in those red blood cells, and excretes cholesterol. It secretes bile salts to break up large fat globules.

The liver stores glycogen (which can be broken down to glucose), fats, minerals such as copper and iron, and many vitamins. It synthesises proteins for blood plasma, including those involved in blood clotting and in transport of some hormones and cholesterol. It activates vitamin D. It is easy to see why one's general health depends on good liver function, as in the joke: "Is life worth living?" "It all depends on the liver!" Bile is stored in the gallbladder, emptying into the duodenum through the Sphincter of Oddi. Bilirubin is yellow and in the intestines is turned brown by bacteria, giving faeces their colour. Some bilirubin gets back into the blood through the intestinal walls and is excreted by the kidneys, making urine yellow.

One's urine is often darker after eating liver or kidneys as they contain bilirubin. Bilirubin accumulation causes jaundice, with yellowing of the skin and the whites of the eyes, and darker urine. Jaundice can be caused by excessive breakdown of red blood cells (e.g., from malaria), from liver malfunction (e.g., a paracetamol overdose), or obstruction of the bile duct by gallstones.

Hepatitis is inflammation of the liver, caused by viruses or harmful substances. Sometimes there is no permanent damage or hepatitis can persist for many years, causing scarring (cirrhosis). Serious hepatitis can cause fatal liver failure or liver cancer. Initial symptoms of viral hepatitis are similar to having flu, such as muscle and joint pains, a high temperature, feeling sick, headaches, and sometimes yellowing of the eyes and skin. Chronic hepatitis causes exhaustion, depression, jaundice and feeling unwell. Hepatitis may have no perceptible symptoms, with the person unaware of it.

Hepatitis A, caused by the hepatitis A virus, is common where sanitation and sewage disposal are poor. Around 350 cases are reported each year in England, usually from faecal contamination of food or water abroad. A vaccine is recommended for travellers to the Indian subcontinent, Africa, Central and South America, and the Far East.

Hepatitis B is caused by the hepatitis B virus which occurs in blood and body fluids, such as semen and vaginal fluids. It can be spread during unprotected sex, by sharing needles for injecting drugs, and from pregnant women to their babies. It is uncommon in Britain and America but more frequent in East Asia and Africa. Most sufferers recover fully but some get cirrhosis or liver cancer. A vaccine is available.

Hepatitis C, from the hepatitis C virus, is the commonest type of viral hepatitis in England, with about 215,000 having chronic hepatitis C, with four million in America and more than 20% of Egyptians. According to Ghany and Gara (2014), 170 to 200 million people worldwide have chronic hepatitis C, causing 30% of cases of liver cirrhosis, 25% of liver cancers and more than 350,000 deaths a year.

The virus occurs mainly in blood, but to a lesser extent in saliva, semen and vaginal fluid, with transmission usually through blood from needle-sharing. It may cause no noticeable symptoms, or be mistaken for flu, so many people are unaware of being infected. One in four sufferers fights off infection and becomes free of the virus. In three out of four sufferers, infection lasts many years, giving chronic hepatitis C which can cause cirrhosis and liver failure. There is no vaccine.

Alcoholic hepatitis occurs with varying severity in a quarter of heavy drinkers, often being unnoticed. An English taxi driver who considered himself a moderate drinker drank about 10 pints (5.7 litres) of beer a weekday night and 30 pints (17 litres) over a weekend!

Cirrhosis of the liver (Greek *kirrhos*, tawny) makes it tawny. In 2000, cirrhosis of the liver killed more men than did Parkinson's disease, and killed more women than did cervical cancer. The liver shrinks and hardens, with abnormal growth of fibrous connective tissue which may cause liver enlargement. There is a failure of normal functions, of varying severity, including death. Prolonged excessive alcohol consumption is the main cause, but cirrhosis can be caused by viruses as described above, by autoimmune attack or metabolic disorders of iron and copper, or drugs.

There are many possible complications, including fluid accumulation (dropsy) in the abdomen, blood coagulation disorders, blood pressure increases in the liver, liver cancer, mental confusion and brain disorders. Symptoms can include weakness, anaemia, weight loss, sickness, impotence, jaundice, and red spider-like veins in the skin. Treatments include stopping alcohol or drug abuse, or using antiviral antibiotics.

One's liver and kidneys get rid of about one unit of alcohol an hour. Drinking large amounts of alcohol at one time (binge drinking) is much worse for health than consuming the same amount over several days, as the former puts more stress on the liver, kidneys, brain and blood circulatory system, as well as problems from drunken behaviour.

12.11. Diet and cancer

A South African study showed 20 times more liver cancer in Blacks than in Whites, although cancer of the colon was ten times more frequent in Whites. The diet of Blacks often included food with fungal spoilage, where moulds produce aflatoxins which can cause liver cancer. The diet of Blacks was usually much higher in fibre than that of Whites, and high-fibre diets reduce cancer of the colon.

Fresh fruit and vegetables are said to be good for reducing cancer, through reducing the effects of free radicals. The amounts of dietary **saturated and unsaturated fats** affect the occurrence of cancer in different parts of the intestines. Animal fats are often saturated, and some plant products, such as coconut flesh, are high in saturated fats. Nearly one third of all cancers are considered diet-related.

12.12. Meat, amino acids and malnutrition

Meat is a great source of minerals, especially easily absorbed iron, amino acids and vitamins, and can counteract anaemia and malnutrition. Vegetarians sometimes do not get enough of amino acids. Adults need about two ounces (60 g) protein a day, with a mixture of proteins to get enough of all 20 essential amino acids required for proteins. Malnutrition is sometimes due to a shortage of amino acids **lysine** and **methionine**, rather than to a lack of food, so breeding high-lysine cereal crops has helped combat malnutrition in some countries. Meat fats can raise blood cholesterol levels. Nine-month-old Areni Manuelyan died of malnutrition after her vegan parents put her on a **fruit-only diet**, thinking that it was healthy. One man lost most of his eyesight after a strict vegan diet. Dietary fats are essential for development.

12.13. Vitamin and mineral deficiencies

Insufficient vitamin levels cause a range of symptoms affecting all parts of the body and general health. Most people on varied diets, including fresh fruit and vegetables, do not need supplements. Excessive vitamin consumption can be harmful — some tablets contain many times the daily recommended dose of some ingredients. Cereals are often supplemented with vitamins and iron.

Vitamin A deficiency causes dry, inflamed eyes and poor night vision. Of the **B complex vitamins**, lack of **thiamine** causes beri-beri, with swelling of the body or numbness and weakness. This became prevalent in parts of Asia where rice is the staple food when people switched to eating polished (white) rice, as thiamine occurs mainly in the husks of brown (unpolished) rice. Lack of **B12** causes pernicious anaemia, with extreme tiredness and dizziness. **Deficiency of vitamin C** causes scurvy, with skin, gum and teeth problems which used to afflict sailors on long voyages before citrus fruits such as limes, lemons and grapefruits were used to combat it.

Vitamin D occurs in some foods and its precursor is made in the skin when exposed to sunshine, with dark skins less able to make it than are pale skins. A shortage can cause rickets, where children's bones do not develop properly. In 2003, the *Lancet* had an article (Wharton and Bishop) stating that hundreds of children a year in Britain had malformed hips and bow legs from rickets, from insufficient vitamin D. Well-meaning middle-class parents were blamed for excessive use of sunscreens on their children, insufficient time outdoors, and faddy diets containing little vitamin D and calcium. **Macrobiotic diets** were bad for children as they exclude dairy products, meat and wheat, all good sources of vitamin D and calcium. The 250,000 vegans in Britain avoid meat and dairy products. **Vitamin K** is needed for the liver to produce several factors involved in blood clotting.

Macrominerals are needed in relatively large quantities, more than 0.1 gram a day: sodium, calcium, magnesium, potassium, chlorine, sulphur and phosphorus. **Microminerals** are needed in lesser amounts, including manganese, zinc, selenium, copper, chromium, cobalt, fluorine and iodine.

Iron deficiency causes anaemia, and 6% of British females have full-blown anaemia, being pale and lethargic. Heavy menstrual bleeding makes women ten times more likely than men to be anaemic, but fad diets can cause anaemia in either sex. The best source of available iron is red meat, but chicken, fish, liver and eggs are also good. The marathon-winning athlete Paula Radcliffe was diagnosed with anaemia and now takes ostrich meat, chocolate and a sprouted grass-seed drink.

12.14. Tea, coffee and caffeine

Caffeine is a stimulant and diuretic, with amounts per cup of about 40 mg (100 mg = 1/284th of an ounce) in tea, 55 mg in instant coffee, 105 mg in filter coffee and about 30 mg in so-called decaffeinated coffee. It can cause mutations, especially in micro-organisms. About 85% of Americans are regular coffee drinkers, averaging 280 mg caffeine a day, equal to two mugs of coffee or four bottles of a caffeinated soft drink.

Tea has many **polyphenols**, which tea producers claim combat adverse effects of its caffeine, decrease blood capillary fragility and reduce inflammations. The tea pluckers of Sri Lanka drink several large flasks of black tea a day and get plenty of fresh air and exercise; they are generally very healthy, especially with respect to heart disease. Compounds in tea, but not coffee, have been shown to inhibit enzymes in the brain which promote Alzheimer's.

While caffeine raises alertness, it also raises blood pressure, increases anxiety and decreases hand steadiness. After we drink tea or coffee, caffeine reaches its highest blood concentration after about 35 minutes, then half is eliminated after 3 to 6 hours. For regular caffeine consumers, getting **caffeine withdrawal symptoms** (reduced alertness, poorer general performance, depression and irritability) takes 24 to 36 hours. As a student, I tried a two-day fast and had headaches which I now attribute to caffeine withdrawal. A 76-year-old man suffered intermittent bouts of sickness for many years. He tried giving up coffee, getting a bad headache, but then his health improved greatly. He had had caffeine intolerance.

Caffeine is combined with aspirin, paracetamol, ibuprofen and codeine in various painkillers and cough medicine. Its stimulant action helps fight drowsiness caused by other ingredients. Much caffeine is used in energy pills such as Pro Plus, and in stimulant drinks. The caffeine content of Red Bull is 80 mg/250 ml. Fizzy drinks such as Coca Cola often contain 30 to 40 mg caffeine per can or bottle. There is little caffeine in white chocolate but milk chocolate contains about 18 mg of caffeine per 100 g and dark chocolate has more.

One can **overdose on caffeine.** A 17-year-old Londoner was rushed to hospital with heart palpitations and fever after she drank seven double expressos. She was hyperventilating, with a really rapid heartbeat. A Danish study showed that drinking more than eight cups of coffee a day doubles the risk of a stillborn baby. In some adults, coffee drinking used to pose no problems but later they find that drinking coffee in the evenings disturbs their sleep, while decaffeinated coffee does not. Neither does tea, which is surprising in view of its caffeine content, unless other ingredients in tea combat caffeine's adverse effects.

12.15. Food and allergies

An **allergy** is an acquired inappropriate specific **immune reaction** to a normally harmless substance (**allergen**), giving **hypersensitivity.** Later exposure to that allergen provokes an immune attack whose effects vary from mild local irritation to a potentially lethal shock to the whole body. Between 1991 and 2004 in Britain, the number of children up to age 14 needing hospital treatment for allergies jumped from 16 to 107 per million. Mere dislikes are not allergies.

Part of the **hypersensitive reaction** involves the release of histamine, which causes dilation of the arterioles, increasing blood flow to the tissues and makes blood capillaries more permeable. Another part is the release of 'substance of anaphylaxis' which causes strong and prolonged contraction of smooth muscle, especially constricting the small airways.

In serious allergic reactions, the entire body goes into an **anaphylactic shock.** A massive loss of blood plasma fluid into the tissues through the capillaries and widespread dilation of the blood vessels can cause a major loss of blood pressure and heart failure; constriction of the airways causes severe breathing difficulties. Such shocks are often fatal unless the sufferer can be quickly injected with a drug such as adrenaline to constrict the blood vessels and open the airways. **Antihistamines** combat some symptoms but are insufficient for a major anaphylactic shock.

Some allergies are very specific. Belinda George, aged 26, had eaten nuts many times with no ill effects. At a party she ate a Brazil nut for the first time and died within two hours from anaphylactic shock. Raya French knew that she had a strong allergic reaction to tomatoes and had nightmares about it. She died aged 37 from anaphylactic shock on opening a tin of spaghetti bolognese. She self-administered adrenalin three times but that did not save her.

Allergies to eggs, milk or peanuts usually develop in young children, but **fish and shellfish allergies** often start in the teens and twenties. **Nut allergies** can start in middle age. A person may eat a certain food for many years, then suddenly become allergic to it. Many individuals are **allergic to shellfish** such as prawns, lobsters and crabs, others to preservatives. In the last 50 years, there has been a large increase in the incidence of **peanut allergy**. One ex-student has this so badly that if someone pours peanuts into a bowl across the room, his throat starts to swell uncomfortably. To eat even one peanut could be lethal.

Allergy tests usually involve panels marked on upper arms or the back. Each panel has a possible allergen, such as birch pollen or dust mite excreta, inoculated on it; developing a red patch in the course of a few hours indicates an allergy.

Personal Account. *I suffer from many allergies* [intolerances?]

The sheer number and variety of things that bring about my allergic reaction are what set me apart. I'm allergic to a range of things, from vinegar and prawns to penicillin, sulphur drugs, high acid foods, preservatives and pure wool next to my skin. I have to be cautious about the foods I eat. If I get the allergic reaction, the antidote is an antihistamine drug which I have to take immediately.

Even though my parents knew about these allergies since I was an infant, my awareness of them was raised when I developed a serious attack at age seven at school. The initial discomfort was followed by unbearable itching and small islands of swelling all over my body. It was made worse by no one being familiar with such a situation, and more severe symptoms developed. I felt faint, disoriented, my vision became blurred and everything seemed to be coloured in shades of green. I was taken home after being given the necessary tablet.

Allergic reactions this serious have occurred three or four times. As the itching starts, I take a tablet and within an hour or so the swelling subsides. I have to be constantly conscious of what I eat and make sure I always carry my tablets.

Though it has sometimes been restrictive, especially at barbecues and parties, it hasn't affected my life immensely. The topic only comes up if I feel the need to mention it, that is, when I get an allergic reaction or have to explain why I cannot eat certain foods. Avoiding the allergens has become almost a reflex action.

Miss Shaznene Hussain, a Sri Lankan aged 18.

About 15% of Britons and Americans say that they have food allergies, but strong allergic reactions only affect about 2%, with dairy products, eggs, shellfish and nuts, especially peanuts, being the main causes. With a real allergy, symptoms start within seconds of eating even small amounts of the triggering food. The lips and tongue tingle and swell, and a really swollen tongue can cause death from suffocation. In severe cases, sufferers should inject themselves with adrenaline.

Food intolerance is less serious than a real allergy, and is often treated with antihistamine tablets. Intolerance causes a delayed response to eating the trigger food, from a few hours to two days. Symptoms vary, including headaches, migraines, tiredness and muscular pains. More severe and faster reactions include itching, vomiting and diarrhoea. The British Allergy Foundation estimates that almost one in two people have food intolerances, including ones to dietary staples like wheat, dairy products and potatoes. The traditional way of identifying a food intolerance is the **exclusion diet**, excluding one type of item from the diet for several weeks to see if the symptoms clear up, but there are quicker immunological tests. Allergies and intolerances, as well as personal preferences, can make hosting meals really difficult with the proliferation of different diets, as I found when entertaining students to dinner. One student said, "My diet is vegan — horror — no, it's simple. Just veg/fruit/nut/pulses/Quorn OK/egg white only."

On undergraduate field trips in the 1960s, we all ate anything we were given. When Imperial College students go on field trips today, the department has the list of names with descriptions such as vegan; no beef; no fish; no cheese; vegetarian (no dairy products, no fish); vegetarian (eats dairy products and fish); normal; allergy to peanuts; no lamb; nothing containing milk; allergic to aspirin; no ham; no bacon or ham; no eggs or fish; gluten free; no red meat; no shellfish, nuts, soya, etc. Many supermarkets stock a 'Freedom From' range of foods for people with real or imaginary food allergies.

Allergic reactions of the skin are covered in Chapter 13. Some people have '**multiple chemical sensitivity**' (MCS), falling ill at the slightest smell of cleaning products or perfume. Nineteen-year-old Nicole Gray was taken to hospital more than 100 times in three years for a **changing set of allergies** including to cheese, maple syrup, pears, face wipes and some soaps. She lives in fear that an unknown allergy would kill her. Boaz Gaventa, aged seven, was allergic to many foods, had asthma and carried an adrenaline 'pen' in case of reactions. His parents fear that a wasp or bee sting could kill him, as they are more dangerous for asthmatics. In Britain, between two and nine people a year die from stings, especially wasp stings.

One acquaintance has strong allergic reactions to coffee fumes and man-made perfumes. Other people say they become weak or sick near mobile phones or other emitters of radio waves or electromagnetic radiation. There is a block of flats in Zurich, Switzerland, for people with multiple chemical sensitivity. The few permitted cleaning and personal hygiene products are displayed in the lobby. About 5,000 Swiss suffer from MCS.

In 2014, the Food Standards Agency proposed laws requiring all outlets which sell prepared or non-packed foods to specify if they contain anything in the European Union's list of 14 'top allergens': eggs, molluscs, crustaceans, celery, milk, fish, tree-nuts, sulphites, soya, sesame, peanuts, mustard, lupins and gluten.

According to the *Canadian Medical Association Journal* news release, April 7, 2015, food allergies can be passed on through blood transfusions. An eight-year-old boy with no history of allergies was given blood products and quickly developed allergic reactions to salmon and peanuts. The blood source was a donor who had acute reactions to peanuts, shellfish and all fish, including salmon. Dr Julia Upton of Toronto said that this situation was very unusual and that allergies were too common to rule out transfusions from people with them. The food allergies could be passed on through immunoglobulin E in blood platelets, but the effects disappear within about five months.

12.16. Research on diets and allergies

Vasso Xyloyiannis did her project on *Student Eating Habits and the Reasons behind the Choice of Diets*, surveying 358 students of varied national origins. Their diets are shown in Table 12.1, where vegan means eating no animal products, vegetarian means eating no meats, although they may eat fish, eggs, dairy products and honey.

There was a higher proportion of vegans and vegetarians amongst females. The number of years of being vegans or vegetarians ranged from 1 to 19 (average 7.6), so most had adopted that diet at school. The frequencies of students eating particular food items were: 79% eating beef, 80% lamb, 81% pork, 84% fish, 92% chicken, 95% eggs and 97% dairy products. Of the vegetarians, 16% ate fish and 6% ate chicken. One reason given by vegetarians for eating eggs and dairy products was that they were obtained without killing animals. There was astonishing diversity in the amount spent on food each week, varying from £4 to £150 (1999 prices). One who spent little on food said she was not interested in eating, having a tin of cold baked beans when hungry.

Table 12.1.   Diet frequencies in undergraduates.

	Vegan	**Vegetarian**	**Normal**	**Total number**
Females	0.6%	12.1%	87.3%	165
Males	0%	5.7%	94.3%	193
Both sexes, number	1	31	326	358

Table 12.2. The reasons for different diets.

Reason:	Religious	Ethical	Medical	Personal	Upbringing	Total number
Vegan		100%				1
Vegetarian	8%	31%	6%	33%	22%	31
Normal	4%	2%	4%	47%	44%	326

The reasons for their diets are shown in Table 12.2. Their comments included:

- *Don't like fish or seafood.*
- *Take multivitamins because college food is poor nutritional quality.*
- *Was vegetarian from 14 till last year but missed chicken too much.*
- *Take calcium tablets as I eat few dairy products due to personal taste.*
- *Believe it is wrong to eat animals.*
- *Believe I am an omnivore, evolved to have all types of food.*
- *Man is an omnivore and therefore should eat all foods. Lack of meat leads to deficiencies.*
- *Do not eat beef because of mad cow disease.*
- *I eat less meat because it is expensive.*
- *I am vegetarian because I do not like meat nor the way animals are pumped full of hormones.*
- *I do not want to eat anything that requires killing. I eat organic free range eggs.*
- *My parents are Indian so I eat Indian food.*
- *I am lactose intolerant so I avoid milk and milky products.*
- *Used to eat beef before BSE.*
- *I became a vegetarian because I did not believe in the ethics of modern farming. Now I could eat meat that has been raised in a manner I find acceptable.*
- *I try to avoid all meats except chicken for ethical reasons and do not drink milk for taste reasons.*
- *I do not eat wheat or aspartame* [an artificial sweetener] *as I am allergic to them.*
- *Am Jewish.*
- *I eat meals that are easy to cook and prepare, although they are not healthy.*

- *I am allergic to eggs. I would buy organic food if I could afford it. I eat genetically modified food because it is so difficult to avoid.*
- *I am allergic to eggs, milk, dairy products, tomatoes, strawberries, fish and nuts, and take vitamins and minerals (iron). My diet is based around my allergies.*
- *Throw up when I drink milk.*
- *Multivitamins and cod liver oil because I don't eat any fruit.*
- *Multivitamins to make up for eating crap.*
- *Do not eat beef because of my religion. I am sick because of lack of exercise.*
- *What I eat is limited by my cooking abilities.*
- *I take iron tablets and multivitamins to replace iron lost in my periods and vitamins I do not get enough of. I take fish oils as I don't eat much fish and my circulation is bad.*
- *No red meat because I am concerned about the way animals are transported across Europe. I do not give up chicken or fish because my diet would be too unbalanced as meat is an important protein source.*
- *I take Sanatogen vitamins and minerals because I don't eat enough fruit and vegetables.*
- *I am Muslim so only eat halal meat and am not bothered about organic food.*
- *Take multivitamins and minerals because of mother's orders.*
- *Have eczema due to dairy products.*
- *I keep away from sweets because of a medical teeth problem.*
- *It is healthy for humans to eat a bit of everything.*
- *Take iron for increased oxygen flow.*
- *I hate organic food.*
- *What is on offer at Safeways. Laziness.*
- *I believe it is unhealthy not to have the full range of foods in your diet. Why else did we evolve to be omnivores??!!*
- *I can't do without meat.*

74% considered that they had a balanced diet and 26% thought that their diet was unbalanced (especially males). Dietary supplements were taken by 47%, of which 91% took vitamins, 44% took minerals and 13% took other supplements. Supplements were taken by 50% of females, 44% of males, 46% of those on normal diets and 53% of vegetarians. Vitamins were often taken as multivitamin tablets, while others took vitamins A, B, B1, B6, C, D or E. Minerals included calcium ('to get better nails'), iodine, iron ('because I'm anaemic'), magnesium, multi-minerals, selenium and zinc. Other supplements included cod liver oil ('to help with joint injuries'), creatine ('I'm trying to put on weight'), *Echinacea purpurea* ('to improve my immune system'), fibre mixes, garlic powder or capsules, *Ginkgo biloba*, ginseng, peppermint oil, Pharmaton and yakult. Some with balanced diets took supplements 'for peace of mind'.

The list of allergies in these 358 students is astonishingly long, with 22% reporting one or more allergies. One was allergic to eggs, milk and other dairy products, tomatoes, strawberries, fish and nuts. Another was allergic to medicines amoxicillin and Actifed, and peach skin. Another was allergic to aspirin and to mixing fish and strawberries in the same meal. Foods causing vomiting included snails and milk, while aspirin made one student 'throw up blood'. One was allergic to broad beans and some drugs: 'loads of stuff, because of a missing enzyme, glucose-6-phosphate-dehydrogenase' (Chapter 12.19). One was allergic to dandelions.

The numbers **allergic to drugs or medicines** were 18 to penicillin, two to amoxicillin, four to aspirin, three to erythromycin, two to arnica, and one each to a particular brand of contraceptive pill, neomycin, Actifed, tetracycline, benzoic acid and other benzenes, larium (malaria drug), Benylin cough medicine, codeine, most types of sunscreen, 'some antibiotics'.

Seven were allergic to milk or dairy products, three to chocolate, three to eggs, three to peanuts and one to nuts in general. One was allergic to fish, two to seafood, two to shellfish, and one each to scampi, to prawns and shrimps, to snails, to wheat. Fruit allergies included three to strawberries and one each to pineapples, kiwi, peach skin and oranges. Vegetable allergies included one each to broad beans, peas and mangetouts (but not peas or beans). Allergies to preservatives included one to monosodium glutamate, one to Chinese food (which often includes that), one to salty food, and one to sausages, presumably because of preservatives.

12.17. Types of diabetes

The main symptoms of **diabetes** are thirst, frequent urination, especially at night, extreme tiredness, weight loss, blurred vision, and there may be genital itching or recurrent thrush infections. There are several types of diabetes. Diabetes UK's estimate is 3.8 million sufferers of full-blown diabetes in Britain, with 738 people being diagnosed a day with type 2 diabetes and 30 with type 1. In America each year more than 50,000 sufferers from diabetes undergo foot or leg amputations after poor blood circulation.

In 2014 it was announced that frequencies of **borderline diabetes** (blood sugar 5.7 to 6.4 millimoles per litre (mmol/L); levels of 6.5 or more signify diabetes) in England had tripled in eight years, with about 15 million people in Britain having this **pre-diabetes**, and four million having diabetes, although many had not been officially diagnosed. Borderline diabetes can lead to complications such as heart disease, stroke, amputations and kidney disease. Vitamin B1 shortage has been linked to vascular problems in diabetics, affecting the kidneys, retina and nerves in the limbs. About 10 to 20% of those with pre-diabetes progress to full diabetes.

The WHO estimates that over 347 million people worldwide have diabetes. Nearly four million a year die from it, especially those in their 60s and 70s. Diabetes is increasing even faster in Latin America and Africa than in America. There are about 40 million diabetics in each of India and China, and 19 million in America. 90% are of type 2. Both sexes are equally affected. Compared to non-sufferers, people with type 2 diabetes are three times more likely to die between the ages of 35 and 54.

Recommendations to diabetics include healthy eating, not smoking, regular exercise, monitoring blood sugar, and regular medical check-ups. There are many blood monitoring devices available for home use, and diabetics should aim for a blood sugar level of four to seven mmol/L of glucose before meals and not more than 10 after meals.

Diabetes is a problem of regulation of blood sugar (glucose) levels by the protein hormone **insulin**. Insulin is produced in the pancreas (located behind and under the stomach) in response to high blood sugar levels (**hyperglycaemia**) and circulates in the blood. It increases glucose uptake by cells, prevents breakdown to glucose of stored **glycogen** (the animal equivalent of starch in plants, acting as the major reserve of carbohydrates, and found mainly in liver and muscle) and stimulates production of stored glycogen from glucose. Glucose is the main digestion product of starches and other sugar sources; when it gets into the blood after a meal it normally causes a rise in blood insulin, followed by a reduction in blood sugar levels.

Symptoms of **hypoglycaemia** (from too much insulin, excessive exercise or missing meals) include headache, aggressiveness, hunger, shaking, feeling faint, dizziness, sweating and a dry mouth, with blood glucose lower than 4 mmol/L. This leads to starvation of glucose in the brain and if severe can cause a **diabetic coma**, with unconsciousness and even death if not treated. Dextrose (glucose) or sugar should be taken immediately. Most diabetics carry something sugary, so that they can take it immediately they feel hypoglycaemia coming.

Symptoms of **hyperglycaemia** are blurred vision, lethargy, increased urination and thirst, weight loss, dehydration, ketones in the urine, a smell of acetone (pear drops) on the breath, and going into a coma. They are caused by too much food, insufficient insulin or reduced physical activity.

Diabetes can be caused by a **lack of insulin production**, or by the body becoming **insensitive to insulin**. Insulin deficiency leads to high blood sugar levels, excessive secretion of glucose in the urine, which is an easy way of diagnosing diabetes, and to depletion of liver glycogen reserves.

Diabetes (Greek *diabetes*, 'running through' and 'syphon'), refers to the increased urine production. Before chemical analysis was available, two forms of diabetes were distinguished by tasting the urine. **Diabetes mellitus** (Latin, *mellis*, honey) gives sweet urine from glucose excretion, while there is no sweetness in urine from diabetes insipidus. **Diabetes insipidus** is a disorder of the pituitary gland, leading to kidney malfunction. The pituitary gland does not make enough **anti-diuretic hormone vasopressin**, with the patients producing up to 35 pints (20 litres) of urine a day instead of the normal 2.6 pints (1½ litres), because the kidneys fail to reabsorb most of the water. Sufferers spend hours drinking and urinating, but diabetes insipidus can be controlled with vasopressin from a nasal spray.

Diabetes mellitus has three main types. **Type 1** (insulin-dependent diabetes; juvenile-onset diabetes) is multifactorial, with a frequency of 2 per 1,000, arising most often in children, but can start in older people. Conservative Home Secretary Theresa May lost two stones in weight (28 lb, 13 kg) and at age 56 was diagnosed with type 1 diabetes in 2013. She needs insulin injections at least twice a day.

Type 1 diabetes involves a lack of insulin secretion. It is frequent in Finland but rare in China. If an identical twin has it, the other has a risk of 35%, compared to 7% for a non-identical twin. The risk to brothers and sisters is 7%, and for a child the risk is 10% from an affected father, 4% from an affected mother. Infections by viruses are one cause, but in most cases environmental triggers are not understood, nor is the genetics. Type 1 diabetics require injected insulin for survival, because their pancreatic β-cells have been largely destroyed in auto-immune attacks. Being a protein, insulin is broken down in the guts and has to be injected, not taken as tablets. It is a big problem for insulin-dependent diabetics to balance injections with diet and exercise, and they have to monitor their blood sugar several times a day.

Type 2 diabetes (non-insulin-dependent diabetes; maturity-onset diabetes) patients have either reduced amounts of insulin, or insulin receptors in the tissues have reduced sensitivity. It arises most often in adults, especially the over-40s, but can affect children. It often appears in under-40s in South Asians and Afro-Caribbeans. In Britain, South Asians have about a five-fold increase in type 2 diabetes compared with white Europeans. It develops more slowly than type 1 and symptoms are less severe.

It is multifactorial, affecting 3 to 7% of adults in most Western countries. Its frequency rose from 4% to 8% in America in the past four decades. If one identical twin has it, the other has nearly 100% chance of getting it, while the risk to non-identical twins and brothers, sisters and children is 10%. The risks are greatest in those with a family history of diabetes and those who are overweight with a sedentary lifestyle.

Type 2 diabetes can arise in fat, normal or thin people. It can often be controlled by diet and exercise, without needing insulin. With a reduced food intake, insulin secretion decreases and the number of insulin receptors often slowly returns to near normal. Severe cases need insulin injections. Drugs are sometimes needed in tablets to stimulate insulin production, to help the body make better use of its insulin, or to slow down glucose absorption from the intestine. There are different subtypes of type 2 diabetes.

According to Dr Pemberton (*The Daily Telegraph*, 21/7/2014), the risk of stroke in newly treated type 2 diabetics is more than double the normal one, and they have a four-fold increase in cardiovascular disease. About a quarter of diabetics have renal damage, leading to kidney failure and the need for dialysis. Diabetics are more likely to have blood vessel damage in the eyes, leading to blindness, and nerve damage leading to ulcers and possible foot amputations. He wrote that type 2 diabetes was progressive, however well controlled. In Australia, after six years, 44% of patients no longer responded to oral medication and needed insulin injections.

The third type of diabetes mellitus is **maturity-onset diabetes of youth**, usually inherited as an autosomal dominant.

Personal Account. *I developed insulin-dependent diabetes when I was 34, and was terrified for my baby during pregnancy*

I am 36 and have insulin-dependent diabetes [type 1]. *I was diagnosed by chance two years ago when I went to register with the GP after my arrival in UK from Sri Lanka for postgraduate studies. I had noticed weight loss, tiredness and extreme thirst, which are characteristic symptoms of diabetes, for a few weeks before I was diagnosed. Even though my blood sugar was very high (19 mMol/L) when I was diagnosed, I was alert and normal. I was told I should have gone into coma with this amount of blood sugar. I was immediately admitted to Accident and Emergency and given insulin. I was advised about the type of food to eat and about the importance of controlling my blood sugar level.*

I have to inject insulin twice a day, before breakfast and before my evening meal and check my blood sugar at least twice a day after two of my three main meals. At the start I found it very difficult to maintain my sugar level, going very low quite often, leading to hypoglycaemia with characteristic symptoms, such as sweating, trembling, and blurry vision. If I get a hypoglycaemia attack, I immediately take three dextrose tablets, followed by a small amount of any carbohydrate. Now I have learnt to adjust my insulin doses so that I do not get hypoglycaemia or hyperglycaemia. Having a snack between meals helps.

The worst experience was when I became pregnant. It is very important to control blood sugar during pregnancy because high levels can cause thickening of foetal heart walls, and low levels can decrease the amount of oxygen available. I was terrified about these problems. Pregnant women with

Eating, Drinking, Diet, Digestion, Liver, Cystic Fibrosis, Diabetes, Allergies,

insulin-dependent diabetes in the UK are referred to special antenatal units where their sugar level is monitored very carefully. When I was three months pregnant I was put on four injections a day, three times before main meals, with fast-acting insulin, and once before bed, a slowing-acting one. I did not have many problems until the eighth month, but after that I started to get hypoglycaemia, especially in the middle of the night, so the bedtime insulin was discontinued. I was told that full-term pregnancy can cause complications in diabetics and labour should be induced at 38 weeks. With all these troubles I had a very healthy baby girl weighing 7 lb (3.18 kg).

Since I am diabetic, her pancreas worked hard to maintain her blood sugar while she was inside. When she was born, her blood sugar was very low and she was immediately given formula milk, and blood sugar was checked five times a day. It became normal when she was two days old. I have gone back to my normal insulin doses.

No one in my family has insulin-dependent diabetes. My grandmother and one of my uncles developed type-2 diabetes in their 60s. I do not understand why I have developed this so-called 'inherited disease'. Now I am worried about my daughter, because there is a 50–60% chance [that is too high] *for the baby of an insulin-dependent diabetic developing diabetes.*

Mathy. Her daughter is now healthy at age 14, but Mathy needs five injections a day.

Personal Account. *I have adult-onset diabetes — and a very unconventional diet*

I am a Type 2 diabetic. Type 2 diabetics have some residual insulin production; they may have to inject insulin, but often control their blood sugar with tablets, diet and exercise.

I am 58 and was diagnosed 4 years ago as a result of tests for an unrelated problem. Most Type 2 diabetics are diagnosed after symptoms have occurred, which means that high blood sugar may have damaged some vital organ. I am assured that all my vital organs are functioning normally or above normal. I do not feel any different from having diabetes, except for having to take care over my diet.

Diabetes often runs in families but no member of my family, up to grandparent and down to second cousin, is diabetic. The causes are not known. It is very prevalent in India and extremely so in parts of Polynesia.

The only effect on me is having to monitor my blood sugar at least twice a day and, as well as taking two insulin-receptor stimulating drugs, I must control my diet to produce acceptable blood sugar levels. If my blood sugar gets too high, I feel tired. Diabetics whose blood sugar level gets too low can fall into comas, but I never suffer from low blood sugar as I do not take insulin.

I follow a low carbohydrate, high protein, diet which means I eat very little bread, potatoes, pasta, sugar or rice and concentrate on meat, fish, eggs, vegetables (not roots) and eat little or no fruit. This diet is the opposite of the received wisdom of the British Diabetic Association, which recommends the current view of a 'healthy diet' — i.e. pasta, fruit, vegetables, limited red meat and eggs — a low fat, high carbohydrate diet. On that diet my blood sugar was poorly controlled. When I changed my diet after reading the views of Dr Bernstein (a diabetic American diabetologist) my blood sugar control improved dramatically. Nothing is totally forbidden, not even sugar. However, sugar and sweet wines have to be taken in strict moderation and are usually avoided, but dry wines in reasonable quantities cause no problems.

I monitor my blood sugar using a drop of blood placed on a test strip that is analysed electronically and produces a reading. A long-term blood sugar test (together with cholesterol, lipids, liver function, etc.) is carried out every 6 months when I see my consultant endocrinologist. A normal range for blood sugar is 4–5.5 mmol/L. My recent reading was 5.9, which places me in the best 5% of diabetics, but has ranged up to 13.

Diabetes, without good blood sugar control, can lead to many unpleasant complications such as circulation problems, which can lead to amputation; retinopathy affecting eyesight and in extreme cases causing blindness; an increased tendency to stroke and heart problems. It is potentially life shortening. Good blood sugar control minimises these risks.

Rob Wilkins

The incidence of diabetes is increasing rapidly. In Britain, there were about 300,000 sufferers in 1950, but estimates in the *British Medical Journal* were for three million in 2010, 6% of the population. Serious complications include kidney failure, coronary heart disease, high blood pressure, nerve damage, blindness and damage to the arteries. Smoking, being overweight and taking little exercise all increase the risks. According to Diabetes UK, diabetes is not caused by eating sweets or the wrong type of food, nor by stress.

In 2015, the Navajo Area Indian Health Service, which covers 300,000 Navajos in Arizona, New Mexico and Utah, announced that 25,000 people there had type 2 diabetes and 75,000 were pre-diabetic. Many now had a sedentary lifestyle. Their traditional natural diet of piñon nuts, wild potatoes, berries, prickly pears and beeweed greens had largely been replaced by fried potatoes, tortillas, cookies, crisps, sugary drinks and Spam, with up to 60% obesity in some age groups, and much heart disease and high blood pressure. Their worried politicians imposed — for the first time in the USA — a 'junk food tax', in the Healthy Diné Nation Act. (information from *The Daily Telegraph*, 6/4/2015).

12.18. The body's energy balance; calories used in different activities

Glucose is a simple sugar which gets into cells with the help of insulin. It is a component of many plant and animal items in food and milk, including starch, glycogen, cellulose, maltose and lactose. It is the main energy-producing compound carried in blood. In the cells' mitochondria, it is oxidised in **aerobic respiration** to carbon dioxide gas and water, with the release of usable energy. If the muscles have worked too hard for enough oxygen to reach them, they undergo **anaerobic respiration**, breaking down glucose without oxygen to lactic acid and less energy than in aerobic respiration, building up an **oxygen debt**. When exercise is lessened, aerobic metabolism is re-established, and the lactic acid is oxidised in the liver.

The brain is fastidious about the blood glucose concentrations it can tolerate, typically 70 to 110 milligrams per 100 millilitres, averaging 100 mg glucose per 100 ml blood plasma, 5.6 mmol/L. The brain needs a continuous controlled amount of glucose, even between meals and at night, yet food intake is intermittent. Excess energy from food is stored for short-term and long-term use, including fasting between meals.

Excess glucose from carbohydrate digestion is stored as **glycogen** in the liver and muscles, but these short-term stores have a capacity of less than one day's energy needs. Once that store is full, excess glucose is converted to fatty acids and glycerol, which combine to form triglycerides which are mainly stored as fats in fat cells, with smaller amounts in muscle.

Excess fatty acids in the blood, from **digestion of dietary fat**, are converted to stored fats. **Excess amino acids**, from **dietary proteins**, are not stored as protein but are converted to fats. The body's fat-storing tissues typically store enough energy for two months. In fasting for more than a day, most energy comes from the breakdown of fats from adipose tissue. The body's main store of protein is in muscle, so using proteins during long fasts gives muscle weakness, leading eventually to death. Amino acids from dietary proteins and from the breakdown of body proteins can be metabolised in the liver to **urea**, excreted by the kidneys.

Some tissues, such as the brain, depend almost entirely on glucose for energy. Others can get energy by oxidising fatty acids, so if glucose is in short supply, those tissues use fatty acids as fuel, leaving glucose for the fussier brain. Body proteins can also be broken down to glucose.

During a meal's digestion, glucose is usually abundant and excesses have to be stored as glycogen. Most meals are fully digested within four hours or less, so there will be fasting states between meals and at night, when the body mobilises its energy stores to maintain blood glucose at the normal level, especially for the brain, whether the body is at rest or active. Skipping breakfast tends to increase, not decrease, weight, as breakfast helps to stabilise blood sugar levels and to regulate appetite.

The young need nutrients for growth, repair and energy, and adults need them for repair and energy. The flow of nutrients into and out of stores is controlled by several hormones, including insulin, glucagon (which has the opposite effects to insulin), epinephrine, cortisol and growth hormones. One gains fat and weight if one's food and drink intake provides more calories than the body uses. Slimming by dieting is the subject of huge numbers of articles and programmes. Viewed cynically, the process often goes like this:

- Feel too fat and overweight.
- See diet advertisement picture of attractive slim model who has probably never tried that diet.
- Try diet, fairly rigorously at first.
- Have a minor weight loss.
- Reward oneself with 'treats'.
- Begin to relapse into old habits.
- Regain one's original weight or more.

A great slimming feat was by an American wrestler, William Cobb, born in 1926. From 1965 he lost 40 stone 10 lb (570 lb, 259 kg) in three years, going from 57 st 4 lb (802 lb, 364 kg) down to 16 st 8 lb (232 lb, 105 kg), while his waist measurement went down from 101 inches (257 cm) to 44 inches (112 cm). A modern example of weight loss and determination is Deborah Dale, age 40, from Staffordshire, England. She placed a £50 bet at odds of 20-1 that she could go from 31 stone (434 lb, 197 kg) down to 19 stone (266 lb, 121 kg). By dieting and exercise she managed it in a year. An extremely rapid weight gain was recorded by an American. Doris James (born in 1929) put on 23 st 3 lb (325 lb, 136 kg) in 12 months before dying in 1965, aged 38 and weighing 48 st 3 lb (675 lb, 306 kg). She was only 5 ft 2 in (157 cm) tall.

Sportsmen frequently have to lose weight temporarily for races or competitions. Terry Biddlecombe (1941–2014) often had to spend hours sweating off excess weight before riding his horse. For one big race, he lost 9 pounds (4.1 kg) in two hours but this weakened him so much that he was almost unconscious on passing the winning post. The course doctor prescribed a pint of Guinness and a tablespoon of salt.

There are many diets. The 'Food Combining' diet, in contradiction to its name, involves not eating proteins, fats and carbohydrates at the same meal. As nearly all foods are a mixture of all three, and because we have evolved a digestive system to cope with such mixtures, there is little reason for that diet. The Atkins diet involves eating as much as one likes of protein and fats, but very little carbohydrate. The 5:2 diet involves calorie restriction for two days a week.

Just eating less, eating a good mixed diet to include plenty of fresh fruit and vegetables, exercising more, and keeping up one's motivation to lose weight, are as effective as elaborate diets. 'Yo-yo dieting', with rapid weight loss followed by regaining weight, followed by severe dieting again, etc., is bad for the health. Setting attainable gradual loss targets is much better, with self-control to keep weight down once the final target is achieved. Many people find it helpful to belong to groups of dieters for mutual support. The typical energy expenditure of average-sized adults who take little exercise is about 2,000 kilocalories (kcals) a day for women, 2,500 for men. Most activities take more calories for men that for women, as men are on average larger. Table 12.3 shows the energy typically used in different activities, with values depending on how hard they are done. There is a huge diversity in energy expenditure by different people in similar activities, with age, sex, weight and metabolic rate involved. Obviously it takes more energy for a heavy person to climb stairs than for a light person. One's **basal metabolic rate** is largely controlled by the amount of **thyroxin** from the thyroid gland; taking regular exercise can increase one's metabolic rate even at rest.

Table 12.3.   Average energy usage in different activities.

Activity	Kilocalories/hour, women	Kilocalories/hour, men
Sleeping	55	75
Lying still, awake	65	90
Eating	70	100
Walking slowly	170	225
Sexual intercourse	240	320
Golfing	260	350
Cycling on the level	260	350
Playing tennis	330	440
Swimming	420	570
Mountain climbing	480	640
Moderate running	510	675
Rowing	700	950
Climbing stairs	930	1,265
Sprinting	1,030	1,500

Table 12.4. Approximate calorific values of types of food.

	Kilocalories per ounce	Kilocalories per gram
Fats	250	9
Alcohol	200	7
Proteins	114	4
Starches	106	3.8
Sugars	106	3.8

It takes about 3,500 kilocalories to put on a pound (0.454 kg) of body fat. My weekly average alcohol consumption of 17 units, at 90 kcals a unit, equates to 1,530 kcal a week, about 9% of my calorie intake. 39 glasses of wine or half pints of beer equate to one pound (454 g) of body fat unless burnt off by exercise. Table 12.4 shows the approximate calorific value of foods, with fats having the highest value and starches and sugars having the lowest.

There was an interesting question (*The Sunday Telegraph*, 10/8/2003) about whether running or walking to work burned the most calories. Robert Matthews answered that someone running at 6 miles per hour (9.7 km/h) burns up about 11 kcals a minute, about three times the rate burned by walking at 2 mph (3.2 km/h). However, it takes three times as long to walk as to run, so running and walking over a fixed distance burn a similar number of calories.

12.19. Glucose-6-phosphate-dehydrogenase (G-6-P-D) enzyme deficiency, giving anaemia from certain drugs or broad beans

G-6-P-D is an enzyme necessary for cellular respiration. A greatly reduced activity for this is rare in Whites, but occurs in about 9% of American Blacks and in some people of Mediterranean origin. About 400 million are affected worldwide, with some malaria resistance. Males with the recessive, defective allele and homozygous recessive females for this sex-linked trait are normal most of the time. They become ill, with the sudden destruction of many red blood cells giving severe anaemia, when they inhale pollen of broad beans (*Vicia faba*) or eat raw beans (**favism**), or take certain drugs such as sulphanilamide or the antimalarial primaquine. They recover when the causative agents are removed. Heterozygous females usually have intermediate enzyme levels.

12.20. The stomach, digestion, ulcers and irritable bowel syndrome

The stomach is a muscular sack starting below the diaphragm. Its upper and middle parts descend almost vertically, then the lower end curves to the right, crossing the midline to empty into the duodenum which goes vertically down to the rest of the small intestine, largely below the waist.

Food from the mouth passes down the oesophagus, taking about eight seconds, too little time for digestion. The oesophagus secretes copious mucus to lubricate the passage of the lump (bolus) of chewed food. The oesophagus goes through the diaphragm, and the gastro-oesophageal sphincter relaxes to let the food through, then closes to prevent the stomach's acid contents flowing back. If that occurs, it causes the pain called heartburn, felt near, but not involving, the heart.

Almost everyone gets **indigestion** at times, from overeating, eating the wrong food, eating too quickly, or from bacterial or viral infections. More persistent indigestion often comes from excess stomach acidity. Anti-acid treatments are widely available.

The stomach has different shapes in different individuals, even when empty. Its size depends partly on how much food it contains. Its internal volume varies from about 0.09 pints (50 ml) when empty to about 1¾ pints (one litre) when full. The empty stomach expands during eating by relaxing the walls, allowing the deep folds to open out. If too much food is eaten, the stomach walls are over-stretched, giving discomfort and bloating. The top of the stomach accumulates gases which may be released upwards in burping. The stomach is separated at its lower end from the duodenum by the pyloric sphincter. Pyloric stenosis was covered in Chapter 8.5.

The stomach stores food, as a meal can be eaten in a few minutes and the narrow-bore small intestine only takes small quantities at a time. The stomach secretes **hydrochloric acid** and **pepsin enzyme** to digest proteins, secreting about 3½ pints (2 litres) of gastric (Greek *gaster*, belly) juices a day. The **muscular churning** of the stomach walls mixes the food with the gastric juices, giving **chyme**, a thick liquid. About every 20 seconds a wave of muscular contractions passes from the top of the stomach towards the bottom, helping to mix the food. The stomach takes roughly **one to four hours to empty** into the duodenum, taking less time for a protein and carbohydrate meal and longer for a high-fat meal. **Emotional factors** can affect digestion.

It seems surprising that the stomach should secrete something as potentially harmful as hydrochloric acid, and why does the protein-digesting enzyme pepsin not digest the cells which produce it? Cells lining the stomach secrete an inactive enzyme precursor called **pepsinogen**. This is converted in the stomach by hydrochloric acid into the active enzyme, pepsin, and the acid gives an optimum pH (level of acidity) for pepsin to digest proteins. The hydrochloric acid kills most, but not all, bacteria, fungi and viruses in the food. A salivary enzyme, **lysozyme**, also helps kill bacteria. The stomach walls are protected from stomach contents by layers of mucus, by almost impermeable outer cell membranes on the gastric cells, and by tight junctions between the lining cells, preventing penetration by hydrochloric acid. With chemical attacks and harsh wear from churning, there is a high rate of replacement of the cells lining the stomach, with the entire lining replaced about every three days.

Secretion of digestive juices is promoted by the arrival of food in the stomach, but thinking about food, seeing and smelling food, and chewing and swallowing all promote gastric secretions. Even if food is not present, **caffeine** and to a lesser degree, **alcohol**, stimulate the secretion of these highly acidic juices. The acid, without food to act on and dilute it, can irritate linings of the stomach and duodenum, and so can aggravate **gastric or duodenal ulcers** (collectively known as **peptic ulcers**; Greek *pepsis*, digestion, as in dyspepsia, indigestion). They usually occur in the stomach, or in the first portion of the duodenum, or occasionally in the lower part of the oesophagus if acidic stomach contents are frequently regurgitated through the gastro-oesophageal sphincter.

In spite of those protective mechanisms, the inner surface of the stomach is sometimes locally damaged, with a peptic ulcer resulting. The classic cause was considered to be **prolonged stress**, from overwork or emotional problems, causing excessive gastric acid secretions. Harassed executives were portrayed as asking each other whether they were in a one-ulcer or a two-ulcer job. Aspirin can break through the mucus barrier lining the stomach, and caffeine and alcohol can cause excessive acid secretions.

We now know that nearly 90% of peptic ulcers are associated with a bacterium, *Helicobacter pylori*, which can persist in the stomach and weaken the gastric mucosal barrier. Ulcers caused by it usually heal after several weeks of treatment with mixed antibiotics. About 40% of the population are infected with this bacterium but only 10% develop a peptic ulcer. Stress and spicy foods can aggravate pain from existing ulcers.

When the gastric mucosal barrier layers are broken, the acid causes the surface cells to release histamine, which causes more acid to be secreted, giving a vicious cycle, enlarging the ulcer. Ulcers can be extremely painful and can result in bleeding, or even perforation of the outer stomach wall, letting stomach contents out into the body cavity — a dangerous **perforated ulcer**. People with ulcers are advised to relax, to eat small amounts of food at any time, to restrict the intake of caffeine, alcohol and spicy foods, and to avoid drinking milk which soothes ulcers but increases acid production.

Irritable bowel syndrome sufferers get stomach cramps and pelvic pain, usually after a meal, with bloating, wind, and constipation or diarrhoea. There is no ulceration and the symptoms are so diverse that it is difficult to treat them all. There are muscle relaxants for cramps and different treatments for wind, constipation and diarrhoea. A long-term bacterial infection is involved in many cases, and sometimes a food intolerance.

12.21. Eating disorders: anorexia, bulimia, pica syndrome

Anorexia nervosa is a disorder in which sufferers eat far too little and lose weight, to a health-threatening and even life-threatening extent. It is particularly common in adolescent girls and young women, especially from middle-class Caucasian families with pressures to be high achievers, with an incidence of about 1 in 100 in such families. The incidence in boys is about 1 in 2,000, but rising. Older people are affected less often. The age of onset is typically the teenage years. It is rife in ballet schools. Contributing factors include perfectionism, anxiety, intestinal illnesses, stress, bereavement and fear of sexual development, so girls starve themselves to retain a more child-like shape. According to the University of North Carolina, genetics accounts for just over half its liability.

Sufferers are often intelligent, knowing that they are harming themselves and distressing their families, but seem unable to eat normally. Often they feel guilty and try to hide the fact that they are not eating much, surreptitiously disposing of food. About 13% of young anorexics have periods of bulimia (see below); about 56% are clinically depressed, 40% have self-induced vomiting, 33% exercise excessively and 8% abuse laxatives.

Anorexia often results from a distorted body self-image, where the sufferer falsely thinks that she is too fat (Chapter 9.6, personal account). There is a persistent, unreasonable dread of fatness. Treatment is with counselling, antidepressants, psychotherapy, joining a self-help group, encouraging a more realistic self-image, and advice on diet and weight. After five years, about half the sufferers recover, and about three quarters after 10 to 15 years. 20 years after developing anorexia, some 10% will have died from suicide or health complications.

The effects of severe weight loss include muscle wasting, lack of menstrual periods, lack of male sex drive, constipation, low blood pressure, increased liability to infection, and poor skin and hair condition. Severe malnutrition results in spells in hospital and intravenous feeding. Death may occur from malnutrition or suicide from depression. Puberty may be delayed or arrested, with growth ceasing. If sufferers recover, puberty is usually normal but delayed.

Bulimia nervosa involves bouts of gross overeating, followed by self-induced vomiting, which reduces the chance of putting on weight. Laxatives and diuretics may be misused to reduce weight gain. These bouts — usually secretive and shame-inducing — may be several times a day. Bulimia often goes undiagnosed because there may be no obvious symptoms. Catherine Scott died at the age of 24 of heart failure brought on by four years of binge-eating and making herself sick.

Bulimics often suffer from depression and bulimia mainly occurs in adolescent girls and young women. Their weight is often normal or low, even if sufferers think they are overweight. According to the MEDLINEplus Medical Encyclopedia, bulimia has an incidence of 3% in the Western population but 20% in college women. My student surveys suggest a much lower incidence at Imperial College London. Treatments are mainly as for anorexia, with about half recovering fully. The condition rarely leads to death as there are few weight problems, unless depression leads to suicide. Frequent vomiting can lead to **erosion of tooth enamel** by regurgitated stomach acid, and to ulcers in the oesophagus.

A much rarer phenomenon is **pica** (Latin for magpie) **syndrome**, a **compulsion to eat strange items.** A Manchester man, Edward Cope, died aged 33 after swallowing 10 buttons, a drawing pin, pieces of chain, bone and foam rubber.

12.22. Taste sensitivity; sweeteners

People have remarkable **differences in sensitivity for particular tastes.** There are chemicals which sharply divide individuals into the **taster and non-taster categories.** Phenylthiocarbamide (PTC) either tastes very bitter or is virtually tasteless, depending on taster's genes, with 'taster' dominant. About 70% of people can taste PTC, varying from 58% for indigenous peoples of Australia and New Guinea to 98% for indigenous peoples of the Americas. Some find sodium benzoate sweet, some salty and some bitter.

Most people enjoy sweet flavours, with natural selection training humans to prefer sweet tastes, as in ripe fruit, to bitter or sour tastes, which can indicate unripeness or toxins. Concerns about weight lead many people to try artificial sweeteners such as saccharine or aspartame in coffee or tea. They are widely used as food additives to reduce the calorie content compared to using sugar. Aspartame (NutraSweet, Canderel) is composed of two amino acids, aspartic acid and phenylalanine, and is about 200 times sweeter than sugar, with almost no calories. The phenylalanine makes it unsafe for people with PKU (see Chapter 12.5).

The main natural sweetener is sucrose (household sugar), although its breakdown products, glucose or fructose, may be used. A teaspoon (4.2 g) of sugar provides 16 kcals. Other sweeteners include glycerine, the milk sugar lactose (sometimes used to sweeten beer, e.g., milk stout, as yeasts do not consume it), honey, molasses, golden syrup and hydrolysed starch.

High levels of acidity may disguise high levels of sweetness. The fizzy soft drink 7UP has 26.5 g of sugar in a 250 ml bottle, with masking by citric acid, lemon and lime juice — 103 kcal. One bottle (20.29 US fl. oz.)

of Coca-Cola contains the equivalent of 15 teaspoons of sugar. For comparison, here are the kilocalories in 250 ml of some fruit juices: mango, 54; guava, 54; lychee, 55; red grape juice, 70. Traditional jams contain more than 60% sugar, and jams spoil more rapidly with less sugar. A typical jam has about 244 kcals/100 g and honey has about 307 kcal/100 g. A Burger King Triple Whopper Meal has 2,040 kcals, more than a woman's recommended daily allowance.

Most sweet wines have 4.5 g or more sugar per 100 ml. The alcohol in alcoholic drinks is ethanol, which provides 7.1 kcals/gram and has a specific gravity of 0.789 g/ml at 25°C. A small glass, 100 ml, of a dry wine of 10% alcohol by volume therefore provides only 56 kcals from alcohol and a few calories from other ingredients. A very large glass, 250 ml (one third of a bottle), of a dry wine of 13% alcohol would provide 182 kcals. One gram of sugar provides 3.9 kcals, so 100 ml of a sweet wine with 4.5 g of sugar in 100 ml and 10% alcohol would contain 18 kcals from sugar and 56 kcals from alcohol, a total of 74 kcals.

References

Ghany, M. G. and Gara, N., Quest for a cure for hepatitis C virus: the end is in sight. *Lancet* (2014) **384**: 381–383.

Lison, M. *et al.*, A polymorphism of the ability to smell urinary metabolites of asparagus. *Br Med J* (1980) **281**: 1676–1678.

Munafò, M. R. *et al.*, Genetic polymorphisms and personality in healthy adults: a systematic review and meta-analysis. *Mol Psychiatry* (2003) **8**: 471–484.

Wharton, B. and Bishop, N., Rickets. *Lancet* (2003) **362**: 1389–1400.

Recommended reading

Corder, R., *The Red Wine Diet.* (2007) Avery Publishing Group.

Plate 1. Normal and aberrant human chromosomes, sperm, blood group determination (Chapters 1, 8, 17, 23).

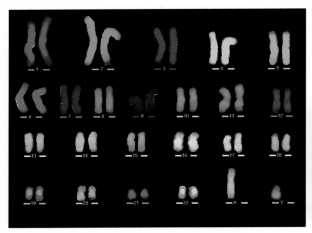

A. Normal male (XY) chromosomes in false colour, arranged in size order. Note the smaller Y than the X, and the smallest, 21.

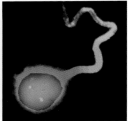

B. Deviant sperm stained with blue nucleus and too many red 21s and green 15s.

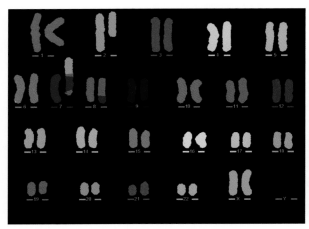

C. Human female (XX) chromosomes, aberrant, showing translocations between numbers 2, 7 and 8.

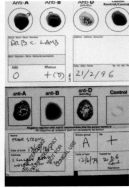

D. Determining blood groups using anti-A, anti-B and anti-Rhesus (D) antibodies.

Plate 2. Multi-racial hybrids (i) (Chapter 2).

A. Edward, Thai/English.

B. Sonia, Japanese/English.

C. Meera, Chinese/Indian.

D. Leo, Japanese/Greek.

Plate 3.   Multiracial hybrids (ii) and racial types (Chapters 2, 12).

A. Yasmine and brother, Swedish/Pakistani.

B. Alexandra,
Colombian/Chinese.

C. Jasmine,
Chinese/Vietnamese.

D. Jasmine showing
the alcohol flush.

E. Hannah,
English/Lebanese.

F. Adrienne,
mixed origin.

G. Ernst, French/Swiss.

H. Talal, Bangladeshi.

Plate 4. Racial types (i) (Chapter 2).

A. Diego, Italian.

B. Tazio, Japanese.

C. Abu, Nigerian.

D. Brazilian.

E. Bernard, English.

F. Chinese.

G. Louisa, Hong Kong (dimples!).

H. Meiting, Singaporean.

I. Ivy, Hong Kong.

Plate 5. Racial types (ii) (Chapter 2).

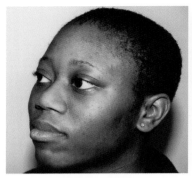

A. Temitope, Nigerian.

B. Chamila, Sri Lankan.

C. Yun, Chinese.

D. Ranjana, Sri Lankan.

E. Katrina, San, South Africa.

F. Argentinian.

G. Brenda, English.

H. English Celtic type.

Plate 6. Racial types (iii) (Chapter 2).

A. Leona,
Hong Kong.

B. Argentinian.

C. Well-groomed
American.

D. Indian minority
dancer.

E. Chilean dancer.

F. Elsa, French
model.

G. Susan,
Philippines.

H. Mark, English.

I. Felicity, Nigerian,
hair in braids.

Plate 7.   Racial types (iv) (Chapters 2, 5, 15).

A. So-Jin, Korean.

B. Korean philosopher.

C. Meriem, Algerian.

D. Italians, thinning hair, wavy on left.

E. Indian beggar girl.

F. Monisha,
Bengali bride.

G. Atacama Desert
Indian, Chile.

H. Japanese lady.

Plate 8. Religion (Chapter 2).

A. Female Jain pilgrims with mouth-masks, India.

B. Young Buddhist monk, Sri Lanka.

C. Turbaned Sikh officials, Golden Temple, Amritsar, India.

D. The sacred lake at the Golden Temple, Amritsar, India, with bathing male pilgrims.

E. A Christian miracle? Walking and preaching after the martyr was beheaded.

Plate 9. Costumes (Chapter 5).

A. Udaipur Maharaja's bagpipe band.

B. Brazilian lady with fruity hat.

C. Hiruni, Sri Lankan girl, dressed for a play.

D. Academic robes Bristol, DSc.

E. French model.

F. French model.

G. Minimal bikini.

H. Lasantha, Sri Lankan bride.

Plate 10. Muslim ladies' coverings (Chapters 2, 5).

A. Charming, open Malaysian.

B. Heavily veiled British.

C. Ratni, British.

D. Bruneian.

E. Turkish.

Plate 11. Beards, moustaches and tattoos (Chapters 5, 15).

A. Dave, Australian winemaker.

B. Maori winery worker, New Zealand.

C. René, French winemaker.

D. Indian attendant, Amer Palace, Jaipur.

E. Shabbir, Pakistani plant breeder.

F. English tourist in India.

Plate 12.   Head and face adornments (Chapters 5, 15).

A. Modern Japanese lady.   B. Peking Opera make-up.   C. Akihiro, Japanese, dyed hair.

D. Chinese Miao minority dancer.   E. Face painting in England.

Plate 13. Hair and nails (Chapter 5, 18).

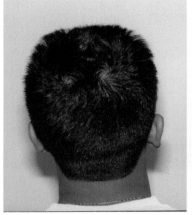

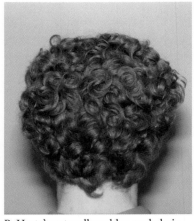

A. Anthony's double crown (two whorls).

B. Hesta's naturally golden curly hair.

C. John's gelled hair.

D. Stephanie's deep black hair, Greek.

E. Mohican haircut.

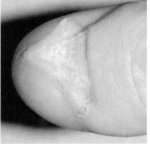

F. Vincent's rainbow dyed hair.

G. Triangular regrown nail after an accident.

Plate 14. Happy relationships (Chapters 9, 10).

A. Jacqueline with great-granddaughter Anna.

B. Chinese wedding couple.

C. Japanese honeymooners in Indian dress at the Taj Mahal.

D. Rasika with her mother.

E. Jenny with her mother.

F. Thesanya, 4, with her sister Buddhima, 6.

G. Good staff-students relations; Bernard with Karen, Xi Yu and Martin, planning a human diversity experiment on whisky drinking.

Plate 15. Groups and special occasions.

A. Military wedding, England.

B. Tall Caspar and girl friend.

C. Family wedding, 1967, England.

D. Colombian wedding, 2007.

E. Family portrait with albino identical twins.

F. Sri Lankan schoolgirls with parasol umbrellas.

Plate 16. Skin troubles (Chapters 11 and 13).

 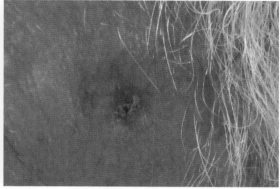

A. Acne in adult Vietnamese. B. Basal cell carcinoma (rodent ulcer).

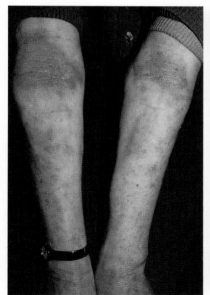 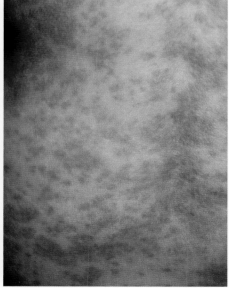

C. Eczema/dermatitis. D. Urticaria (hives), over most of the lady's back, from a reaction to amoxicillin antibiotic.

Plate 17. Skin: vitiligo and freckles (Chapters 1, 13).

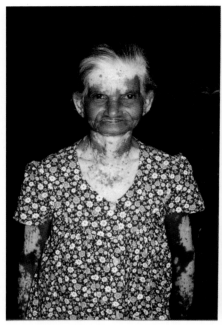

A. Dramatic vitiligo in Sri Lanka, Nandawathie, aged 74.

B. Symmetrical vitiligo.

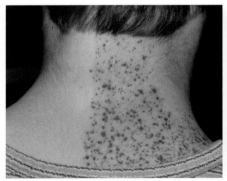

C. Segmental freckles, very rare. The lady is a mosaic for two genetically different skin types.

D. Indian field worker with regular skin colouring.

Plate 18. Eyes (Chapters 15 and 18.5).

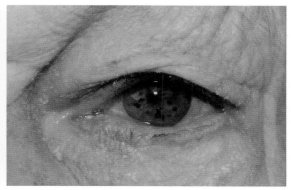

A. A blind Cambodian beggar using her foot and withered arm.

B. Jacqueline, colour flecks in eye, not related to her cataract operation.

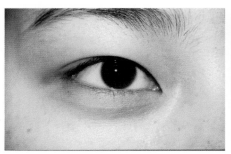

C. Lucy, very dark Chinese eye.

D. Corinna, English, eye make-up.

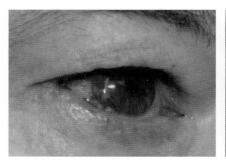

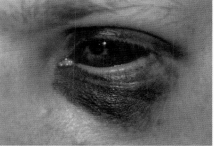

E. Spontaneous bleeding, haematoma.

F. Black eye from a botched eye operation.

Plate 19. Head and face (Chapters 1, 15, 23).

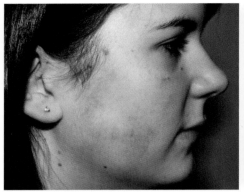

A. Eleri, permanent blush.

B. Australian waitress with face and ear piercings and bleached hair.

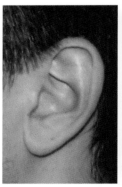

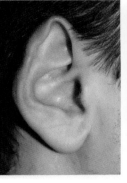

C and D. James with differently shaped ears.

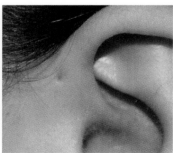

E. Hanna, Down syndrome face.

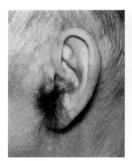

F. Hairy ear.

G. Natural hole in head near the ear.

H. American with hair thinning at temples.

Plate 20. Childhood (Chapter 21).

A. Sashini, 9 months, breast-fed.

B. Wartime English baby.

C. Young Chinese dancer.

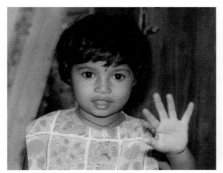

D. Shashi, innocent curiosity of childhood.

E. Sarah, 6; David, 8 with freckles.

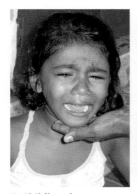

F. Childhood tears.

G. Malintha, 15, adolescent.

H. Thapthi, 20, young adult.

Chapter 13

Skin, Skin Colour and Disorders

13.1. Introduction

The skin is our largest, most visible organ, and part of our identity. There is enormous diversity in skin colours, textures and markings, with **genetic** and **environmental** causes including climate, diet, accidents, age, diseases, life-style and treatments such as make-up, moisturisers, tanning or skin-lightening lotions. Our skins reflect emotions, signalling to other people, as in *purple with rage* or *white with fear*.

There are **physical, psychological and social consequences** of skin differences. Visible disorders can ruin people's lives, with embarrassment causing social withdrawal and low self-esteem. A lady's beautiful skin may cause pride, adoration or envy, while skin colour can be a trigger for racially-motivated attacks.

Skin is essential, with its complete loss normally being lethal, but there is one example of survival: **the woman who shed her skin, outside and in.** Sarah Yeargain, a young American, lost all her skin from a **whole-body allergic reaction** to Bactrim, an antibiotic for a sinus infection. It caused 'toxic epidermal necrolysis', starting with minor swelling and discolouration on her face, followed by blistering. Skin was then lost from her face and entire body, and even from her internal organs. She lost the membranes covering her mouth, throat and eyes.

Fluids and salts leaked out from raw surfaces. Sarah was sedated for several weeks to reduce the chance of a heart attack from the pain, and was given a drug to induce amnesia of the trauma. She looked bright red. The doctors in San Diego covered her with an artificial skin called TransCyte.

She eventually began to regrow her own skin and later went home, with 100% recovery expected in six months. The trauma left her emotionally frail, unable to look at photos of herself skinless.

Blushing shows embarrassment but can be attractive. Some people hardly ever blush while others blush easily. Eleri in Plate 19A has a natural blush low down on both cheeks. **Severe blushing** and the fear of it (erythrophobia) tend to run in families. The cheeks can go dark red, with the colour spreading to the ears, neck and chest, with feelings of being on fire, causing great embarrassment. Emotional stress causes the sufferer's sympathetic nervous system to open wide the superficial blood vessels, giving strong blushing. Swedish surgeons pioneered an operation to cut the nerves to small blood vessels of the face, with 94% success.

There are more than 2,000 skin diseases and disorders. Some are described below. In Britain, about 15% of doctor consultations are for skin conditions, especially inflammatory ones such as psoriasis, eczema/dermatitis, urticaria, acne and skin cancer.

Many people are deeply conscious of their **skin colour** and of that of others. It can be white, pink, yellow (natural or from jaundice), sallow, olive, coffee-coloured, red, brown, mahogany, black, or anything in between (Plates 2 to 20). In the 1950s, some British landladies did not want anyone 'looking too foreign' as lodgers and took only ones of 'a light-coffee colour or paler'.

Many individuals are unhappy with their colour. In the words of a satirical South African song: *All de white men wanna go black, and de black men wanna go white!* In countries where skins are dark, many women want to look pale, as shown by advertisements claiming that the offered bride is 'fair' or 'light complexioned'. Many of my female friends in Sri Lanka use creams claiming to make their skins 'whiter and more beautiful', but only act as moisturisers. One Sri Lankan baby girl was heavily criticised by female relatives for being 'much too dark'.

Women — and increasingly, men — spend huge sums on skin creams and lotions, deodorants, anti-ageing and anti-wrinkle preparations, tanning lotions, sun-protection cream, Botox, lipstick and other cosmetics, and cosmetic surgery: see Chapter 5. Novelists try to make villains seem evil by describing them as pock-marked, swarthy, or having an ugly scar or broken nose, which is unfortunate for people with those features naturally.

Our skins and their **convenient inlets and outlets** are vital for taking in food and excreting urine, faeces and sweat — and making and having babies! Our skins are essential for our senses, with nerve endings for heat, pressure/touch and pain. The skin and its glands produce tears, sweat (and body odours after bacteria attack it) and pheromones (including sex pheromones — volatile chemicals which can influence the behaviour of other individuals of that species).

The skin is our barrier against the external environment, dehydration, UV light, diseases and parasites; it keeps our internal organs internal. We walk on our skin, breathe through it and talk through it. Its **texture and sensitivity** vary in parts of our bodies, as confirmed by touching our soles, cheeks and eyelids. Few people would relish being called 'horny-handed sons of the soil', although manual labour is a respectable occupation (Plates 4F, 17D). Protection is important. Gardeners wear heavy gloves for dealing with roses, brambles and nettles. Bee-keepers wear protective outfits when opening hives. People with eczema are advised to wear gloves for washing-up, and doctors, dentists, nurses and lab workers use gloves as protection against infection or harmful chemicals.

The **topic of skin colour is a landmine**, bedevilled by **political correctness** and **hypersensitivity**. In some social circles even mentioning skin colour is deemed a serious error. What phrases are acceptable for describing people with dark skins — especially in America — change in mysterious ways. The term, 'a person of colour', is illogical as pink, yellow and red are also colours. In my view, simple descriptive colour words require no euphemisms.

One can change one's skin colour to a certain extent, darkening by sun-tanning, sun beds or fake-tanning lotions, lightening by bleaching. In the Middle East, the husband often has dark skin but the wife (or wives) has pale olive skin from avoiding sun exposure.

In India, it is a compliment to say that someone is 'gora', meaning fair, and Indians (men as well as women) spend more than £200 million a year on skin-lightening products, in spite of doubts about their effectiveness. In 2007, Shah Rukh Khan, a Bollywood actor, agreed to promote 'Fair and Handsome', a skin-lightening cream. The president of the All-India Democratic Women's Association accused him of racism, saying, "It is downright racist to denigrate dark skins." Indian traditions equate fairer skins with high caste, good looks and eligibility for marriage, as shown by advertisements ranking a potential bride's 'porcelain skin' as more desirable than a university degree.

Local skin colour changes are achieved by **tattoos**, placing non-fading pigments below the skin layers which are shed. See Plate 11B for Maori tattoos. Visible tattoos make a statement, and some employers are reluctant to employ people with them. Some tattoos are in private parts while others are flaunted. Tattooing is a matter of fashion in certain circles. When celebrities such as Madonna and the Beckhams have tattoos, many youngsters slavishly follow suit. Actresses Melanie Griffith and Angelina Jolie had tattoos with the names of their then husbands, Antonio Banderas and Billy Bob Thornton. Angelina is now married to Brad Pitt, having divorced Mr Thornton. Half of those tattooed regret it later. Laser treatment for tattoo removal is expensive and may not work completely. **Transfers** are more easily removed. In the Indian subcontinent, so-called henna tattoos are common, with a bride often having extensive ones on her hands, arms and feet. Their use has spread to females in other countries.

Sweating, producing cooling by evaporation, is a major mechanism for reducing body temperature. Men sweat about twice as much as women. There is an X-linked recessive gene causing a **lack of sweat glands**. Males carrying this defective gene are sufferers, unable to lose heat through sweating, while females with one good gene and one bad one are **mosaics**, with some skin patches lacking sweat glands and others having them. **Hyperhidrosis** causes excessive and embarrassing sweating. It affects about 3% of the American population and is genetic. The American Food and Drug Administration approved the use of Botox toxin as a treatment. There is an operation to disconnect the sympathetic nerves which control sweating of the arms and hands, and many antiperspirant sprays.

Although individuals differ widely, rough averages for an adult male, per day, are 38 g of sebum secreted, one litre of sweat produced (depositing 10 g salt on the skin), and about one billion skin flakes are shed. For skin cancer, see Chapter 11.11.

13.2. Skin structure, function and smell

In an average adult, the skin weighs about nine pounds (4 kg) and covers an area of two and a half square yards (2 square metres). Skin thickness varies from 1/50th inch (0.5 mm) on the eyelids to 1/5th inch (0.5 cm) on the upper back.

Our skin separates us from the outside world, giving us protection from mechanical damage, micro-organisms and other parasites. There are specialised regions over the eyes, middle ear, the olfactory region of the nose, etc. Through skin we **sense** light, sound, smell, taste, touch and temperature. It is a **biochemical factory**; with sunlight it photosynthesises ergosterol, which is converted by the liver and kidneys into **vitamin D**, essential for healthy bones and teeth.

Exposed skin gives us clues about an individual's ancestry, age, health and sun-exposure. If the face is largely covered as in some Muslim women, where the niqab leaves only the eyes visible (Plate 10B), that reveals religion and customs. The importance of skin condition as a sex-attractant and confidence-booster is shown by the multi-billion pound skin-care industry.

There is a fatty layer beneath the skin, then the lower three quarters of the skin's thickness is the **dermis**. Above that and separated from it by the **basement membrane** is the **epidermis**, including the layer we see. The dermis is well supplied with blood and lymph vessels, nerves and sense receptors. **Fibroblast cells** synthesise a connective tissue matrix with a network of collagen fibres that give the skin elasticity and strength, while **mast cells** are involved with immune reactions, inflammation, dilation of local blood vessels and defences against parasites. The dermis includes hair follicles, sweat glands and sebaceous glands which make fat-rich **sebum** discharged into hair follicles, helping to protect the skin against drying out.

There are huge differences between societies in attitudes to **visible or easily smelled sweat**. It is natural to sweat from heat or exercise, and people in hot, humid climates often sweat profusely. Except at the gym or in sports, observable sweat on clothing, and any sweaty, smelly, body-odour smells, are often looked on with horror in so-called 'polite society', but treated as natural and acceptable in some more realistic societies.

While the sweat glands function in cooling the skin and body, and open directly on to the epidermis, the dermis also includes the smell-producing **apocrine glands** which open into the hair follicles. They are found especially in the armpits (up to 100 per square centimetre), the pubic region, around the anus and the mouth. Women have more than men and Orientals have fewer than Europeans or Africans. The fluid from the apocrine glands is initially odourless but within six hours skin bacteria produce from it about 15 volatile chemicals (organic fatty acids and odorous steroids such as androstadienone) which constitute our natural body odour, which most of us rigorously supress by washing, using deodorants, antiperspirants and/or masking perfumes. Armpit hair provides a huge surface area for the bacteria and for smell dispersal. The most abundant component of human body smells is 3-methyl-2-hexanoic acid, described as having an aroma which is 'urinous, musky, sweaty, burnt, or woody'.

There is a large genetic difference between races in body odours, which are milder in Orientals and fiercer in Africans and Europeans (see Stoddart, 2015). Compared with others, most Orientals have weakly developed armpit scent glands and have a gene inhibiting apocrine secretions reaching the skin's surface. Stoddart's fascinating book gives details of 'sweaty T-shirt' experiments on the effects and likeability of human odours in relation to sex, age, attraction and HLA genes, and discusses whether human sex pheromones actually exist. The apocrine glands are formed before birth but only function under the influence of sex hormones at puberty. Our **body smells** are also influenced by diet, especially garlic, onions, chilies, spices, asparagus, strong cheeses, cabbage, celery, milk and lamb, so as different groups tend to have different diets, so they tend to have different smells.

The lowest part of the epidermis is the **basal layer** which includes upwardly branching **melanocyte cells** which synthesise **melanin pigments** which give skin its main colours. As surface layers are continually worn away, they are replaced by cell divisions in the basal layer, with cells taking about 30 days to migrate to the top. **Skin cancers** are caused by unregulated cell divisions.

The main cell type of the lower epidermis is the **keratinocyte**, which synthesises **keratin**, a fibrous sulphur-rich protein also found in nails, hair, calluses and warts. By the time cells reach the surface, they are flattened and dead, making up the **horny layer** which is thickest over the palms and soles. People who go barefoot on hot, hard ground develop protective thick pads of horny layer, as do those doing abrasive manual labour.

Sweat provides nutrients for **skin bacteria and fungi**. We normally have about 10,000 bacteria per square centimetre (0.16 square inches), especially *Staphylococcus epidermidis*. We have lesser amounts of *Staphylococcus aureus*, which can live harmlessly on the skin or cause boils.

13.3. The effects of age on skin

A baby's skin is thin, easily irritated and sunburns easily. 'Nappy rash' is common from contact with urine and faeces, or from infections with the fungus *Candida*. Temporary patches of eczema may be triggered by environmental irritants.

Puberty can cause problems as **sex hormones** stimulate the sebaceous glands to secrete oily sebum. Its overproduction can block pores and follicles, giving bacterial growth beneath the blockages, with acne, blackheads and whiteheads resulting. It can give greasy hair. The same hormones can change hair follicles in the groin and armpits from producing the small soft vellus hairs to giving the puberty-associated **body hair**. Its amount differs widely between individuals and races. Some men's bodies are largely covered by thick dark hair, which can look primitive.

Between the **ages of 30 to 50**, less sebum is produced and the skin becomes drier. Fewer elastic fibres are made, the lower layers retain less fat and water, so the skin may appear less full, with creases and wrinkles occurring, especially 'crow's feet' at the outer corners of the eyes. In **old age**, the skin gets drier, thinner and more wrinkled. The amount of the elastic fibre, collagen, in the skin decreases by about 1% a year in adults, and even more in post-menopausal women.

Wounds heal more slowly as there are fewer fibroblast cells to divide. Many billions of pounds are spent each year on anti-ageing preparations, with treatments being more expensive than effective. Smoking and exposure to UV from sun or sun beds age the skin faster. A German group found that smoking activates genes for an enzyme which breaks down collagen, so smokers tend to have less elastic, more wrinkly, skin.

13.4. The sense of touch

Our most **touch-sensitive parts** are the fingers (especially the tip of the index fingers) and tongue, followed by the lips. **Sensitivity** varies between individuals, especially in relation to callusing. People who grow up going barefooted in hot climates with spiky vegetation, like South African runner Zola Budd, develop hard, insensitive skin on their feet, whereas the Briton with soft feet feels the pebbles badly on a stony beach.

The dermis and epidermis contain **sense organs**. The most frequent are free nerve endings, fanning out as naked nerve fibres which can detect heat, cold, pain, light touch or heavy pressure. There are specialised nerve-ending bodies, some with specific sense abilities, such as light or heavy touch, slow- or quick-change motion sensors, and vibration sensors. We can learn from touch about an object's weight, temperature, roughness, wetness and hardness.

Shaking hands is usually a socially acceptable way of touching someone on meeting. A 'social kiss' on meeting a member of the opposite sex is normal in some societies, but would be frowned upon as immoral in others. In China, a social kiss from a complete stranger is bad behaviour. In some societies, men embrace each other with a hug or kiss on meeting. Romantic kissing is an example of highly pleasurable and sometimes prolonged touching. It is no evolutionary accident that the hands, lips and genitals are well supplied with sense organs!

The **brain interprets signals** from nerves in the skin. There can be **phantom feelings** where the brain perceives apparent sensations even in limbs which no longer exist, following amputation. 'Pins and needles' is a sharp tingling sensation not caused by pointed objects, but occurs when a limb is recovering from numbness.

13.5. Skin colour, tanning and sunburn

The much-branched **melanocytes**, mainly responsible for skin colour, make up about 5% of all skin cells. They contain small bodies called **melanosomes** in which the enzyme tyrosinase synthesises **melanins** from the amino acid tyrosine. Exposure to UV light speeds up melanin production, giving sun-tanning. In Orientals and Caucasians the melanosomes are grouped inside a membrane in complexes of three or more melanosomes, but in Blacks, the melanosomes are larger, separate, and occur throughout the thickness of the epidermis. Melanosomes contain red or yellow **pheomelanin** pigments, or black or brown **eumelanin**. Each melanocyte can make either type of pigment or a mixture, giving our human diversity in skin colour.

Research in Sweden showed that eumelanin efficiently converted harmful UV into heat but pheomelanin did not. This explains why people with red hair and pale skins, who predominantly have pheomelanin, burn easily in the sun (*The Daily Telegraph*, 29/9/2014).

Surprisingly, the **number of melanocytes** is similar in all races. Differences in skin colour depend on the type, dispersion and distribution of melanin particles in the melanosomes. It is interesting to compare the **upper and lower surfaces of hands** of people with dark skins. Often the dark skin on the back of the hands contrasts with a pink or light-brown palm, with a clear meeting of the different colours.

The **red haemoglobin in our blood** affects skin colour, and so do pigments like bilirubin, a product of red blood-cell breakdown in the liver. Someone with **jaundice** from liver disease looks distinctly yellow. The **dilation and contraction of superficial blood vessels** in the skin can give dramatic temporary changes in white people's faces. If they become overheated, expansion of blood vessels to give cooling can make them red in the face. Similarly, shock or prolonged cold can make their faces very white. The face colour of people with dark skins can also change temporarily from blood vessel expansion or contraction, to a surprisingly strong extent. Severe anaemia and strong emotions can change skin colours. Prolonged exposure to the elements can leave people ruddy faced and weather-beaten, with small surface veins on the nose and cheeks.

The **number of genes** controlling skin, hair and eye colour includes more than 150 alleles at more than 50 loci. In general, the children of a man and a woman of different skin colours tend to be intermediate between their parents in colour. Conversely, two parents of intermediate colour can have more extreme children, so humans will never all end up the same colour. Skin colour was so important that there were special words such as 'mulatto', with one white parent and one black, 'quadroon', with one-quarter black ancestry and 'octaroon' with one-eighth black ancestry.

There is a traditional **limerick about skin colour**. It is politically and genetically incorrect, but fun:

> *There was a young fellow named Starkey*
> *Who had an affair with a darkie;*
> *The result of his sins*
> *Was quadruplets, not twins —*
> *One white, one black and two khaki!*

This wrongly assumes that there is only one locus involved, with two alleles, one giving white, one giving black, with no dominance, so having one copy of each gives khaki. Even if that were true, *white × black* would only give khaki, and the expected ratio of one white, one black and two khaki offspring would apply to *khaki × khaki* matings, not *white × black*.

Usually **dark hair is dominant** to pale. Two blonds usually have blond children, sometimes red-haired children, but seldom have dark-haired children. Two dark-haired parents, if heterozygous for the relevant genes, will most often have dark-haired children but may also have red- or pale-haired ones. Red hair is controlled by a recessive gene. Early greying of the hair is usually caused by a dominant gene. The kinky/woolly hair of most Africans (Plates 4C, 5A) is dominant, so most Negroid × Caucasoid children have kinky hair. Two straight-haired parents usually have straight-haired children. Waviness and curliness can be strong, medium or low. Hair can be thick, medium or thin, and the number of head hairs per square centimetre is variable, often low in Orientals.

There are various **patterns of baldness** in men, affecting the temples (Plates 7D, 19H), top and back of the head, with genetic controls of those patterns and the age when baldness starts. Baldness depends on male sex hormones, so eunuchs castrated in childhood do not go bald.

Depictions of **different skin colours** go back to about 1300 BC in Egypt, where the wall of the tomb of Sethos I shows a relatively fair-skinned Libyan, a very dark Nubian, and an Oriental and an Egyptian with intermediate colouring. The Egyptians in Chapter 18, Photo 1, appear to show sex differences in skin colour. Skin colours range from albino pale (Plate 15E) through the pale pinky-white of untanned Caucasians, to the sallow buff-beige/yellow of many Orientals, to the darker olive of many Mediterraneans and some North Africans, to the browns of Indians and Pakistanis, to the browns and blacks of Africans (Plates 2–20), with some really black skins in Equatorial Africa. Twins studies gave an estimate of 72% of skin colour variation being due to genes and 28% being environmental, on average.

Compared with pale skins, thick, dark skins protect against sunburn about 12 times and against skin cancer about 500 to 1,000 times. The highest incidence of **UV-induced skin cancer** is in white Australians, who traditionally like outdoor activities. Pale skins are much better at

Table 13.1. The six main skin types with respect to sun tanning and sun burning.

Skin type	Sun burning	Sun tanning	Complexion, example
I	Burns easily, even on a cloudy day	Never or very little	Pale, Celtic type (Plate 5I)
II	Burns, but less easily	Light	Fair/pink, many English
III	Sometimes burns	Moderate	Fair to pale brown, many French
IV	Rarely burns	Easily	Yellow, olive or light brown, Oriental or Mediterranean
V	Rarely burns	Tans deeply	Dark brown, Indian, lighter Afro-Caribbean
VI	Rarely burns	Tans deeply	Black, darker Afro-Caribbean

photosynthesising vitamin D than are dark skins. It is accepted that humans evolved in Africa and were fairly dark, but after migration to less sunny regions there must have been natural selection for lighter skins in response to vitamin D requirements (Chapter 2.8). In Britain and other less sunny climates, there have been many cases of **rickets** (softened bones) among female Muslims who wear garments covering nearly all their skin (Plate 10B), so they cannot make enough vitamin D from sun exposure.

Dermatologists recognise six main skin types, as shown in Table 13.1. People of one race may differ in skin type, depending on their genes and where they have lived.

My student Sheila Figueiredo investigated the effects of sunshine on skins of different colour in more than 1,130 individuals, covering many countries, races, ages and skin colours. People aged 16 to 20 tanned fastest and had the longest-lasting tans, while those over 50 tanned slowest and had tans disappearing fastest. Whites tanned faster than Blacks or Browns, but those with olive skins tanned fastest and had tans lasting longest. Strong sun exposure actually resulted in lighter skins in some Blacks, due to **sun-bleaching**: their covered skin areas were measurably darker than their sun-exposed areas.

The lighter-coloured skins were predictably the **most likely to burn** in the sun; the order of most burning was white, yellow, olive, brown, black. On the whole, people with white, pink or olive skins **wanted to be darker**,

Table 13.2. The amount of sunburn in people of different skin colours on exposure to strong sun.

Amount of sunburn	Skin colour			
	White	**Yellow**	**Brown**	**Black**
None	16%	31%	48%	62%
A little	60%	52%	42%	31%
A lot	24%	17%	9%	7%

while those with yellow, brown or black skin generally wanted to be lighter, especially among females.

Table 13.2 shows the **huge diversity in sunburn responses** even in people with the same skin colour. All four skin-colour types had some people with no sunburn on strong sun exposure, some with a little and some with a lot. Although black pigment gives some protection from sunburn, these results show that with strong sun exposure, 31% of people with black skins get a little sunburn and an unexpectedly high 7% of people with **black skins get a lot of sunburn** (burning, reddening, flaking of skin, as opposed to tanning).

Albinos (Plate 15E) have very white skin, white to straw hair and pink to blue eyes, and sunburn very easily as they lack protective melanin. Their frequency is about 1 in 20,000. Albinism is usually due to an autosomal recessive gene, and 1 in about 70 people is an unaffected carrier. There are several types. See Chapter 15.4 for a personal account. One very striking type, *OCA2*, is most common in Africans. Sufferers produce some pigment, often having coffee-coloured skins and orange-red hair, and typically do not tan.

For unknown reasons, there are about 200,000 albinos in Tanzania, 1 in 1,429 births. Many die from sun-induced skin tumours. More sinisterly, there are deliberate killings of albinos in Tanzania, Kenya, the Democratic Republic of Congo, Swaziland and Burundi. There is an African superstition that body parts of albinos transmit magical powers, with witch doctors using them in rituals and potions claimed to bring prosperity. Hair, arms, legs, skin, eyes, genitals and blood have been used. Not only have albinos been persecuted, killed and dismembered, but even their graves have been dug up. Albinos have resorted to sealing graves with concrete.

Older Tanzanian albino women with red eyes were murdered as witches. The killings continue, with few prosecutions. In some parts of Africa, albinos are ostracised as they are presumed to be cursed and bring bad luck.

Freckles (Plate 20E) are small brownish spots on the face, arms and elsewhere. Their colour becomes more intense with strong sun, so they show much more in summer. Having freckles usually behaves as an autosomal dominant, so a freckled person usually has at least one freckled parent. They are most noticeable on individuals with fair skins and reddish hair. Our research on skin colour mentioned above showed that 31% of females had freckles compared with 21% of males, averaged over all racial groups. Their incidence was 37% in Whites, 22% in Yellows, 14% in Browns and only 3% in Blacks, but freckles are hardest to detect on black skins.

Plate 17C shows a fascinating and extremely rare (fewer than one in ten million people) case of **segmental freckles**. The young English lady has them in very clearly defined areas, such as part of the back of the neck, along one surface of her right arm, part of the right-hand side of her face, etc.

Red to copper-gold hair (Plate 5I) is usually an autosomal recessive trait, as is the often associated pale, freckled skin type. Notable redheads include Queen Boadicea (who fought the Roman invaders in about 60 AD), Queen Elizabeth I, Oliver Cromwell, Nell Gwynne, Sir Winston Churchill and Prince Harry.

Liver spots are flat brown patches on sun-exposed skin, especially the face and backs of the hands, from middle age onwards. They are usually left untreated unless malignant.

There are many inherited skin conditions. The **Ehlers-Danlos syndrome** has a frequency of about 1 in 150,000, giving lax joints and a **highly elastic, easily bruised skin** which can be stretched several inches (about 8 cm) but returns to normal when released. Some sufferers have a reduced life expectancy from fragile blood vessels, organ rupture and poor wound healing.

Xeroderma pigmentosum (XP) is caused by an autosomal recessive gene which results in an inability to repair damage to DNA caused by UV light. There is much freckling of the skin, with death usually happening before the age of 30 from multiple skin melanomas. There is no cure. Sufferers are strongly advised to stay out of the sun, wear strong sun-block, all-enveloping protective clothing and large hats. It involves both sexes equally and all races. Its incidence is 1 in 250,000 in America, but about six times that in the Japanese.

There are **less dramatic conditions involving excessive sensitivity to light**, some to sunlight and some to the energy-saving **compact fluorescent lights** (CFLs) which have been imposed on Europeans in the name of 'being green'. According to the European Commission Scientific Committee on Emerging and Newly Identified Health Risks in 2008, CFLs pose health risks from the ultraviolet and blue light emitted, which can cause **cataracts**. Their light at distances of less than 20 cm (8 inches) could lead to UV exposures approaching the limit to protect workers from skin and retinal damage. About 15% of Europeans suffer from **polymorphous light eruption**, with red itchy skin eruptions caused by UV and by CFL bulbs. **Chronic actinic dermatitis** shows as inflamed skin from sunlight or artificial light, especially CFLs. It affects 1 in 6,000 Scots.

Flicker from fluorescent tubes can cause seizures in sufferers from **photosensitive epilepsy**. About 5 to 20% of the population suffer from **excess sensitivity to light** which can cause headaches and eye damage. A minority have **photophobia**, a morbid fear of light, with obvious consequences.

13.6. Some skin disorders

Many sufferers use **camouflage preparations** to hide skin blemishes, but exact colour matching can be difficult. Such preparations may be greasy, rub off easily or not be waterproof. Skin cancer is covered in Chapter 11.11.

In a study of skin conditions of 745 undergraduates and postgraduates at Imperial College London, Connie Chen found the following incidences: eczema, 15%; acne, 22%; urticaria (hives), 4% (8% in females, 1% in males); psoriasis, 2%. It is hard to diagnose some disorders, say giving pink weals, in dark-skinned people.

The impact of skin conditions on an individual's life is not directly related to their severity. Someone with physically mild symptoms may have more difficulty coping than those with severe symptoms. Reactions such as avoiding being in public, avoiding skin exposure through sports or exercise, and avoiding intimate relations, can cause further physical and mental problems, including overeating and drinking too much, with increased frequencies of depression, type 2 diabetes, obesity and cardiovascular disease. There is more unemployment in people with visible skin conditions, and sufferers who have major flare-ups of unsightly conditions often need time off work to cope with them. Many sufferers fear to go out, worried that others will think that their rashes mean they are dirty and infectious.

Cara Delevingne is a British supermodel. According to speakers at the *Medicine and Me: Psoriasis* meeting at the Royal Society of Medicine in 2013, she has psoriasis, with visible flare-ups sometimes restricting her modelling. Other celebrity sufferers are Britney Spears and Kim Kardashian. About 90% of people with a family member affected by psoriasis say that their own lives are also affected, e.g., by effects on social life, holiday plans or time needed off employment during flare-ups.

Some skin disorders have associated medical problems. People with **atopy** have a genetically increased sensitivity to environmental allergens. Sufferers are more likely to get one or more of **asthma, hay fever** and **atopic dermatitis**. In atopic dermatitis (atopic eczema) there is itchy chronic inflammatory skin disease, where scratching results in skin thickening and marking. Secondary bacterial infections give exudations and crusting. Atopic dermatitis occurs in about 25% of children, especially in the first year of life, but disappears in half of sufferers by the age of 10. Triggers may be detergents, dust and dust-mite excreta.

The commonest form of **psoriasis** makes patches of skin silvery-red and scaly, flaking off, with epidermal thickening, small abscesses and dilated upper blood capillaries. There may be strong itching. It affects roughly 3% of the population, men and women equally, with about 1.8 million sufferers in the UK. It is less frequent in Orientals and West Africans. On black skins, it shows as violet patches with grey scales. The commonest form, **plaque psoriasis**, can affect any part of the body, especially the scalp, elbows, lower back, navel and knees. Cells only take 2 to 6 days instead of 21 to 30 days to pass from the bottom to the top of the epidermis. Symptoms of severe psoriasis overlap with eczema.

Psoriasis has a strong multifactorial (see Chapter 8.4) **genetic component**. About half of its heritability has been attributed to the gene *PSORS1* in the major histocompatibility region of chromosome 6, controlling the HLA CW-6 antigen. Of those with severe psoriasis, 40% have a family history of it. The life-time risk is 3 to 4% if there is no family history of it, 28% if one parent is affected and 65% if both parents are affected. It often has an auto-immune basis. Most onsets happen between 15 and 40.

Psoriasis's **environmental triggering factors** are poorly understood, although stress is a common one, causing flare-ups covering up to 80% of the body with bloody-looking red plaques. Other factors include wounds, infections, starting or stopping over-the-counter drugs, and smoking and alcohol. Some students suffer from psoriasis under exam stress. The severity often comes and goes. One sufferer went from having one small affected patch to 80% of his body affected within a week, from reaction to sun-protection lotion, with open skin weeping blood.

The **psychological effects of psoriasis** can be severe. According to the (British) Psoriasis Association, 10% of sufferers contemplate suicide, 33% experience depression and anxiety, often taking anti-depressants, 40% experience humiliation, and 20% report being rejected and stigmatised. Social isolation often causes compensatory over-eating and drinking. Of those with moderate to severe psoriasis, one third has alcohol problems. Many sufferers develop **psoriatic arthritis** with stiffness, pain and swelling of the joints, which can lead to deformity as well as severe discomfort and aching.

Many treatments for psoriasis are similar to ones for vitiligo, including narrow band UV light (311–313 nm, three times a week), PUVA, topical steroids, immunosuppressive drugs such as calcineurin inhibitors, vitamin D analogues, plus emollients and bathing with coal tar products. Some treatments are messy and involve being swathed in bandages to stop staining from spreading. Expensive protein biological agents are also used, costing about £12,000 a year.

Eczema is another condition with inflammation of the skin, with swelling, very small raised papules, and sometimes exudations. The chronic itching (pruritus) results in damage from scratching, often with bleeding and secondary infections. Outbreaks arise from internal or external factors such as irritant dermatitis and **allergic contact dermatitis** (Plate 16C). Irritants include acids, alkalis and detergents. Some occupations are prone to cause eczema, such as hairdressing with its strong chemicals or mechanics in contact with oils and petroleum products. Cement workers with chromate sensitivity can get severe hand eczema which persists even after they cease working with cement.

Eczema allergies may be to hair dyes, plastics, some medicines, rubber additives, nickel in jewellery, etc. A common form affecting infants and children is **atopic eczema**, linked with susceptibility to asthma and hay fever; it has a complex genetic background, with triggering environmental factors. The frequency of sufferers is as high as 1 in 5 in children, reducing to 1 in 12 adults. Its frequency has risen four-fold in 60 years. **Seborrhoeic dermatitis**, with excess sebum from sweat glands, can be caused by yeast infections, with reddening and scaling.

Personal account (by a male of Indian ancestry). *I have severe recurring eczema*

I have suffered with eczema since I was two weeks old. My childhood memories are full of episodes of extreme flare-ups. When I was dropped off at my new middle-school-to-be to take the entrance tests, I was referred to as an 'alien' by a boy there. My face was red with atopic eczema, a wet rash, weeping and generally horrible. Circumstances like this affected my confidence greatly; the looks I received from others when at school spoke volumes. The fact that I had quite a severe form of eczema made me very focussed in my studies, which is where I tried to excel, given that sports were difficult as sweating aggravated the condition.

My social life was restricted since I didn't feel like going out when my face was inflamed. It was easy to become paranoid that people would be staring, because one becomes so wrapped up with worrying about how one's face looks.

On the day of an A-level exam, I woke up to find the side of my face stuck to the pillow; my skin and scalp had been weeping. Peeling my face from the pillow was excruciating, as was washing my face — this was a daily torture when my skin was weeping and sore. I am still amazed, that with the looks of shock (from teachers and students alike) at my raw skin when I entered the exam hall, I managed to concentrate and pass the exam!

I have tried every treatment from traditional steroids to allergy tests, special diets, light therapy, Chinese Herbal Medicine, Ayurvedic medicine;

all helped to a certain extent but then side effects meant changing to another treatment which was frustrating. I manage my condition now with elements from each therapy. My eczema flares up when emotions are heightened (sad or happy), and there is little I can do to calm my skin down.

At university, my skin began to improve slightly, and I began to understand some triggers of my condition, such as environmental pollutants, extremes of temperature, moisturising products and diet. It is a case of trial and error to find what suits my skin in different seasons. I usually know now when my skin is about to flare up: there are tell-tale signs. The products I use to moisturise in the summer differ from those in the winter. My eczema is unusual in that it changes from a dry form to an atopic form in different climates.

Atopic eczema is when the skin is 'wet', red, swollen and often weeping. Dry eczema is when the skin is dry, flaky, and if not moisturised can crack and bleed, which can result in opportunistic infections. Atopic eczema requires a soothing, cooling gel application, whereas dry eczema requires rich moisturisers.

During the last months of university, the eczema in my scalp was irritated. Within days, a few lumps appeared on my head. They grew very rapidly; in 24 hours they were fully developed and excruciatingly painful. Since there is no stretch in the scalp, these lumps (boils) resulted in pain, headaches and nausea. They recurred four times in six weeks.

The treatment consisted of repeated antibiotics and high-dose painkillers, since the boils were due to Staphylococcus, *a normal skin bacterium which can cause boils and other infections in breached skin. These boils were large and caused hair loss and scarring. This was worrying since I had four courses of antibiotics back-to-back; antibiotic resistance and impaired immunity were possible. The episodes of boils, like eczema, were unpredictable, and controlled my life for their duration.*

I thank my strong family for unconditional love and support through trying times. This allowed me to carry through the bad days and go on to achieve in other spheres.

Acne (Plate 16A) is a common disorder affecting 90% of the population at some stage, with a peak at 18 years. Three in four teenagers have acne at some point, sometimes briefly, sometimes for years, and about one in five adults is affected at ages 25 to 44. Sex hormones at puberty start an increased production of oily sebum from the hair follicles' sebaceous glands. The follicles can become blocked and infected with bacteria such as *Propionibacterium acnes*.

The greasy skin develops closed **whiteheads**, open infected **blackheads**, small papules or larger spots/zits/pustules/pimples, which may itch or be painful as well as unsightly, causing anxiety, distress and embarrassment. The pus from each spot usually clears up in a few days, but inflammation, itching and scratching can lead to permanent scarring. Sufferers are extremely self-conscious, desperately trying various treatments. Many sufferers use covering make-up. The condition often clears up spontaneously in the early twenties. In dark skins it can cause persistent extremely dark marks.

Urticaria (hives) results in skin weals, small reddish flattened swellings due to blood vessel dilation in response to histamine released by mast cells in the dermis. **Dermographism** is where the weals result from scratch marks. **Acute urticaria** is an immune response triggered by some plants, raw fruits such as strawberries, by nuts, shellfish or antibiotics. The bad urticaria shown in Plate 16D affected the whole of the lady's back, stopping at her knees. The antibiotic amoxicillin was the cause. **Chronic urticaria** lasts for weeks or years and usually has no known cause. **Contact urticaria** comes from direct stimulation of the skin by the allergy-causing stimulus, not from its ingestion.

Vitiligo (Plate 17A, B) is a series of related disorders giving white skin patches, affecting about 1 person in 100. People can acquire it at any age. It is not contagious and cannot be caught. Vitiligo shows up most on dark skins. In some sufferers it develops symmetrically; in others, it is irregular. It may spread rapidly or slowly, or stabilise, or some repigmentation may occur. The face, neck, hands, arms, knees, legs and genital area are most affected. The comedian Bob Monkhouse needed a lot of make-up to hide severe vitiligo. He said that below his underpants was "a riot of polka dots and moonbeams."

I developed vitiligo when I was 50 and the genital area is affected all year round. In winter, no symptoms are visible on my exposed skin, but by summer my hands are mainly white or pink, contrasting with light brown forearms. About one third of sufferers has an affected near-relative, showing that vitiligo has a genetic component. Some predisposing genes have been identified.

As well as a genetic predisposition, some **triggering event** seems necessary for vitiligo. Possibilities include stress, hormonal changes as in pregnancy and the menopause, grazes, emotional trauma, and there are associations with thyroid disorders and diabetes. Treatments which sometimes work include steroid creams, psoralens plus UV light, narrow band UVB, immune suppressors and skin grafts. Most forms of vitiligo probably have an **auto-immune cause** with an attack on melanosome production.

Children with the white patches are often horribly teased and mocked by other children, and teenagers, adolescents and adults are very conscious of being different, not wanting others to know that they have vitiligo. They may avoid sports, especially swimming, and revealing clothes. One white man told me that when he was at school and went into the swimming pool, all the other boys got out, as if he had something contagious. This caused severe problems of self-esteem. In some Asian and African cultures, women with vitiligo are not considered marriageable, possibly from false associations with leprosy and contagious conditions.

The old Sri Lankan lady (Plate 17A) with extensive vitiligo was feared by the other villagers who wrongly believed that it was contagious. Her skin is a **superb example of diversity within one person**, with patches of white, pink, coffee, brown and very dark brown.

I know three English ladies with almost **100% vitiligo**, with virtually no pigmented skin. They look very pale and one might not suspect vitiligo as the tell-tale patchiness is absent. When one of them developed some local repigmentation, she considered permanent bleaching of the repigmented areas to avoid visible patchiness.

The Vitiligo Society magazine, *Dispatches*, has many harrowing accounts of how vitiligo has ruined peoples' lives, restricting socialising, destroying self-esteem and causing them to stay out of the sun (and out of the fun). The encouraging accounts are ones where people eventually accept their vitiligo and learn to live with it.

Here are some extracts:

- *I am in constant fear of being 'found out' and avoid anyone seeing my patches at all times. I cannot bear to look at them myself as I see them being ugly and unattractive.*
- *I don't know what it is like to wear a t-shirt without fake tan to cover the patches, or a cute sleeveless dress or a pair of shorts in summer. … I sometimes get the odd glimpse as people try to figure out why my face is black and my hand is white.*
- *I used to cry at night and say that I wish I was dead as I did not want to be different to everyone else. … I was bullied. I had trouble with people staring and making comments during swimming lessons.*
- *I resigned myself to never getting married or having children.*

Darcel de Vlugt, is **a black girl who turned completely white with vitiligo.** Her parents came from Trinidad and Darcel called herself black. Her parents noticed white spots on her forehead and forearms when she was five. At seven, white vitiligo patches had appeared very visibly on her legs and white spots elsewhere.

At 12, when her body was 80% white, she tried laser treatment of the white patches, but it didn't work. By 17, she was entirely white. She was viciously bullied as a teenager, being called 'spotty' and 'Dalmatian'. At a sleepover, someone spread the rumour that anyone sleeping in the same room would catch the condition and die. Being so self-conscious, she missed out on many teenage activities. Darcel felt she had lost her black identity along with her skin colour. Her comfort was to think about singer Michael Jackson, who appeared to have the same colour change. She believed he had vitiligo, possibly with skin bleaching to make it look even. He always wore lots of make-up and would not be seen in shorts. Darcel now lives in London. Her skin is so pale that even on a dull summer's day she wears factor 100 sun protection to avoid burning.

Birthmarks may be present at birth or develop later, varying in colour, texture and extent. Although they seldom do physical harm, they can cause people to feel repellent to others, with severe social and psychological consequences.

Port wine stains are dark red or purple patches on the face or upper body, typically on only one side. They are caused by enlarged blood vessels, giving large irregular areas of splotchy deep flushing. In children they tend to be pink and flat, becoming darker, more raised and blotchy. Treatment with argon lasers usually gives marked lightening. Lasers also work fairly well on **thread veins** which are small purple broken blood vessels, particularly on the legs and face, or injections can shrivel the veins.

Strawberry marks usually appear as small red spots during the first month after birth, soon growing to give red swellings of spongy blood vessels, affecting at least 1 in 5 babies, but disappearing spontaneously in more than half by age seven.

Moles are areas of pigmented skin, often raised, up to 5 mm (1/4 inch) across. Whites may have up to 400, averaging 30, reducing with age. They are removed only if large and awkwardly placed. Having many moles is associated with increased risks of malignant melanoma.

A study at King's College London (published in 2007) of more than 1,800 female twins showed that people with more than 100 moles had longer telomeres (specialised ends of chromosomes) than those with fewer than 25. Longer telomeres are associated with **longer life spans**, and the difference between the two groups was the equivalent of seven years extra life for the more moley individuals. Identical twins tend to have similar numbers of moles, while non-identical twins differ more, showing a genetic element.

Scars can result from wounds. Some fade, others persist. Visible scars in prominent places can cause loss of self-esteem and shyness. Large scars which extend beyond the original wound, by overgrowth of scar tissue, are called **keloids**. They are often raised, sometimes 5 mm or more above the skin surface and may be reddish or flesh-coloured. They can come from wounds, operations, body piercing or tattooing, and are most common in Afro-Caribbeans. One young Indian lady has a zigzag scar on her forehead, from a fall. I find it likable and interesting rather than disfiguring.

Rosacea gives reddened skin on the nose, cheeks, forehead and chin, and the eyes are often affected. It affects about 1 in 100, typically starting in middle age. Papules (small lesions), pustules (small blisters) and surface small blood vessels may follow, with irregular thickening of the skin and enlarged pores. Enlarged, blotchy red noses are typical. **Red noses** may come from excess drinking, but **rosacea** is separate. It is made worse by temperature extremes, wind and smoking. Food triggers differ widely between individuals, including chocolate, coffee, tea, spicy foods, alcohol and citrus fruits. Treatments may include antibiotics, special skin care products, sunscreens and avoiding triggering factors.

When the skin falls off — Epidermolysis Bullosa. Epidermolysis Bullosa (EB) is a family of inherited disorders causing blistering and shearing off of skin even from mild friction. 'Epidermolysis' means breakdown of skin, and 'bullae' are blisters. The frequency is about 1 in 60,000, taking the three main types together. The most common form, **EB Simplex**, affects about 70% of sufferers. It is the mildest, giving blistering of hands and feet, especially from rubbing shoes. It is caused by a dominant gene, so most sufferers have an affected parent. The skin dissolution occurs above the basement membrane, and if mild does not require medical intervention, although one form gives blistering all over the body.

Junctional EB is much more serious, with dissolution of the skin through the basement membrane. Severe blistering in the pharynx and oesophagus, between the mouth and the stomach, causes death in half the sufferers within the first two years from malnutrition and anaemia. The other half have a milder, non-lethal form. About 10% of EB sufferers have Junctional EB, caused by a recessive gene.

Dystrophic EB involves dissolution of the skin more deeply, under the basement membrane. 'Dystrophy' means a progressive weakening or wastage. The blisters heal with much scarring, which can lead to fusion of fingers and toes, joint problems, contraction of the mouth and narrowing of the oesophagus. About 20% of EB sufferers are dystrophic, with an incidence of about 1 in 300,000 at birth. The form caused by a dominant mutation may be mild in childhood but with increasing severity later with scarring, wastage and fusions. The recessive types often lead to major handicaps, a fairly short and painful existence, with a strong chance of getting squamous cell cancer before 35. About 40% of sufferers from dystrophic EB die from that.

Dystrophic EB became better known in Britain in 2004 from a Channel 4 documentary, *The boy whose skin fell off*, about a brave sufferer, Johnny Kennedy, who died of cancer in 2003, aged 36. At birth, he had no skin on his right leg from the knee down, and damage to his vocal chords gave him a high-pitched childish voice. It was painful to see film of his agony as the blood-soaked bandages were eased from his almost skinless body, 70% of which was covered in weeping open sores, as was the top of his head. His hands and feet were completely bandaged; he was confined to a wheelchair and needed 24-hour care, being unable to turn over in bed or to feed himself. He had the recessive form, with unaffected carrier parents.

Johnny and his mother agreed that it would have been better if he had been aborted before birth, to save him a life of extreme pain, discomfort and frustration. His northern sense of humour helped sustain him in his last year when he knew he was dying of cancer. He was really active in arranging his own funeral, flying and taking part in publicity events about EB. In his final week, he was taken to meet the Prime Minister's wife and persuaded her to support the charity DebRA UK (www.debra.org.uk) which works for sufferers. There is now prenatal diagnosis for this distressing condition.

'**Black men's beard rash**' (pseudofolliculitis barbae) is a face and neck rash from ingrowing beard hairs, and is fairly common in African men who shave. Shaving stiff, curly beard hairs can leave sharp tips; if the hairs grow sideways or downwards, the tips can penetrate the skin, causing infections. There can be small lesions or infected blisters, with excess black pigment produced.

Skin ulcers show as inflamed lesions of variable size and depth, with a loss of surface tissues. **Varicose ulcers** arise when non-return valves in large leg veins lose function, so blood stagnates and pools, causing varicose veins which can become ulcerated and painful. Such leg ulcers may be several centimetres across and very difficult to heal. They are common in older women and may be associated with diabetes, rheumatoid arthritis and syphilis, and can ruin their last years. Treatment is with elevation of the limb, compression and frequently changed dressings to keep ulcers from infection.

Bedsores/pressure sores/decubitus ulcers occur after prolonged pressure on the skin, where bones push by gravity against the skin, as in bedridden patients who are not turned often enough.

Scurvy comes from a deficiency of vitamin C and is easily cured. It results in lethargy, depression, spots, spongy gums and bleeding. Advanced cases have open wounds, loss of teeth, jaundice, and eventually fever and death. It was common and often lethal in the days of long voyages by sailing ship. It was described by Hippocrates (about 460 BC to 380 BC). In 1753, Royal Navy surgeon James Lind proved that scurvy could be prevented by eating **citrus fruits**.

13.7. Contagious diseases affecting the skin

Contagious skin diseases can be caused by insects such as mites, and by micro-organisms. Although not counting as diseases, **blood-sucking insects** like mosquitos, fleas and bedbugs can cause pain, blisters and skin irritation from their bites, which can produce allergic reactions.

Scabies is caused by a mite, *Sarcoptes scabiei hominis*. It burrows into skin, giving intense itching and inflammation between the fingers, under nails, on the buttocks, trunk, male genitals and elsewhere. The mites eat their way into the epidermis and females lay eggs in the burrows. The larvae travel to the surface and mature in shallow skin pockets to become adults. Treatment is with insecticidal lotions or creams. Some individuals have allergic reactions to the female mite, with very itchy red rashes.

Scabies is contagious and spreads though skin-to-skin contact, sex, shared clothing, towels or bedding. It is endemic where there is a high population density and limited medical care, including much of Africa, South and Central America, India and Southeast Asia. In developed countries, outbreaks occur in crowded schools, nurseries and nursing homes.

Lice affect the skin, with extremely common (much feared by parents) outbreaks in school children, causing itching. The **body louse** (*Pediculus humanus corporis*) and **head louse** (*Pediculus humanus capitis*) affect different parts. Body lice attach eggs to clothes, whereas head lice attach eggs to the base of hairs. There are about ten million cases of head lice a year in America and one million in Britain. The lice are found mainly behind the ears and neck. There is racial diversity, with head lice infestations being rare in African Americans as their hair shape and width are less suitable than in other races.

Nits are yellow to white oval **lice eggs**, about 0.3 by 0.8 mm (0.01 to 0.03 inches). They are cemented to the base of a hair shaft, as they need body heat for incubation. They can be seen on the body, in the hair, or in seams of clothing, hatching in seven to ten days into nymphs which feed on blood and mature in about 10 days. The **adult** body louse has six legs and is tan to greyish-white. To live, lice must feed on blood; if separated from their host and the host's body temperature, they die fairly quickly.

In films, **to suggest poverty**, mothers in poor countries are often shown patiently examining their children's heads for lice and nits, but lice are no respecters of country, class or wealth, as British middle-class mothers know. Treatment is by thorough combing to remove lice and nits with special fine-toothed combs, and insecticidal shampoos.

The **pubic or crab louse** (*Pthirus pubis*) is found in pubic hair and other hairy parts. It feeds voraciously on blood. At about 1.7 mm (0.07 inches) long, it is smaller than body and head lice, and has a roundish body with large claws on the hind legs. The eggs are usually laid on hairs in the genital and perianal regions and take a week to hatch, followed by three nymphal stages, which take about two weeks to become adults. The itching can be very strong, especially when there is hypersensitivity to louse saliva. About 2% of the world's population are infected.

Infection is by close contact, largely through sexual intercourse, or from shared towels, clothing or beds. Crab lice die quickly away from the body, so infection from toilet seats is unlikely. Contrary to folk lore, they cannot jump. A graffito in a Bristol University toilet was incorrect: *It's no use standing on the seat, the crabs in here can jump ten feet!*

Tinea diseases are **fungal**, caused by species of *Microsporum*, *Trichophyton* and *Epidermophyton*. **Athlete's foot, tinea pedis**, is extremely common. It can be transmitted from infected skin particles in gyms, swimming pool surrounds, hotels, and particularly from damp changing room floors. It attacks the skin between the toes, causing fissures, whiteness (extra blackness in dark skins) and erosion, with itching and if severe, burning sensations. **Scalp ringworm, tinea capitis**, infects the scalp and hair follicles, causing hairs to break, with itchy reddened bald patches, mainly in children. It is fungal, not caused by worms, and can leave scars. In Britain it is commonest in school children of Afro-Caribbean ancestry. **Crutch rot/jock itch, tinea cruris**, causes red areas on the inner thighs, crutch and groin areas, which can become itchy. All these tineas can be treated with fungicides but may be persistent.

Impetigo is a highly contagious common skin infection causing weeping sores and blisters. When a boy of 14 in our class came to school with his face covered in sores, buff-coloured crusty deposits and white antiseptic cream, he looked hideous and we were afraid to go near him. It mainly affects children. Bullous impetigo causes large, painless, fluid-filled blisters. Non-bullous impetigo is more contagious; its sores quickly rupture, leaving a yellow-brown crust. It is caused by the **bacterium** *Staphylococcus aureus* or by Group A streptococci such as *Streptococcus pyogenes*. The incubation period is only one to three days. Each year about 140 million people catch it, about 2% of the world's population.

Chicken pox is another very infectious disease mainly affecting children. The red pustules on any part of the body are very itchy and if scratched can cause pock marks. The spots become fluid-filled blisters which crust over, forming scabs. Chicken pox is caused by the **herpes varicella-zoster virus** with an incubation period of 14 to 17 days. It is so common that about 90% of adults in Britain have some immunity to it, but that may not be complete, so about 13% of adults catching it have had it before.

The chicken pox virus can persist for many years in a dormant state, especially in nerves. It can escape the immune system and break out to cause **shingles**, particularly in the over-50s. About 3 in 1,000 adults get it a year in Britain. It normally affects only one side of the body, following one peripheral nerve, causing red skin rashes and much nerve pain. It can have dangerous complications, including ones affecting the eyes. A vaccine for the elderly was introduced in 2013 in Britain.

Warts are small, raised, rough, hard lumps on the skin, usually on hands or feet. They are mildly contagious but not cancerous. Warts are caused by the **papilloma virus** which makes the skin produce excessive amounts of a hard protein, keratin, in the epidermis. Most warts eventually disappear without treatment but that can take years. Some people claim to cure warts by 'magic', but salicylic acid, other chemicals and cryotherapy (freezing) are the medical treatments. **Verrucas** are warts of the soles of the feet, often caught in changing rooms.

For **leprosy**, see Chapter 11.6.

Further information about skin. A good source of information and illustrations is NHS Choices, www.nhs.uk/Conditions. For many skin (and other) disorders, there are **national patient support groups,** easily found on the Internet. Sufferers often feel isolated and can find it extremely helpful to talk with others having the same condition at meetings of patient support groups, nearly all of which have informative websites, e.g., www.vitiligosociety.org.uk for the Vitiligo Society. The British medical school curriculum is so crowded that many qualified doctors have done little dermatology, often only one week, so seeing a specialist is best. For more than 500 colour photos of skin conditions, see *Ethnic Dermatology: Clinical Problems and Skin Pigmentation,* by C. B. Archer (2nd edition, 2008, informa healthcare). The **British Association of Dermatologists** (www. bad.org) publishes excellent free, downloadable patient information leaflets on skin conditions.

Reference

Stoddart, M., *Adam's Nose, and the Making of Humankind.* (2015) Imperial College Press, London.

Chapter 14

The Skeleton, Muscles, Osteoporosis, ME, Motor Neurone Disease, Muscular Dystrophy

14.1. The skeleton; rib diversity

Even the skeleton and its musculature show diversity. Roughly 1 in 200 individuals has 13 pairs of ribs instead of 12. Some people are asymmetric, with 12 ribs on one side and 11 or 13 on the other, often without realising it. Bifurcated ribs, with an end split in two, have an incidence of 1%. **Rib variations** are more common in females and on the right side. Variants include incomplete development of the third and fourth ribs, or the absence of ribs eight, nine and ten. The twelfth rib may be small or long.

In a small adult weighing 7st 12 lb (110 pounds, 50 kg), the skeleton weighs about 1st 1.5 lb (15.5 pounds, 7 kg), 14% of body weight. The longest bone is the femur (thigh bone) and the smallest is the ear's stapes, about $1/5^{th}$ inch (5 mm) long.

Of the adult's 206 bones, 23 are in the neck and skull, 3 in each ear, 26 in the spinal column, 25 in the chest, 32 in each arm and hand, and 31 in each hip, leg and foot. Bone counts depend on how one treats fused bones; three are fused in each hip bone, five in the sacrum. Below the sacrum is the small triangular coccyx, consisting of four fused rudimentary vertebrae corresponding to the larger and more numerous bones in an animal's tail. It supposedly resembles a cuckoo's beak (Greek, cuckoo, *kokkyx*). The cranium is made of eight bones fused at wavy ridges called sutures, while the facial skeleton has 14 fused bones. The wavy joints, like interlocking jigsaw pieces, are stronger than straight joints.

There are more than 100 **movable joints** between bones, with hinge joints at elbow and knee, ball-and-socket joints in shoulder and hip, saddle joints in the base of the thumb and gliding joints between adjacent bones in the feet and palms.

Bones serve for **support, protection,** like the rib cage around the lungs and heart, and for the **attachment of tendons from muscles. Red blood cells** and many white blood cells are made in bone marrow, especially in the skull, ribs, backbone, breastbone, and in the upper parts of the humerus (upper arm), and femur.

Bone is five times stronger than steel, resists compression twice as well as granite and resists stretching four times better than concrete. It is penetrated by nerves and blood vessels, and contains many living cells. The matrix around those is about 35% protein, especially collagen; the rest is largely mineral deposits of calcium phosphate and related compounds. **Cartilage** has less mineralisation and is more flexible, so our ears and nose can bend. Bands of cartilage hold our windpipe in shape.

Loss of minerals in **older people** makes their bones more liable to break or fracture, with hip, leg and arm or wrist fractures being common. Various biphosphonate drugs show promise in strengthening bones. In some sports, broken noses are frequent, as in boxing and rugby, while champion jockey Terry Biddlecombe suffered 47 broken bones in his riding career. **Ballet dancers** have enormous stresses on muscles, ligaments, tendons and joints. Ballet Rambert expects 10% of its dancers to be off with injuries at any time.

14.2. Brittle bone disease and osteoporosis

Brittle bone disease causes extreme bone fragility, leading to fractures before and afterbirth. There are genetically different forms with a combined frequency of 1 in 20,000, with about 3,000 sufferers in the UK. A typical sufferer is Laura Wiggins who had more than 150 fractures by age 10, including breaking her neck turning over in bed.

Mutations in collagen genes cause diseases known as *Osteogenesis imperfecta* (Oi), **imperfect bone production.** The abundant **type I collagen** molecules are long, helical and thin, aggregating into fibres which give mechanical strength to skin, bone, tendons and ligaments. The collagen defects in Oi give bone and skull defects, and often hernias; intelligence is usually normal. Undermineralisation leads to repeated bone fractures.

Osteoporosis mainly affects the elderly, with demineralisation and weakening of bones, causing 'Dowager's hump' on the back and a big increase in fractures. 80% of sufferers are women. Osteoporosis affects 10 million Americans, with 34 million having the lesser condition of **osteopenia**, with a risk of fractures. At least eight **loci influence osteoporosis**. Brothers and sisters of a sufferer are six times more likely to have low bone density than the general population. The spine weakens with age, becoming more curved and fragile. Regular exercise reduces the risk of fractures. About 40% of female ballet dancers get osteoporosis.

Scoliosis is a sideways curvature of the spine, occurring most often during the growth spurt before puberty. Two per cent of adolescents, 90% of them girls, develop this, with 12% needing an operation. Scoliosis is often not noticed by parents unless the curve reaches 25 or more degrees. Claire Gracie had a severe curvature of 57 degrees but it was only spotted when she was 14 by her hairdresser.

My scoliosis is mild (Photo 14.1), but it can be disabling. Usually no treatment is necessary. Shakespeare described King Richard III as a 'hunch-backed toad'. His skeleton, excavated in 2012, showed that he was not a hunchback but he had marked scoliosis and a twist in his spine.

Lower back pain is extremely common in the West. In one study, 55% of 15-year-olds were found to suffer from back pain at times, from slouching and carrying heavy school bags slung over one shoulder.

14.3. Muscles

Muscles, 'red meat', make up about 45% of weight in males, 35% in females. We have about **640 different muscles** attached to the skeleton by tendons of tough connective tissue, and can control most of them consciously, as in raising a hand. That is **voluntary muscle**, with a different structure from **involuntary muscle**, such as that in your gut walls, which you cannot move by thought.

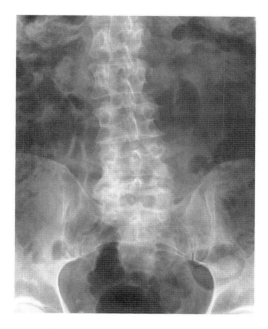

Photo 14.1. Mild scoliosis (sideways curvature) above the pelvis.

Our largest muscles are the **gluteus maximus** ones, forming the contours of our bottom and weighing about 2¼ lb (1 kg) each. Bodybuilders try for large biceps and triceps and well-defined pectoral and abdominal muscles ('six-pack'). **Muscle development** is influenced by genes, exercise and drugs like anabolic steroids.

Muscle wastage ('use them or lose them') is a serious problem in bedridden patients, those in wheel chairs, and when limbs are immobilised in plaster. In 1991 the space shuttle Columbia spent nine days in orbit, during which the seven astronauts lost 25% of their weight-bearing muscles' mass because it requires less effort to move with little gravity.

Some muscles are specialised for **power**, like those of the jaws, legs, arms, hands and feet, others for accurate alignment, as in the six muscles controlling the direction of each eye. Vocal cord muscles produce large changes in pitch from extremely small changes in length.

Many muscle disorders are known, including **ME (myalgic encephalomyelitis, chronic fatigue syndrome)**, which has 250,000 sufferers in the UK.

Personal account. *I have ME*

On a bad day I get up feeling more dead than alive, and after an hour need to lie down till lunch. After another 2 or 3 hours lying down I can manage to play easy computer games or watch TV. On a good day a couple of hours on the bed after lunch are sufficient, but my activities are restricted. I can stand up for only a few minutes, and walk about 30 metres because of foot pain. I tire very rapidly during physical activity, and can't concentrate for more than 40 minutes. Things which I could do in an hour or two now take days or weeks. I could not live unaided, relying on my wife for most things. ME's severity varies greatly from person to person, and over time. Some are confined to a wheelchair or bed.

My ME started 11 years ago, in my early fifties, creeping up gradually over a year or more. I thought my fatigue resulted from stresses at work, but being signed off didn't make much difference. A psychiatrist diagnosed depression. I only discovered it was ME when I talked to fellow sufferers at a local support group. As there was no prospect of recovery, I took retirement. Being cut off from friends at work, and spending so much time lying down, I had almost no social life, and succumbed to low self-esteem and depression. The depression was treatable but ME persisted. To keep depression at bay I need to keep as active as I can, without overdoing it and causing ME to worsen. Thankfully, I rebuilt my social life with help from the local MIND centre.

What is ME? No one knows. The confusion is seen in the plethora of names: Myalgic Encephalomyelitis (ME), Chronic Fatigue Syndrome (CFS), Post Viral Fatigue Syndrome (PVSF), Fibromyalgia... The 'experts' don't even agree whether it is physical or psychological. There is no satisfactory test. The main features are feeling tired most of the time; becoming rapidly exhausted after physical or mental activity; muscular aches and pains; disturbance of sleep patterns and lack of refreshing sleep. There is no cure although medication helps with side-effects.

ME is increasingly recognised in teenagers. Many recover spontaneously after three-to-five years, not always completely. ME may last decades. In spite of considerable research, its cause is a mystery.

David Ockendon. He used to be a great walker.

The Fibromyalgia Network provided this information about **Fibromyalgia Syndrome (FMS)**, which overlaps with **chronic fatigue syndrome/ME**:

'FMS is a widespread musculoskeletal pain and fatigue disorder of unknown cause, with the name meaning pain in the muscles, ligaments and tendons. It used to be called fibrositis, implying muscle inflammation but the muscles are not inflamed. Aching all over is common, with twitching or burning sensations in the muscles. It can affect people at any age, with more women than men sufferers.

The symptoms often resemble bad flu, with great muscle pain and almost no energy or motivation. Fatigue and lethargy vary from mild to incapacitating, giving difficulty in concentrating. Frequent bursts of brain activity give disturbed and unrefreshing sleep. Other common but variable symptoms include irritable bowel syndrome, chronic headaches and increased sensitivity to certain smells and foods, bright lights, noise and medications. Depression, anxiety and over-exertion can lead to aggravated symptoms. ME and FBS are difficult to treat. They can ruin lives and lead to the loss of the capacity to work, and to diminished social lives but do not usually kill sufferers.'

In contrast, **motor neurone disease** is fatal, killing about 1,000 sufferers a year out of 5,000 affected people in the UK. There is no known cause, cure or treatment, with the disease worsening until death. There is a **progressive degeneration of nerves to the muscles**, causing weakness, lack of control and muscle wasting, starting with the arms and/or legs. Chewing, swallowing and speaking become difficult. The mind, senses, bladder and bowel are usually unaffected. Its highest incidence is in the 50–70s. The number developing the disease in a year is about 2 per 100,000.

The actor David Niven (1910–1983) fell ill in 1982 with a virulent form of motor neurone disease, killing him within a year. Stephen Hawking, born in 1942, is a famous astrophysicist. He has a slower-developing form but is now almost entirely paralysed.

Muscular dystrophies are inherited disorders, usually eventually lethal, with progressive muscle wasting and weakness. They initially affect the hips, pelvic area, thighs and shoulders, progressing to the heart and breathing muscles. Pneumonia is often the final killer. The commonest type, **Duchenne muscular dystrophy**, has an incidence of three in 10,000 males. Because it is a **sex-linked recessive condition** (see Chapter 8.3), it rarely occurs in females, although 7% of female carriers have some muscle weakness. Affected boys are usually confined to a wheelchair by the age of 10, with death before 20 or 30. About 20% have some mental retardation. The very large gene has been cloned and the function of dystrophin protein is understood; gene therapy is being tried. Pre-implantation and prenatal genetic diagnoses are available.

In 2014, Tania Clarence, 42, pleaded guilty to manslaughter of Olivia, 4, and three-year-old twins Ben and Max, claiming diminished responsibility. All three children suffered from **type 2 spinal muscular atrophy**, a genetic disorder giving little control over movement and a drastically shortened life expectancy. Tania killed her children to prevent them suffering later. People have very diverse opinions on the **ethics of 'mercy killings'**.

Chapter 15

Head, Face, Eyes, Ears, Sight, Hearing, Smell, Taste

15.1. Introduction

People's faces are all so different. They give us impressions of personality, attractiveness, sex, mood and possible racial origin. A friendly smile, a hostile scowl, a bored look — all tell their story. One rare but striking **within-person example of diversity** is having one blue eye and one brown eye. Although this could be due to a mutation in development, the usual cause is that migrating pigment cells from the neural crest in the embryo reach one eye but not the other, in a genetically brown-eyed baby.

In 2008, a baby girl called Lall was born in India with **two faces**, and was said to be a reincarnation of the Hindu god Lord Ganesha. A more prosaic explanation is that she had cranio-facial duplication due to a mutation in a gene called *sonic hedgehog.*

Many people **alter their faces**, making cosmetic changes to improve their image (Chapter 5), or have facial adornments (Plates 11, 12). Some Oriental men have sparse facial hair and many Turks and Greeks grow large moustaches. Large beards, 'at least a clenched fist in length', were worn by the Taliban in Afghanistan. In ancient Egypt, female pharaohs had a small false beard strapped to their chin as a symbol of royal authority.

15.2. Head size and shape

Head shape, independent of size, is measured by the **cephalic index**: its breadth is divided by its length, and multiplied by 100. My head breadth is 16 cm and its length is 19 cm, so my cephalic index is $(16/19) \times 100 = 84$. The technical terms are **dolichocephalic** for long, narrow heads, with values from 70 to 75, **mesocephalic** for intermediate types, and **brachycephalic** for broad or round heads, with values of 80 to 90.

Average cephalic index values include: Australian Aboriginals, 72; South African Bushmen, 75; Ituri Pygmies, 76.5, Eskimos, 78; Iranians, Armenians and Assyrians, 80; Japanese, 81; Eastern Chinese, 82; Germans, 82.5; Hawaiians, 84; Norwegian Lapps, 85. People in colder climates tend to have rounder heads than those in hot regions, with long, narrow heads best for heat dissipation and round ones best for heat conservation. On a cold day, heat lost from the head by a fully-clothed person can be 80% of total heat loss.

Ashley Montagu's classification of ethnic groups (Montagu, 1963) describes **Caucasoids** (including Baltic, Nordic, Irano-Afghan, Alpine, Mediterranean) as largely but not exclusively brachycephalic, white, narrow nosed, with mainly wavy hair, **Australoids** (including Veddah, Australian Aboriginals) as mesocephalic, brown, moderately broad nosed, with mainly curly hair, **Negroids** (including Negro, Negrito, Negrillo, Melanesian) as largely dolichocephalic, black, broad nosed, with mainly woolly hair, and **Mongoloids** (including Mongol, Japanese-Korean, Eskimo, Amerindian, Indo-Malay) as largely but not exclusively brachycephalic, yellow-brown, moderately broad nosed, with mainly straight hair. His ethnic groups diagram is a circle, so that some types are on the borderline between his four major ethnic groups, with Bushmen between Negroid and Mongoloid, Polynesian between Mongoloid and Caucasian, and Indo-Dravidian between Caucasoid and Australoid.

Head shapes can change. Boas in 1911 published his *Report on Changes of the Bodily Form of Descendants of Immigrants*. He compared immigrants raised abroad with their children raised in America. The children were larger and heavier (at a particular age) than their parents, mainly due to better nutrition. Surprisingly, head shape was one of the more changeable characters, with American-born children tending to have longer, narrower heads than their parents. Children lying for long times on cradle boards can develop a flat back of the head. **Cranial deformation** by applying pressure to babies' or children's skulls has been practised by many American and South American tribes but is not inherited.

Head size and head shape are independently determined genetically. Broad heads are generally dominant to long. Genes affecting head shape are segregating in many populations, so that although Whites are generally brachycephalic, all types occur in white populations, with Scandinavians having a high frequency of dolichocephalic forms and people in Alpine villages showing a high frequency of brachycephalic forms.

For **head width**, the average difference between members of identical pairs of twins is 2.8 mm, while for non-identical twins (of the same sex) it is 4.2 mm. This difference shows that there is a strong genetic element for this polygenic, multifactorial character (Chapter 1.3).

15.3. Head and spinal cord defects, including spina bifida

A serious defect is **Apert's syndrome**, an autosomal dominant condition with a frequency at birth of 1 in 10,000. The head is flattened from back to front with a high vault, broad forehead, a wide space between the eyes, and webbing or fusion of fingers and toes. Mental handicap is mild in 31% of sufferers and severe in 7%. The effects are usually so severe that sufferers tend not to reproduce, with most cases due to new mutations.

Hydrocephalus ('water on the brain') is a progressive enlargement of the cerebral part of the head from an excessive accumulation of fluid which forces brain tissues against the skull, damaging them. Its incidence varies from 0.4 per 1,000 to 2 per 1,000 live births. It mainly arises in children but can occur in adults. If untreated, it has about a 55% death rate, with survivors having mental and physical handicaps. **Symptoms** depend on the extent of obstruction of cerebrospinal fluid circulation in the brain and the age when it develops. Once the skull bones fuse at about five years of age, the head cannot expand much. The condition can cause irritability, muscle spasms, delayed development, lethargy, incontinence, headaches, vomiting, poor co-ordination and mental aberrations.

Hydrocephalus can result from infections (including meningitis and encephalitis), tumours, birth injuries, haemorrhages and incomplete closure of the spinal canal. It can be genetic as an autosomal recessive or an X-linked recessive when most sufferers are boys with strangely flexed thumbs. Treatment is by surgery to remove the obstruction, or insertion of a drainage channel.

Anencephaly ('without a head') involves the absence of the skull and the cerebral part of the brain and is lethal. The frequency at birth is about 4 in 10,000, but many cases abort spontaneously. It can be diagnosed by ultrasound early in pregnancy and parents often opt for a termination. The neural tube in the foetus which should form the brain and spinal cord fails to develop properly. In the USA, the federal government requires grain products such as flour to be supplemented with folic acid, which reduces the incidence of neural tube defects.

Macrocephaly ('large head') means having an abnormally large head. There may be no adverse symptoms, as in benign familial macrocephaly, caused by an autosomal dominant gene. Other forms may have mental handicap or progressive dementia.

Microcephaly ('small head') means having an abnormally small head, with mental retardation. The immediate cause is usually failure of the brain to grow normally, since a baby's skull growth is driven by brain expansion. The causes include chromosomal disorders such as Down syndrome (Plate 19E), autosomal recessive genes, congenital infections with rubella, toxoplasmosis or cytomegalovirus, the mother having untreated phenylketonuria (Chapter 12), severe malnutrition, or maternal drug abuse.

Spina bifida ('cleft spine') involves incomplete closure in the spinal column. The mildest form, **spina bifida occulta** (in medicine, *occult* means 'not obvious on inspection'), has an opening in one or more vertebrae (usually lumbar 5 or sacral 1 or 2) without damage to the spinal cord. About 40% of Americans have this mild form and most are unaware of it. The other two forms have clear symptoms and are called **spina bifida manifesta**, with a combined frequency of 1 in a 1,000 live births. Of these, 4% have the **meningocele** form, with an intact spinal cord but the protective membranes (meninges) around it protrude through an opening in the vertebrae in a sack called a meningocele. This can be repaired surgically with little damage to the nerves.

The other 96% have the worst type, the **myelomeningocele** form. Part of the spinal cord protrudes through the lower back, sometimes covered with skin, sometimes open, exposing nerves and tissue. There is usually paralysis of lower regions, and loss of bowel and bladder control. About 80% of children with this bad form have hydrocephalus (see above), needing a drainage tube inserted into the brain. In patients with closed lesions, 60% survive to 5 years, with one third each with severe, mild and no mental handicap.

Treatments often involve surgery to repair the open cord. A series of operations on the back is often needed throughout childhood. Bladder and bowel training may be attempted, or a catheter may be inserted for urine. Crutches, braces and wheelchairs are needed if the legs are paralysed. Some determined sufferers manage to have good, active careers, including the British orchestral conductor, Jeffrey Tate, who conducts sitting down. He is really glad that his handicap did not result in his mother terminating her pregnancy!

The **incidence of neural tube defects** varies. For Africa, the USA and Mongolia, about 1 in 1,000 live births is affected; in the 1970s, the figure for south-east England was 3 in 1,000, and a high 7 in 1,000 in Ireland, although the frequency has fallen since then from folic acid supplementation of maternal diets. The risk of a mother with an affected baby of having another affected baby is 1 in 30, dropping to 1 in 100 with folic acid before and during early pregnancy. The risk for the offspring of an affected person is about 1 in 30. Neural tube defects are generally multifactorial but may occur by chance without genetic predisposition.

Prenatal diagnosis for neural tube defects is done by ultrasound scanning and amniotic fluid tests, detecting 100% of anencephaly and 98% of open spina bifida. Testing for anencephaly can be done at 11 weeks gestation and for open neural tube defects at 17 weeks. Testing the pregnant woman's blood for raised levels of **alpha-fetoprotein** at 17 weeks gestation is frequently done because foetuses with neural tube defects tend to leak it into the mother's blood. Unfortunately the test is not specific. 85% of pregnancies involving neural tube defects are terminated.

Well known sufferers from neural tube defects include Tanni Grey-Thompson, Welsh Paralympic athlete, member of House of Lords; Karin Muraszko, chair of Department of Neurosurgery at University of Michigan, the first woman appointed to such a position in the USA; Chandre Oram, a man who has a tail due to spina bifida, and Billy Bridges, Canadian Paralympic ice-sledge hockey and wheelchair basketball player. Some sufferers have excellent arm mobility and strength, and not all have learning difficulties.

15.4. The face

The face consists of the forehead, eyebrows, eyes, nose, cheeks, lips, jaws, mouth and chin. The dimensions of each part are variable between individuals and sometimes between racial groups, and so are their relative placements in the face (see Plates).

The **forehead** varies in hairline position, width, height, curvature, bulges, slope, lines, and prominence of crests above the eyes. These traits show continuous variation, with larger values often dominant to smaller values.

The **eyebrows** may be thick, thin, curving, straight, separate over each eye, or more or less continuous, with many women plucking out central hairs. They often pluck or shave eyebrows, replacing them with artificial dark lines. Really bushy eyebrows are considered very masculine. The eyebrow hairs nearest the nose are usually the most vertical. Eyebrow hair normally parallels head hair in colour but brows may retain colour long after head hair has gone white. Eyebrows divert sweat running down the forehead from entering the eyes, and act as sunshades.

People suffering from **total loss of body hair** find that **without eyebrows** the salty stinging sweat from the forehead runs straight into their eyes, making them red and sore. They frequently have to wipe rain from their eyes. **Botox injections** to reduce lines on the forehead diminish the role of eyebrows in signalling feelings. A woman complained that her Botoxed friend was not able to raise or lower her eyebrows, so the woman could no longer deduce her friend's mood.

The **eyelids** are important as closable shields. In normal light they are open, but in strong light can be partly closed. They help greatly with sleep when fully closed. People often 'screw up their eyes' by partial closure if they are having trouble focussing. **Eyelashes** help to keep dirt out of the eyes and give some shade. They act as touch sensors, allowing the eyes to be closed if the eyelashes are brushed. In many cultures, darkening female eyelashes with mascara is common. Genes control eyelash length, thickness and curvature: long tend to be dominant to short.

The **cheekbones** are sometimes an attractive feature of women but those of men are rarely commented on. High cheekbones on Japanese and Chinese ladies (Plates 12A, D) can look splendid; many Whites use make-up to accentuate their cheekbones.

The **mouth** is the feeding orifice, the gateway to the alimentary canal. It is crucial for speech, eating and drinking, and for **social and emotional signalling** — the smile, the grimace, the laugh. A tentative half-smile might be answered by a disdainful look, by no response, by a half smile or a full smile, all full of meaning.

Female lips receive all kinds of lipstick, outliner, gloss, etc. The skin on the lips is thin and in Whites has few pigment cells, with many small blood vessels providing the natural pink or red colour. In people with darker skins, there is more lip pigment and the red is less noticeable, but the lips are often thicker, especially in Negroids (Plate 5A). The lips are muscular, closing as a sphincter to enable food to be swallowed. Lips have many nerve endings and are very sensitive to touch and temperature.

Given their role in **sex-attraction and kissing**, lips are an **erogenous zone**. In women, a small nose, big eyes and voluptuous lips are attractive. A woman's attractiveness is linked to her hormones during development. While testosterone makes a man's face look mature, a woman's oestrogen helps to keep a youthful facial structure during young adulthood. Higher levels of oestrogen give a woman bigger eyes and fuller lips (Plate 12D), suggesting health and fertility. Lipstick makes a woman's lips more eye-catching. Kissing usually involves the exchange of millions of bacteria from the lip surface and saliva. **Cleft lip and palate** are covered in Chapter 8.5.

15.5. The eyes; sight, colour blindness, cataracts and other eye problems

The eyes are usually the centre of attention as they reveal so much about mood and personality. Humans have for thousands of years used **eye make-up** (Plates 12A, 18D). Eye differences include size, shape, distance apart, colour, how high up the face they are and whether they squint. Temporary differences depend on emotion and physical conditions, including tiredness or alertness, boredom or interest, and exposure to weather. The eyes can be open wide with surprise or interest, or narrowed in bright light or driving wind, or with suspicion or concentration. One alarming temporary change is a spontaneous **haematoma**, where a broken blood vessel within the eye results in the white being coloured a fresh red with blood (Plate 18E). The colour fades and disappears without treatment within a week or so. Black eyes (Plate 18F) come from accidents.

The degree of **pupil opening** is controlled by light, narrowing in high intensities, enlarging when it is darker, but emotion and mood have effects too. An enlarged pupil can indicate a state of love and passion, and look inviting. Some women — particularly in the Middle East — apply lotions to enlarge their pupils on special occasions. In bright light, the pupil may be only 1/25th inch (0.1 cm) across, expanding to 1/3rd inch (0.8 cm) in dull conditions.

Look at an object directly in front of you with your head level and motionless. Move your hands around to establish your angle of vision in up/down and side-to-side directions. Can you see behind you to any extent? Can you see further upwards or downwards?

Try with one eye at a time. A complete circle is 360 degrees.

One's **field of vision** is usually about one third of a circle, 120° up and down, with more downwards than upwards. With both eyes open, we can see about 200° from side to side. That is more than a straight line, 180°, so we can see a bit behind us, but nowhere near as much as can a rabbit or zebra. Of that 200°, the middle 120° can be seen by both eyes, so we have **binocular vision** and good **distance perception** there, while each eye has a region of monocular vision of about 40° to the side, with much poorer distance perception.

Vision is so central to human activities that about 70% of our body's sense receptors are in our retinas, with about 120 million rods per eye and 6 million cones. Our brains can detect about 10 million different colours.

A few babies are born with a **single large eye** in the centre of the forehead. They usually have multiple abnormalities and die before or soon after birth. The condition is called the **cyclopean eye**, after Cyclops, one of a mythical race of one-eyed mountain giants from Sicily.

A dramatic **eye colour variation** is **albinism** (Plate 15E), with no pigment in the iris or at the back of the eye in the normally black choroid coat, so the eyes look pinkish in bright light from reflections off blood vessels within the eye.

Personal account. *I am a rare kind of albino*

Albinism is genetic, caused by a congenital lack of melanin pigment in the skin and eyes. Albinos generally have very light hair and skin with pink or very light blue eyes. It has two main effects, a very light-sensitive skin which burns easily, and impaired sight, from being partially sighted to almost blind. My brother and I have the rarest (1 in a million) of the ten types of oculocutaneous [eye, skin] albinism. We have bad eye problems but have a little melanin in our hair and skin.

The main issue for albinos is the loss of sight, not directly caused by lack of pigment, but by the eye's need of protection from light, which it does in two ways. These are by movement (nystagmus) and astigmatism (changing shape — to rugby ball from round ball) — reducing the amount of light hitting the retina. Because of the movement there is a lack of clarity — everything is blurred. Most albinos wear dark glasses on light days. I find it much easier to see at night, although the vision does not improve in clarity.

The other issue is skin. Many albinos have very poor tolerance to sunlight and will invariably suffer skin damage. Cancer of the skin is usually squamous cell carcinoma and vitamin D deficiency is more prevalent than normal, both due to the lack of melanin.

Life is more complicated. The obvious issues are going out in the sun without good protection (some cannot go out) and not being able to see to drive, but albinos have even greater issues. The biggest for me is that I communicate differently. I am unable to make eye contact unless very close, due to the lack of clarity. This makes it difficult for me to form relationships. When I can make eye contact it is extremely powerful. Imagine walking into a large room with many people; I do not know who wants to talk to me until I speak to them. Reading is slow and uses more energy than for sighted friends.

Other senses are heightened to compensate, especially hearing, which I use as much as sight for crossing roads, and proprioception, where my feet have much greater sensitivity to the ground. My nervous system is much more attuned to my surroundings such that I have greater spacial awareness and very fast reflexes.

John Hemington

Amongst non-albinos, **colours in the iris** surrounding the pupil vary from very pale blue to almost black (Plates 6H, 18C). The **iris** (Greek for rainbow) is a muscular diaphragm of which the central opening is the pupil. Although this opening is the main determiner of how much light penetrates the eye, iris pigmentation is important, with darker eyes giving more protection in strong light. Skin and eye colours are both from melanin pigments which can be reddish, yellowish, brownish or blackish. Pigment particles vary greatly in size.

The iris has a front layer of cells which may or may not be pigmented, and a double layer of pigment cells at the back (except in albinos). If pigment is present only in the rear layers, the eyes are **blue**. Small amounts of pigment the front layer give **grey** or **green** eyes, while a lot of pigment in both parts of the iris gives **brown** or **black** eyes. There are no blue or green eye pigments. The diversity of colours is caused by the amount, distribution and light-scattering properties of eye pigments and tissues. One can **change one's eyes' appearance** with coloured contact lenses, some with odd patterns.

Light of shorter wavelengths (green and blue) is scattered more than light of long wavelengths (orange and red) in tissues, or by dust in the air or water and particles in the sea. Red light entering the eye is not scattered much but shorter blue wavelengths are scattered in all directions, some coming back out of the iris to an onlooker, with the black choroid coat at the back of the eye favouring reflection rather than transmission of light. Onlookers see blueness in the eye, sky or sea, even though none of these has blue pigment. If you look closely at someone's eyes, you will probably see a mixture of blue, green, yellow and brown, with various sized flecks of colour (Plate 18B). Except for albinos, babies are born with bluish eyes, just having the light-scattering effects. In a few weeks, those who will be brown-eyed start making brown pigments which will largely hide the blue.

The **genetics of normal eye colour** is not well understood. It is inherited: identical twins have identical eye colours in 99.6% of cases, while eye colours are identical in only 28% of non-identical twin pairs. A simple explanation, of one locus with a dominant gene (allele) for brown and a recessive gene for blue, often works but it cannot explain many cases where two blue-eyed parents have a brown eyed child. Pedigree analysis of 832 families from Copenhagen identified a gene (*BEY*, brown eye) on chromosome 15 with a dominant allele giving brown and a recessive one giving blue, and another gene (locus), *GEY*, green eye, on chromosome 19, with a dominant allele giving green eyes and a recessive one giving blue; this gene tended to segregate with one for hair colour. Bright light over long periods can cause fading. **Darker sectors** in light-coloured eyes are probably due to mutations during development.

The **sclera** is the tough white layer round the eyeball. The colour of the 'white of the eye' is normally white, with visible small blood vessels (Plate 18B), but it can go yellow with diseases, especially jaundice. The **sclera** is transparent over the centre of the eye, where it forms the **cornea**.

The cornea's curvature is part of the focussing system; if it is not spherical, it causes **astigmatism**, with blurring of the image. The thin, delicate, transparent membrane which covers the sclera and inner eyelid surfaces is the **conjunctiva**. In **conjunctivitis**, this can be inflamed by grit or bacterial infections to give reddening, watering, soreness and itchy irritation. **Allergic conjunctivitis** is caused by an allergic reaction to tree, grass or other pollens, cat or dog skin or hair, cosmetics, smoke, air pollution or other allergens. Conjunctivitis treatments may be with antibiotics, antihistamines or other anti-inflammatories. Lubricating drops ('artificial tears') may help.

Uncontrolled eye movements (nystagmus) may be up and down, from side to side, or rotary. They are rapid, involuntary, and may be inborn or arise from disease or brain injury. Congenital nystagmus is usually mild. Often affected people are unaware of their eye movements and vision is normal unless movements are large. I noticed that a friend had obvious, rapid, unceasing, side-to-side movements of both eyes. I was astonished to find that she did not know she had it. It had not been noticed by her husband, children, friends, or various opticians. Her eyesight was unaffected.

A **squint (strabismus, cross-eyes)** is where the two eyes point in different directions. Some people can deliberately 'cross their eyes'. Squints affect about 4% of American children, boys and girls equally, and can arise in adults. The brain fuses separate images from the two eyes into a three-dimensional image, allowing perception of distance. In half of children with a squint, the brain ignores the image of the weaker eye, resulting in effectively monocular vision and poor depth perception. This is treated with a patch over the 'good' eye, to strengthen the brain's image from the 'bad' eye. If this is done at a pre-school age, it is usually successful and the patch can be discarded. Squints can be treated by surgery on one or more of the six muscles controlling eye movements.

Because one knows the approximate size of most objects, one can judge their distance from their apparent size even without **binocular vision**. Drivers who do not have binocular vision judge the distance between their car and the one in front from its apparent size. If they have one good eye, they can pass the eye part of the British driving test. I have one short-sighted eye and one normal-sighted one, and only discovered that I did not have true binocular vision when I mistook a distant flight of birds for mosquitoes at an open window. Without consciously doing so, I use my left eye for distance vision and my right eye for reading. Unlike many in their seventies, I do not need glasses for reading or for distance vision.

Short-sightedness (nearsightedness, myopia) is where someone sees nearby objects clearly but distant objects look blurred, because the eyeball is too long or the lens is too strong. *Myopia* (Greek *myein*, shut, *ops*, eye) often causes short-sighted people to screw up their eyes when trying to see distant objects. Males and females are equally affected, and there is a genetic component (mainly recessive) as short-sightedness often runs in families. Glasses or contact lenses treat most cases satisfactorily. **Reshaping the cornea** by laser is a popular effective treatment.

Short-sightedness often develops in fast-growing children when the physical length of the eye exceeds its optical length. During growth, different strength glasses may be required at different ages. Medical opinion is usually that neither reading nor watching TV has much effect on the development of short-sightedness, but popular opinion often differs. A Taiwanese student told me that most Taiwanese of university age have to wear glasses or contact lenses; this was attributed to the very long hours of study of books and computer screens in the competitive process of getting into university.

Problems with focussing can cause headaches and tiredness. Children need regular eye tests because not being able to see the board clearly in class retards progress. Children can be very cruel to those who are different: many children are hurt by taunts such as "four-eyes" when they have to wear glasses. There is a saying that *Men never make passes at girls who wear glasses*, but that is untrue.

Longsightedness (far-sightedness, presbyopia, hyperopia) is where someone sees distant objects clearly but cannot focus on close objects, because the eyeball is too short or the lens is too weak. The lens's elasticity is important for focussing close objects and is slowly lost during ageing. From their mid-forties onwards, many people hold reading matter further and further away in order to focus it, often complaining that the print is too small nowadays. The condition is easily corrected with glasses or contact lenses. It can be congenital, usually caused by a dominant gene.

Many older people have **bifocal** or **multifocal** glasses, with the bottom section for reading or close vision and the rest for distance vision. People with contact lenses sometimes have one lens for distance vision for one eye and one for close vision in the other, with some loss of distance perception.

Focussing depends on the ciliary muscle ring, attached to the lens by suspensory ligaments. When these muscles are relaxed, they form a larger circle round the eye, so these ligaments are tight, pulling the elastic lens into a flattened disc. This acts as a weak lens, capable of focussing the more or less parallel rays of light from distant objects on the retina. The light rays from close objects are diverging when they reach the eye and need a stronger lens to focus them on the retina. This is achieved by the ring of ciliary muscles contracting, making a smaller circle and slackening the sensory ligaments attached to the lens. The lens is elastic, tending when relaxed to be more spherical than flat, so when the ligaments relax the lens becomes more rounded, making a stronger lens, capable of focussing the divergent light rays from close objects onto the retina.

The retina's light-sensitive cells are cylindrical **rods** and conical **cones**. Light passes through layers of nerve cells before reaching them. The rods have only the light-sensitive pigment **rhodopsin** which absorbs visible wavelengths of light, so providing vision in different shades of grey. The retina contains about 120 million rods per eye but only about 6 million cones. The rods are particularly used at night and in dim light, giving high sensitivity to low light, but not very distinct vision, in shades of grey. Perception of colour is much diminished in poor light, but one can still see shapes. The cones need brighter light to give high-resolution **colour vision**. By adaptation of the pupil size and within the rods, the eye can cope with about a million-fold difference in light intensity.

There are three types of cone, with slightly different **opsin pigments**, so red cones are mainly sensitive to red light (longer wavelengths), green cones to green light (medium wavelengths) and blue cones to blue light (shorter wavelengths). The absorption ranges of the three types of cone overlap and the brain perceives an **enormous range of different colours** from the relative amounts of stimulation of red, green and blue cones. Yellow is seen when there is equal stimulation of the red and green cones, with little or no stimulation of the blue cones. Green is perceived when there is mainly stimulation of the green cones, and lesser stimulation of red and blue cones. We can generally detect light between 380 nanometres (a nm is one thousand-millionth of a metre), violet, and 750 nm, red.

The light-sensitive pigments consist of an opsin combined with retinene, a derivative of vitamin A, so a deficiency of **vitamin A** causes **night blindness**. Even fairly small reductions in the amount of rhodopsin in the rods make them unable to sense poor light.

Blue colour blindness is rare, with its gene on chromosome 7 so it is not sex-linked. **Red-green colour blindness** was covered in Chapter 8.3; this sex-linked recessive character is inherited on the X chromosome only, so sufferers are much more often male than female. In White populations, about 8% of males are sufferers and about 0.6% of women. The condition is rarer in Negroids and Orientals.

Even in people classified as red-green colour-blind, there is **enormous diversity in colour perception**. The genes (loci) for red and for green colour vision occur very close to each other on the X-chromosome, with their products having 96% identity in amino acid sequence and much DNA homology. The two genes occasionally crossover with each other, generating deletions, duplications and/or hybrid genes. About one quarter of red-green colour blind people have one or both genes missing completely. Sufferers can have good, partial or no red function, combined with good, partial or no green function, all easily detected by test charts. Colour blindness cannot be treated and disbars sufferers from colour-sensitive jobs, e.g., the armed forces where colour-coding of products is used extensively.

Although **cataracts** can be congenital or appear in young children, they typically develop in older people. Cataracts are the occurrence of cloudiness and opacity in the lens. They develop progressively, with increasing visual problems, often simultaneously in both eyes. A slight yellowish clouding of the lens is common in those over 60. In the USA, significant visual impairment from cataracts affects about half of those aged 65 to 74, and nearly three quarters of the over-75s. Metabolic disorders such as galactosaemia (Chapter 12.4) or infections with German measles can cause cataracts in infants.

1 in 250 babies has a congenital **cataract**. Its cause is genetic in about 32% of cases, usually dominant. If both parents are normal, the chance of a second child being affected is 1 in 10. The chance of cataracts in adults is increased by smoking and exposure to bright light (which contains UV) unless eye protection is worn. **Airline pilots** are three times more likely than others to get the most common form of cataract, in the centre of the lens. Cosmic radiation was blamed as its intensity is 100 times greater at 36,000 feet (11,000 m) than at ground level.

Cataracts cause loss of visual clarity; they may cause sensitivity to glare and cause halos to be seen around lights, and poor night vision. They are easily detected during routine eye examinations. **Cataract surgery** is very successful, with the patient able to leave hospital a few hours after the operation, although eye drops for lubrication and for antibiotic purposes are used for several weeks. The original lens is replaced by a plastic one. As a friend reported with delight after her cataract operations, the new lens often means that patients no longer need glasses for distance vision, if they did before, although reading glasses are usually required in older people.

Later-life cataracts running in families are usually dominant when genetic. This account from a Hong Kong student is typical:

> *We have a genetic problem in my family. My grandmother, uncles, aunts and even my father are sufferers from cataracts. That is nothing shameful, but something which I worry about for my father and myself. Grandmother is blind in one eye already and so is my uncle, but fortunately my aunt had a successful eye operation.*

Blindness in one or both eyes (Plate 18A) may be complete, partial, congenital or acquired. Its causes include accidents, untreated cataracts, glaucoma, diabetes and infections such as 'river blindness' (onchocerciasis). In poorer parts of the world, an insufficiency of vitamin A is the major cause of blindness, easily cured by better diet or vitamin supplements. Blindness is socially isolating and dangerous, with many repercussions for education, work, mobility, shopping and leisure, but some blind people have been remarkably successful, including the American, **Helen Keller** (1880–1968), who was blind and deaf before the age of two but became a famous author and lecturer. She remarked that while blindness separates people from things, deafness separates people from people. There have been many very able blind musicians.

Blindness has many disadvantages beyond the obvious one of impaired safety. David Blunkett was blind from birth yet rose from humble beginnings to be Secretary of State for Education and Employment, and later Home Secretary, in Labour governments from 1997 until 2004. He could read only documents in Braille. In an interview (*The Daily Telegraph*, 2/3/2015), he blamed the failure of his first marriage and his upsetting of Cabinet colleagues, such as Gordon Brown and Jack Straw, on his inability to see their facial expressions and body language. Non-verbal communication can be crucial to understanding the feelings of others.

People with sensory handicaps such as blindness may develop other senses such as better hearing. An osteopath I know has poor sight but an excellent sense of touch which is very useful in feeling through skin to find out what is happening with the patient's bones, muscles, ligaments and tendons. Blind people often manage to get around remarkably well, using built-up knowledge, sticks for prodding and sounding out the way ahead, and listening to sound clues. In the **Great Smog** of 1952 in London, blind people helped sighted people home. One man in Notting Hill repeatedly led people to their homes from the tube station as he had a good knowledge of the area.

Macular degeneration leads to poor central vision, leaving peripheral vision intact. The macula is the most sensitive part of the retina, central at the back of the eye where fine detail is perceived. It may degenerate with age, leading to an inability to read or drive, although light is still sensed in peripheral vision. Nearly all sufferers are over 50, with 15% of those aged 75 having symptoms. There is a genetic component. It is more common in smokers, and Whites have the highest incidence. It has no real treatment once symptoms occur, although injections into the eye, dietary zinc supplements and laser treatment can help in the early stages. Stem cell injections are being tried.

Here are some student accounts of eye problems from one intake of about 100:

- *There is a lazy left eye in myself, mother and mother's mother. It is lazy in focusing, so all three need glasses. My sight is all right for distance, not for close up.*
- *My little sister has extra skin over the white of her eye right up to the iris. It has fine hairs, causing trouble, irritation, and needs removing.*
- *I have a collarboma in my left eye, affecting the shape of pupil. I had it from birth. It affects my vision which is partly blacked out, with little detail perceived. My right eye is OK but short sighted.* [A **coloboma** is a congenital hole in an eye structure, usually the iris. Very small ones have no effect but large ones can cause blindness. It affects about 1 in 10,000 births.]
- *A friend has one blue eye, one green eye.*
- *I am red-green colour-blind and wear glasses.*

- *I have keratoconus in my left eye — a genetic eye condition. My dad has it. It causes a misshapen cornea. It is not visible but causes astigmatism and blurring. My eyesight not good, but my right eye has almost perfect sight. I will need contact lenses. It got worse last month — and gets worse until 21 or 22, then is stable.* [**Keratoconus** involves the cornea becoming thin and irregularly (cone) shaped, which prevents focussing correctly and causes visual distortion. These symptoms usually appear in the late teens or late 20s and may progress for 10–20 years, with increasing distortion of vision.]
- *Grandmother and aunt have glaucoma; mother may, but it's too early to know.*

Glaucoma is the third most common cause of blindness in America and involves increased fluid pressure within the eye from reduced drainage of fluid (aqueous humor). This high internal pressure reduces the blood supply to the optic nerve and retina, so nerve cells progressively die. Peripheral vision deteriorates first, then central vision. If untreated, it can cause blindness. It is detected by a simple, non-invasive, pressure test, as done in **routine eye screenings**. These are strongly advised as they can pick up a number of medical problems, including non-sight-related ones, such as the retinal blood vessels showing up circulation defects. Glaucoma affects 600,000 people in Britain, mainly the over-40s.

There are four types of glaucoma. The commonest by far is **chronic angle glaucoma**, affecting about 1 in 75 people over 40. Nearly 1 in 4 of these cases is undetected, with progressive damage to the optic nerve, giving blindness within 25 years if untreated. The small fluid-outlet channels in the eye narrow with time in this form of glaucoma, with no obvious symptoms, so the progressive loss of peripheral vision is unnoticed perhaps for years, and is irreversible. African-Americans are affected four times as often as European-Americans. The risk to a child or brother or sister of an affected person is 1 in 10, and children in families with a history of glaucoma have twice the normal risk. Treatment by medication or surgery is effective but cannot restore vision already lost. Laser treatment of the drainage areas may be sufficient, or surgery can create new drainage channels.

People's **choices of glasses** — frames and lenses — show incredible diversity. If one's vision is not good enough, does one struggle on without glasses out of vanity, or go for glasses or contact lenses? Does one need only a small, discreet pair of reading glasses, to be surreptitiously fished out when required? The need for them could be a sign of ageing.

15.6. The nose and sense of smell

The nose functions in **breathing** and in our **sense of smell**. Our **olfactory epithelium** is inside the nose, on the roof of the nasal cavity, near the brain, communicating directly with the brain's **olfactory bulb**. The epithelium on each side of the central nasal septum is about one third of an inch square (2.5 cm^2). Each of these patches has about 16 million sensory cells, with a life of about 30 days each. There is a slight **decline in the sense of smell with age**, about one per cent of sensory cells not being replaced each year. Practice and close attention to smells can overcome much of this decline. I am a National Wine Judge and was worried when I read that after age 55 the sense of smell declines more rapidly, but having won blind wine-tasting competitions in my sixties and seventies, I am less worried. See Chapter 13.2 for the role of skin in smell. For a detailed account of the mechanisms of smell, the role of smell in human evolution, in courtship, and in social functions at different periods, see Stoddart (2015).

It is amazing how different people are in their **sensitivities to particular odours**. About 10% of the population cannot taste or smell a flavour and odour called 'mouse' from bacterial spoilage in wines, while others find it obvious and revolting. The ability to detect it may depend on the pH (acid/alkali balance) of one's saliva. We can detect some chemicals with extremely high sensitivity, even at 1 part per billion for some smells. We can each recognise more than 10,000 smells, although our vocabulary to describe them is poor. Smells can be evocative and play on our emotions, which is why perfumery is such a huge industry.

According to wine-writer Victoria Moore (*Telegraph Weekend*, 29/3/2014), we each have about **400 different smell-receptor-types**. Because of variability in the 800 different odorant-receptor genes, everyone is unable to smell specific scents. For example, 40% of the population cannot smell violets because the molecule responsible is only received by one receptor, *OR5A1*.

A friend has **anosmia**, with almost no sense of smell. Another is strongly interested in food and drink but lost her sense of smell, which in turn affects taste. She had several nose operations but none restored her sense of smell. When I had a really bad cold, I could not even smell coffee. Human diversity includes **hyposmia**, a reduced sense of smell, **phantosmia** giving olfactory hallucinations and **parosmia** giving a distorted sense of smell, so that roast chicken might smell of rank sewage and seem inedible.

A straight nose is sometimes called 'Greek', while a 'Roman' nose is convex, curving down to the tip, while 'turned up' (*retroussé*) noses are concave. 'Snub noses' are short and turned up. Semitic noses are often convex, almost hawk-like. An old joke deals with the question as to whether there is a 'Jewish nose': *And God said unto Moses, "All men shall have long noses, excepting Aaron, who shall have a square'un."* The Chinese and Japanese, with short noses, sometimes refer to Europeans as 'big noses' or 'long noses'.

The **tips of noses** may be sharp, broad, bulbous, furrowed, rounded, square, turned-up, etc. **Nostrils** may be pear-shaped, slit-like or circular. The **nasal septum** divides the two nostrils from each other, with the **philtrum** being the double-ridged structure going from the centre of the nose to the upper lip, often getting red when one has a cold. The sides of the lower edge of the nose are the **wings**, which may be broad and flaring as in many Negroids, or narrow.

15.7. Jaws, teeth, lips, tongues, chins, taste

The upper **teeth** usually extend beyond the lower teeth, but some people have the lower jaw longer than the upper one. In the Spanish Hapsburgs, the latter condition persisted for six centuries, giving a protruding Hapsburg lip.

There is much visible diversity in teeth size, colour, spacing, regularity, position, the relative size of incisors, canines, premolars and molars, the amount and colour (grey, white or gold) of visible fillings, etc. Teeth are important to one's appearance, with much money spent on cosmetic dentistry. Teeth are vital in eating, which can be difficult if teeth are misplaced so as to impair the **effectiveness of biting and chewing**. Helen had her back teeth meeting too soon when she bit, so that her upper and lower front teeth could not meet to give a proper bite (Photo 15.1). She found it very difficult to bite off bits of meat, so she was vegetarian and only ate foods her teeth could cope with.

Many children, especially teenage girls, wear **braces on their teeth** (Photo 15.2) to improve alignment and regularity, more for appearance

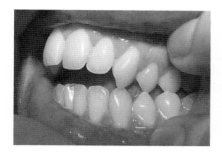
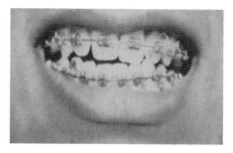

Photo 15.1. Helen, Front teeth not meeting.

Photo 15.2. Nuzreth, Braces to straighten teeth.

than function. A tendency to **tooth decay** has a genetic component but what people eat and how well and often they clean their teeth are more important.

Montagu (1963) quotes data on diversity in the **upper lateral incisor teeth**, the teeth next to the central pair in the upper jaw, normally with one lateral incisor on each side. In Whites, these teeth are missing in 1 in 40 people, much reduced in size in 1 in 40, slightly reduced in 1 in 6, rotated in 1 in 25, crowded in 1 in 13 and duplicated in 1 in 250. Compared with Whites, the Chinese and Japanese have fewer missing lateral incisors (1 in 667 and 1 in 91, respectively) but degenerate ones are more common, 1 in 14 and 1 in 21. Missing ones are rare in Negroids.

There are many abnormalities of the 20 milk teeth and the 32 mature teeth, including persistence of milk teeth, and variation in whether the wisdom teeth penetrate the gum. The normal adult set is 8 incisors (front teeth, wide, flat, biting tops), 4 canines (pointed tops), 16 molars and pre-molars (with grooved, broad, chewing tops), with 4 extra molars (wisdom teeth). Here are some irregularities reported by students:

- *I have a third set of teeth, growing after my permanent teeth. They were removed at age 13/14 as they were pushing out my main set — no one else has this in the family, and there is one more coming through now, in mid-gum.*
- *I had an extra tooth (fang).*
- *One adult tooth is missing (I still have baby tooth but there's nothing under it).*

- *Missing eye teeth.*
- *Hereditary lack of eye teeth (just never had them!).*

Ashik Gavai in Mumbai was 17, with a painful swelling on his right lower jaw and dentists chiselled out **232 small pearl-like teeth**. His family was relieved that it was not a tumour.

Bad teeth were the prime cause of admission of young people to hospital in Britain in 2013–14. There were 25,812 admissions for bad teeth in five-to-nine-year-olds (tonsillitis came second). Some needed all their first teeth removed, while removal of four to eight teeth was most common. Sugary and fizzy drinks and fruit juices were blamed.

Missing or extracted teeth used to be replaced by dentures or bridges. Denture plates often caused embarrassment by becoming loose and by having to be taken out at night. After year 2000, **dental implants** became widely available at a cost of about £2,500 a tooth. A rod of titanium is inserted into the jawbone. When the cuts have healed and the implant has settled in, which can take several months, a porcelain crown is fitted onto the rod.

Tooth whitening is very common for actors, models, celebrities and some politicians. **Bleaching agents** such as hydrogen peroxide are put in fitted trays, with several sessions required. The brightest white teeth are obtained by sticking on **porcelain veneers** after removing some enamel. Veneers are expensive but can cover gaps and irregularities. Genes determine one's basic tooth colour, although red wine, betel chews, tea, coffee and other products can stain teeth. Smoking or chewing tobacco can make teeth yellow or brown. Race does not affect tooth colour but darker skins make teeth look whiter (Plate 4C). Most people have ivory-coloured teeth, although dentists recognise 28 colours.

About 65% of the population can **roll their tongue**, bringing the sides of the tongue up to make a U-shape, and 35% cannot. The ability is conferred by a dominant gene. The tongue is very important in eating, to manipulate and compress the food ball before it is swallowed, and in speech. Tongue defects affect eating and speech.

The **chin**, a protruding part of the lower jaw, may be prominent (genetically dominant), extending beyond the vertical plane of the face, or normal, roughly in the plane of the face (probably dominant), or receding (genetically recessive). Narrow chins are recessive to broad. In fiction, the hero is often portrayed as having a strong, determined, jutting chin, while diffident, weak characters are ascribed receding chins, but this link between chin and character has not been established. A chin may be rounded, dimpled or have a definite cleft, depending on the shape of the ends of the lower jaws.

What we think of as our **sense of taste** is a combination of perception by taste buds in the mouth and the detection of smells by the olfactory area in the nose (see above). As our breath goes from and to our nose to and from the lungs, it passes the back of the mouth. By this channel, odours get from food and drink to the olfactory epithelium. We have about 9,000 taste buds in the mouth, nearly all on the tongue but also on the rear of the mouth's roof and at the back of the throat. The five major tastes we detect are sweetness, sourness, saltiness, bitterness and meaty umami.

The **number of taste buds** on the tongue varies from about 5,000 to about 12,000. Taste buds are too small to see, about 1/500th inch (0.05 mm) high and wide, but one can see small raised papillae. The taste buds occur on the sides and near the base of those papillae.

Each taste bud has 25 to 40 tasting cells, the hairy tips of which do the taste sensing, sending nerve impulses to the brain. We are particularly **sensitive** to bitter tastes and can taste the bitter compound quinine (used in some tonic drinks and as an antimalarial) at about 1 part in 2 million. By trying a group of people with very dilute solutions of sugar, salt, citric acid (or lemon juice) and quinine, one finds amazing **differences in sensitivity** to particular tastes. Our sense of taste helps us to choose good nutritive foods (e.g., sweet ripe fruits, iron- and protein-packed meat) and to avoid poisonous berries and other foods (many have bitter alkaloids), sour under-ripe fruits, or food which could be poisonous because of bacterial decay such as smelly, off-tasting rotting meat or fish.

15.8. The ears and hearing problems

The visible **external ear** is a sound-gathering device. It shows huge differences even within families, with continuous variation in shape and size; large ears are usually incompletely dominant to small. Some people can waggle their ears, together or independently. Humans can **locate the direction of sound** approximately. They cannot work out its distance, unlike with binocular vision.

Nearly all ears are **rolled** at the top but the rolling may stop before half way down, or go most or all of the way down. **Earlobes**, at the bottom of the ear, may join the side of the head at a wide variety of angles. The lobe may hang down then rise to join the head, in the free lobe, controlled by a dominant gene. It may join the head at right angles, in the attached lobe, or may run downwards (a decurrent lobe) on the head, with little true fleshy lobe.

The ear may be **asymmetrical**. A friend has one decurrent lobe and one hanging lobe. Plate 19C, D, shows a man with one 'pixie' ear, pointed at the top, and one rounded. An ear, nose and throat surgeon told me that her commonest ear operations were removing parts of the outer ear in cases of **cancer**, and operating to correct '**jug ears**' ('bat ears'), where they stick out. She could remove up to one third of an outer ear without most people realising that that person then had ears of different sizes, because they did not see the two ears in side view simultaneously.

Females and some males have **pierced** (Plate 19B) **or multiply pierced ears**, with piercings through different parts including the lobe, the rim and the tragus (the small prominence in front of the entrance to the ear). A foreign woman had 14 piercings in each ear, each piercing filled with a jewel or a gold ring of increasing size going down the ear. She also had a piercing and ring in one eyebrow, a piercing and gold object on each side of the mouth, and another piercing and silver object below her lower lip. The objects near the mouth must get in way during kissing, unless removed.

Ear deformities are frequent, with parts missing, shortened or asymmetric from birth, accidents or sports, e.g., boxers' thick 'cabbage ear'. One strange phenomenon is the occurrence of holes (preauricular ear pits or sinuses) in front of the ears (Plate 19G). In one intake of 110 students, two females showed those holes on both sides of their head, but had never noticed them on each other. This is how a student described it in a friend's family:

> *His family has a small red hole in front of the ear, of unknown depth — on his half-sister, stepfather and stepfather's mother — it's been in the family a long time. It is always on both sides of the head and sometimes has pus. Someone unrelated at school has it too, and so has one of the second year students.*

The external ear canal, about an inch (2.5 cm) long, leads to the **ear drum**, a thin, flexible membrane about the size of a front tooth. Three small ear bones transmit vibrations to the spiral, snail-like, cochlea of the inner ear, and amplify the signal about 22 times. Those vibrations are converted to electrical nerve signals in the Organ of Corti. The cochlea is about ¼ inch (0.6 cm) wide and 2/5 inch (1 cm) high. We hear sounds in the loudness range of 1 to 120 decibels, with the latter painfully loud.

A common cause of impaired hearing, especially in older people, is **excess wax** in the external ear canal, deadening sound reception. This is easily solved by softening the wax over a few days with a proprietary fluid or warmed olive oil, followed by syringing with warm water. The restoration of full hearing by such a simple procedure seems little short of the miraculous to those affected.

When young, we can usually hear **sound waves** ranging in frequency from 30 Hertz (waves per second) to 16,000 Hertz, but high-frequency sound perception reduces with age. **The elderly** often hear nothing above 8,000 Hertz. Even musicians with sensitive hearing are restricted to 20 to 20,000 Hertz. The range we hear best is 30 to 3,000 Hertz. The singing voice ranges from about 60 Hertz to 1,000 Hertz, while the piano goes from about 30 Hertz to 5,000 Hertz.

Congenital deafness affects 1 in 1,000 babies and is genetic in half those cases, usually from autosomal recessive genes. The risk of a second affected child is 1 in 6, and the risk in a child of a congenitally deaf parent married to a hearing spouse is 1 in 20. **Joanne Milne** was born deaf. In 2014, at 40, she received **cochlear implants** for each ear. These small electronic devices send impulses to the auditory nerve. When Joanne's implants were switched on, she wept with emotion on hearing sounds for the first time. She was reported to be 'drunk on birdsong and traffic and the sound of lights being switched on', and to be delighted at hearing music and human voices. A further advance in 2014 was the use by an Australian team of cochlear implants plus gene therapy to regrow some of the surviving auditory nerves. Cochlear implants alone give enough hearing for understanding speech but not for musical appreciation.

The most frequent cause of American children visiting paediatricians, with 25 million visits a year, is **fluid in the middle ear**, sometimes with pain or fever. The fluid may be thin and watery, but in **glue ear** it is thick. The cause may be ear infection, or malfunction of the Eustachian tube, which drains the middle ear into the back of the throat. That tube allows the pressure to be equalised between the outside air and inside the ear (behind the ear drum), as is necessary when swallowing or in changes of altitude or air pressure. When that tube is blocked even partly, fluid can accumulate and bacteria can multiply.

Allergies, pollutants, cigarette smoke, infections (including colds) can all cause **inflammation and swelling of the Eustachian tube** and hence fluid in the middle ear. Problems are most frequent in young children. The symptoms may be mild or acute, with some loss of hearing and a feeling of fullness in the ear. The ear drum may be red and painful from infection. The trouble often clears spontaneously within weeks, or antibiotics may be tried. If hearing loss is significant after three months, insertion of pressure-relieving **grommets** may be used.

In Britain, about 20 million people are deaf or hard of hearing, with **age-related hearing loss** affecting about 1 in 4 of those aged 65 to 75, and about three-quarters of those over 75. It is progressive, starting with high-frequency sounds and can be socially isolating. Even if affected people can hear if one person is speaking to them, understanding speech often becomes difficult when there is background noise, such as others talking or music playing. Hearing aids help to some extent but magnify background noise.

There are many causes of **complete or partial deafness**. Minor losses of the ability to hear high-frequency sounds occur in most people from 20 onwards. Some definite loss of hearing due to **nerve deafness** affects 1 out of 5 people by 55; it increases in occurrence and severity thereafter. Nerve deafness is caused by declining nerve function in the ear, and is irreversible. **Conductive deafness** is caused by damage to the ear drum or to the three sound-conducting bones, or by problems with ear fluid, and may be treatable.

Mishearings can be bad or amusing. In Cuba in 1963, one of Castro's aides asked, "Is there an economist in the room?" Che Guevara misheard the word as 'communist', raised his hand, and was promptly made governor of the Bank of Cuba!

Infections with mumps, measles, meningitis and other diseases can impair hearing, temporarily or permanently. One source of congenital deafness is **German measles (rubella)** (Chapter 11.1). Prolonged exposure to loud noises causes nerve damage and hearing loss; health and safety rules are now much stricter about ear protection. Drills, guns, bombs and other explosions may cause deafness. Most of Nelson's sailors on HMS Victory must have been deaf from their own cannon fire. Instructions were given them by a penetrating whistle. Some **medications**, including aspirin, chloroquin and some antibiotics, can impair hearing in sensitive people.

When 53, French chef Jean-Christophe Novelli disclosed that he had age-related and noise-related severe hearing loss. He blamed this on 30 years of working in noisy kitchens, with the clanging of stainless steel, the clatter of washing up and the stream of yelled instructions. "Chefs work in very noisy environments," he said. "You don't speak, you scream." A hearing aid transformed his life.

Tinnitus, a persistent ringing sound in the ears, develops spontaneously, especially with age, is common and difficult to treat. The apparent sounds vary and can be loud or soft, continuous or intermittent. They include ringing, whistling, buzzing, humming and roaring. If moderate or severe, it can affect hearing, concentration, sleep and temper. Hearing aids to increase external sounds may help, and some sufferers use low-level sounds, say of music or a clock ticking, or directly in the ear from a 'tinnitus suppressor', to mask the sounds.

15.9. Hair

Our hair shows much diversity (Plate 13 and others) in **colour, length, thickness and condition**, and in the **amount** of head hair, from a full head to gleaming total baldness, or a close, all-over crop (Plate 17B).

Our hair offers us many **choices in controlling its appearance**, by cutting, bleaching or dying it with any of the huge range of colours available (black, chestnut, strawberry blonde, green, red, mauve, blue, striped (Plate 13F), etc.), or with highlights, lowlights or streaks, and **how we arrange it**. It can be plaited, fringed, pony-tailed, Mohican styled with a central crest (Plate 13E), dreadlocked, centre-parted, gelled into spikes (Plate 13C), brushed upright or in many other ways. It varies in **waviness**, from dead straight to wavy (Plate 7D), curly (Plate 13B) and crinkled (Plate 5A).

There is some control by hair-straightening, waving or curling, including 'permanent waving' which lasts a few weeks. Straight hair is round in cross-section while naturally curly hair has a much flatter cross-section.

Afro-Caribbean hair (Plates 4C, 5A) is often woolly, kinky or very tightly curled, and can get frizzy and tangled, breaking easily. Chemical straightening is common, as is setting wet hair on large rollers to dry under tension, or hot metal combs may be passed repeatedly through the tight curls. Hair braiding (Plate 6I) and hair extensions are popular.

We have **two main types** of hair, the short, fine, often light-coloured **vellus hairs** over most of our body's surface, and longer, thicker **terminal hairs** on the scalp, in eyebrows, eyelashes, genital area and armpits. At puberty, males develop beard and moustache hairs and usually have hairier legs and bodies than do women. Our external surfaces lacking hair are the soles, palms, nipples, lips and the end of the penis.

People with a full head of hair have about 100,000 terminal hairs on the scalp, with about 70 falling out and replaced each day. Head hairs typically grow about half an inch (12 mm) a month, 5.7 inches (14.5 cm) a year. Some hairs will be in the resting, mature stage and others will be growing, so one or two months after a haircut, the ends will be of uneven length.

Each terminal hair **shaft** consists of an inner cortex of keratinocyte cells and contains melanin granules. The surrounding cuticle consists of a layer of thin cells also containing keratin. **Keratinocyte cells** produce large amounts of the protein keratin which makes up most of the epidermis and nails. The cuticle cells overlap each other, giving flexibility and some waterproofing. One reason why dyed hair looks different from naturally pigmented hair is that dyes often colour the outer part, while natural pigments are in the centre.

Each hair grows from a **follicle** in the dermis (Chapter 13.2). The follicle has a blood supply for nutrition and sensory nerve endings, plus a sebaceous gland producing oily sebum which lubricates the hair. The amount of sebum determines whether someone's hair is greasy, medium or dry. On the scalp, hairs usually grow for two to five years, followed by a resting period of about two weeks when a new hair bulb is often formed. The old hair is eventually shed.

In a fatty lower layer of skin, the hypodermis, there is a small involuntary muscle, the **arrector pili**, attached to the bottom of the follicle. When we are cold, or occasionally when we are frightened, these muscles contract, raising the hair and a small bump of skin, the **goose pimple**, which helps to reduce heat loss.

Hair colour (Plates 2–20) is controlled by at least four different genes (loci) affecting melanin production in the follicles' melanocytes. Eumelanin gives shades of brown and black, and phaomelanin gives yellows, browns and reds. True blondes have almost no hair pigments (there is no white pigment), while combinations of these two types of pigment give all the other colours in different shades and intensities of yellow, brown, red and black, with black predominating in Africa and Asia.

Hair colour can **change with age.** I had blond hair until I was three, brown hair for most of my childhood and adolescence, then my hair started going white from about age 25 (Plates 20B, 15C, 14G). My mother, brother and sister also went white early, showing a familial genetic condition. Some children with red or sandy hair develop darker shades such as brown before they are adults. With age, different hairs start showing different amount of pigment, with progressive increases in the number of white hairs to give an overall grey effect, with a mixture of white and dark hairs ('salt and pepper'), not uniform greyness. There are genetic controls of the age at which greying begins, with most people starting to go grey in their mid-30s to mid-40s. Africans tend to go grey later than Europeans.

Hair colour is important to **people's self-image.** Many women hate the ageing implications of white or grey hair, and regularly dye it. Most of my female friends in Sri Lanka who start to go grey dye their hair black. People often do not use mirrors to look at the top of their heads, so pale hair at the roots of dyed hair often goes unnoticed by its producer.

More than half the women in Britain **colour their hair** at home, in addition to those with professional colouring. Colour treatments often have adverse effects on hair condition, damaging and weakening the cuticle, giving split ends and hair breakage. Hair colouring agents have been used for thousands of years, including henna (red), walnut (yellow-brown) and indigo (blue-black, often mixed with henna).

Age-associated hair loss is mainly genetically controlled, although diet, climate and psychological factors such as stress can have effects. It affects men more than women because it requires testosterone, which women usually produce only in small amounts. Pattern baldness affects individuals in a diversity of ways. One type of **male pattern baldness** is hair loss starting at the temples (Plate 7D) and progressing over the top of the head, merging with the bald patch on the back of the head, but often leaving the lower back and sides of the head with hair. In females, there can be thinning over the whole of the scalp, with the loss of many hairs and the remaining ones being thin and straggly.

Hair loss is individual in timing, form and psychological effects. It can lead to serious loss of self-esteem, with feelings of looking old and unattractive. Wigs (toupees) are one solution and are made of human hair, which tends to lose colour with time, of silk, or of man-made fibres. Some kinds of hair loss can be slowed by application of Minoxidil to the scalp, or by taking Finasteride tablets. Some treatments work by combating testosterone in the scalp. Hair transplants put small areas of skin with hair from the lower back of the head on to the top and front areas. In men, testosterone and the appropriate genes can cause **excessive nasal hair and ear hair growth** (Plate 19F).

References

Montagu, A. *Human Heredity, Second Revised Edition.* (1963) Signet, New York.
Stoddart, M., *Adam's Nose, and the Making of Humankind.* (2015) Imperial College Press, London.

Chapter 16

The Heart; Heart Attacks, Strokes, High Blood Pressure

16.1. Introduction

Even for the heart, there is diversity in disease and placement. In Britain, early deaths from heart disease amongst Indians, Bangladeshis, Pakistanis and Sri Lankans are 50% higher than average, and they suffer more fatal strokes. Surprisingly, children with heart defects may be long lived. For example, music composer Roy Douglas had recurrent heart troubles when young, but lived to be 107. Kirsty Howard was born in England with her **heart back to front** and all her organs misplaced. She lived. While having a heart back to front is extremely rare, having the **heart on the right-hand** side occurs in 1 in 20,000 births. Sometimes other organs are on the wrong side as well. Disorders include almost **random arrangements** of the heart, lungs, liver, spleen and stomach.

Charli Southern was born in Australia in 2004, seven weeks prematurely, with her **heart exposed** where the chest failed to close. At six weeks she survived an operation putting her heart in the right place, needing later surgery to reconstruct her breast bone.

Naseem Hasni was born by Caesarean section in Florida in 2006 with his **heart protruding** from his chest with nearly all of it visible, beating normally. Surgeons wrapped it in Gore-Tex, used some of his skin to form a membrane around it, then eased it into his chest. Protective plastic was fitted over the gap. Later there was grafting of rib pieces to form a breastbone.

At rest, the heart pumps about 12,670 pints (7,200 litres) of blood a day, more under exercise. Blood goes through arteries (Photo 16.1), arterioles and capillaries, then back through venules and veins to the heart, at a rate of 8.8 pints (5 litres) per minute at rest. Its speed of flow is highest in the aorta, 20 inches per second, 1.1 miles an hour (50 cm per second; 1.8 kmph), and slowest through capillaries, one thousandth of that. The largest blood vessels have a diameter of 1 to 1.2 inches (2.5 to 3 cm), 3,000 times that of the smallest capillary, 1/3,200 inch (0.007 mm). About 75% of blood is in veins, 20% in arteries and 5% in capillaries.

About 35 pints (20 litres) of fluid a day get out of the capillaries and bathe the tissues; about 30 pints (17 litres) are reabsorbed into capillaries. The five pints (three litres) not reabsorbed are returned to the heart by the **lymphatic system**. This is an extensive system of collection vessels, with small lymphatic capillaries penetrating the tissues, then forming larger lymph vessels, eventually draining into the vein into the right atrium. **Lymph** does not carry oxygen, is a very pale yellow and tastes salty.

The lymphatic system has **many valves**, allowing flows only towards the heart. The lymph vessels get squeezed by surrounding muscles, and distended lymph vessels trigger their own contraction. Openings into the lymphatic system are bigger than those into blood capillaries, so large molecules such as blood plasma proteins and even whole bacteria or viruses can enter. It would be disastrous if harmful microbes could use lymph openings to get into the blood but lymph passes through **lymph nodes** where special cells can destroy micro-organisms. The lymphatic system has colourless **lymphocytes** for immune defence.

Lymphocytes come from precursor cells in red bone marrow, but most new ones are produced by division of existing ones in lymphoid tissues such as the lymph nodes, thymus gland, tonsils and spleen. Lymphocytes circulate in lymph and blood and go into the tissues. An adult has about two thousand thousand million lymphocytes.

16.2. The heart and circulatory system; congenital heart diseases

The **heart** is about the size of a fist, under the breastbone, weighing from about 8 to 14 oz (250 to 390 g). It is about 4¾ by 3½ inches (12 by 9 cm), forcing blood into a network of blood vessels about 60,000 miles (97,000 km) to 100,000 miles (160,000 km) long, enough to circle the world five times. The heart generates enough force to propel a fountain of blood six feet (1.83 m) into the air. In a typical lifetime, it beats about two and half billion times without external maintenance, lubrication or the need to repair valves which open and close 4,000 times an hour.

The heart is a **double pump**, with non-return valves to ensure a one-way flow of blood and to allow a build-up of blood pressure to send the blood round two independent pathways. It has four chambers, with a thick muscular septum between left and right sides. As Valentine cards testify, the heart is often considered the seat of the emotions, especially love, but people with artificial hearts still feel a full range of emotions.

Dark purple-brown blood, low in oxygen and with increased carbon dioxide, returns from the head into the superior (meaning above) vena cava, and from the body into the inferior (down below, not substandard) vena cava into the right atrium, thence through the right atrio-ventricular valve into the bottom right chamber, the **right ventricle**. This muscular chamber pumps blood at a rate of about eight and a half pints (five litres) a minute through the non-return pulmonary valve into the low-resistance system of blood capillaries in **the lungs** where carbon dioxide is excreted and oxygen is absorbed into haemoglobin.

Bright red oxygenated blood returns from the lungs to the left atrium at the top of the heart, then passes through the left atrio-ventricular valve into the most muscular chamber, the **left ventricle**. This pushes blood through the aortic valve into the aorta on top of the heart and out into the **head, trunk and limbs**. Although the left ventricle pumps blood more powerfully than the right one, it pumps the same volume.

The heart-beat cycle consists of **systole** (contraction and emptying) and **diastole** (relaxation and filling); the ventricles fill and expand as the atria contract. A typical resting rate is **65 to 75 beats a minute**, compared with 9 for the grey whale and 1,200 for a hovering humming bird.

The **pressure on contraction** in the left ventricle is about 120 mm of mercury (Hg), compared with only 20 mm in the right ventricle which pumps blood to the lungs. (1 millimetre of mercury = 133.3 pascals, or 1/760 of standard atmospheric pressure. 1 atmosphere = 14.7 pounds per square inch.) Each heart beat pumps about 70 ml of blood from the right

Table 16.1.   The distribution of blood at rest and under moderate exercise, from the left side of the heart.

Body part	Percentage of flow received at rest; total flow, 5 litres per minute	Percentage of flow received during exercise; total flow, 12.5 litres per minute	Change in total flow volume with exercise, %
Brain	13%	5%	None
Skin	9%	14%	up 370%
Heart muscle	3%	4.4%	up 367%
Skeletal muscles	15%	64%	up 1066%
Digestive system	21%	3.7%	down 56%
Liver* (from heart, not guts)	6%	1%	down 56%
Kidneys	20%	4.4%	down 45%
Bone	5%	1.4%	down 30%
Other	8%	2.2%	down 30%

*The liver receives blood directly through the hepatic artery, and from the digestive system through the hepatic portal vein.

side and 70 ml from the left, averaging 4,627,893 pints (2.6 million litres) a year. Fortunately, our heart muscles never tire.

With hard exercises, the amount pumped goes up from 8½ pints (5 litres) to 35 to 44 pints (20 to 25 litres) a minute, with athletes sometimes reaching 70 pints (40 litres). The two ways of increasing heart output are raising the heart rate and the amount of blood pumped per beat. The proportion of blood going to different parts of the body changes from rest to exercise, as shown in Table 16.1.

At rest, the digestive tract and kidneys receive about one fifth of the heart's output each, with the brain and skeletal muscles receiving about 14% each. Under moderate exercise, the skeletal muscles (e.g., in arms, legs, chest and abdomen), receive a massive 64% of the output, supplying them with oxygen and nutrients and removing carbon dioxide. The skin's share increases to 14%, permitting heat energy from aerobic respiration to be dissipated. The heart's muscles get a larger increase, nearly 370%. Organs less involved in exercise get a decreased proportion, including the digestive tract, liver, kidneys and bones.

The right-hand column of Table 16.1 is for the **absolute change in blood flow** during exercise, not relative to other organs. Thus the **blood flow to the brain** is about 650 ml a minute at rest and under exercise, hence the right-hand column records no change, while the **brain's proportion** goes down from 13% to 5% during moderate exercise.

There are **congenital heart disorders** with a combined frequency of about 1 in 50 live births, some resulting in early death. They include failure to close the foetal artery from the pulmonary artery to the aortic artery, defective heart valves, holes in the heart's interior or exterior walls, problems with heart-beat regulation, an enlarged heart, and rheumatic heart disease.

The **main heart problems in newborn babies** result from the **failure of gaps in the foetal heart to close**, so some deoxygenated blood mixes with oxygenated blood and is sent round the body. Affected babies often have a bluish tinge. In the foetus, the oxygen comes via the placenta, with the lungs non-functional until birth. The foetus has two devices keeping blood from being pumped into its unexpanded lungs. The **oval window** allows blood to pass from the right atrium to the left atrium, and blood from the pulmonary artery passes through the **ductus arteriosus** into the aorta. In **patent ductus arteriosus**, the ductus arteriosus does not close fully after birth. Failure to close the oval window results in an **atrial septal defect**. The commonest heart defect in babies is the **hole in the heart**, where the partition between the ventricles has not grown completely, allowing deoxygenated blood to pass from the right ventricle to the left ventricle and hence be pumped around the body, leaving it deficient in oxygen. If any of these defects is minor, the decision might be taken to see if it cures itself. All are treatable by surgery.

16.3. Heart attacks, artery diseases, strokes, TIAs, heart murmurs and deep-vein thrombosis

Disorders of the heart and circulatory systems are widespread, especially in older people. In Britain, about 800 people a day die of heart disease, accounting for 30% of all deaths of those under the age of 65, especially diabetics. Seven million Britons are prescribed **statins** to reduce cholesterol levels and the chance of heart attacks and strokes, especially in the elderly. The **mass prescription of statins**, which are cheap, is controversial. Some doctors want everyone over 50 to take statins, to reduce the 270,000 heart attacks a year in Britain, half of them fatal.

In 2002, cardiovascular diseases were the leading cause of death in America: more than 61 million Americans had them, including high blood pressure, coronary heart disease, stroke and congestive heart failure. High blood pressure affects 50 million Americans, and nearly 5 million have had a stroke. In Britain, there are more than 100,000 cases of stroke a year, with 60,000 deaths.

Heart problems are usually diagnosed by a doctor listening to the heart with a stethoscope, or from electrical wave tracings from an electro-cardiogram (ECG) from electrodes attached around the heart region and to arms and legs, or diagnosis is from chest pains (angina) or difficulty with activities such as climbing stairs.

The **normal heart rate** of 60 to 100 beats per minute is controlled by the **sinoatrial node** in the right atrium. If that fails, or if nerves from it to the rest of the heart are damaged causing a **heart block**, there are other parts which can initiate heart beats, but at a lower rate. If the rate falls below 30 beats a minute, sufferers can lead only a sedentary life or may become comatose. A battery-driven **artificial pacemaker** can be inserted to stimulate the heart electrically to beat at a more normal rate, when such people can lead active lives. The pacemaker is not a pump but is an electronic device which initiates and regulates heart contractions. More than 40,000 patients in England had pacemakers fitted in 2012–13.

Abnormalities of heart rate include going **too fast**, over 100 beats a minute at rest (**tachycardia**), or **too slowly**, fewer than 60 beats a minute (**bradycardia**). People with anxiety or various medical conditions may have faster rates, while athletes usually have a slower rate.

The ECG can detect **abnormalities in rhythm**. In complete heart block when the nerve fibres from the sinoatrial node to the ventricles are damaged, the atria contract at the normal rate of 70 to 80 beats per minute, but the ventricles, using their own fallback beat-initiation systems, contract at only 30 beats a minute or fewer. This causes many problems, reducing heart efficiency by at least a quarter.

In 2001, Adam Griffin, born five weeks prematurely in England, had potentially lethal tachycardia from six days after birth. His heart beats sometimes rose to the dangerous rate of 300 a minute. To prevent damage, doctors gave him an injection to stop his heart, then they restarted it at a more normal rhythm of 90 beats a minute. They had to do this more than 25 times in his first 15 weeks since birth and hoped that the attacks would stop as Adam grew older. In terms of heart beats as indicators of life, he had been 'killed' and brought back to life many times in only four months.

A student had a low heart rate, about 45 beats a minute, but was able to participate in "all high thrill action sports", calling himself "a total adrenaline junkie".

An **inadequate blood supply to the heart** is **myocardial ischaemia** (Greek *mys, myos,* muscle; *kardia,* heart; *ischein,* to restrict; *haima,* blood). A **heart attack**, an **acute myocardial infarction**, is due to the death of some heart muscles from a blockage of their blood supply.

The heart requires blood from its small **coronary arteries** (Photo 16.1) to bring oxygen and nutrients, and **coronary veins** to take away wastes. Their health is crucial. With some blockage, the blood supply for the heart muscles may be sufficient at rest but insufficient during exercise, causing heart failure then.

The most frequent cause of death in America is **coronary artery disease**. An inadequate blood supply to the heart has three main causes: vascular spasm, and plaques or clots in blood vessels. In **vascular spasm**, a sudden constriction occurs in the heart's blood vessels, caused by factors

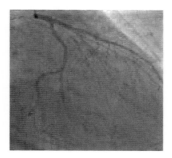

Photo 16.1. Coronary arteries seen in the coronary angiogram.

such as anxiety, physical exertion or exposure to cold. This restricts the blood supply temporarily and does not often do much damage.

Atherosclerotic plaques start as small non-cancerous tumours of blood vessels' muscles. As the tumours enlarge, cholesterol and other fat-derived chemicals accumulate there, producing a growing plaque in the blood vessel, the walls of which are normally elastic and can expand in the region of the plaque, allowing an almost normal flow. As the condition progresses, calcium salts are deposited on the plaque and surrounding blood vessel walls, making them inelastic, causing '**hardening of the arteries**'. They become less able to distend, so the plaques are more serious. An **atheroma** is a cyst with porridge-like contents.

Personal account. *I had worrying heart investigations*

I did not think I had heart problems but during a routine medical examination I had a test on an exercise bicycle while attached to various electrical leads. The technician asked me to stop pedalling and called the doctor. She said that one **electrical trace showed a heart problem**. *I was given aspirin to thin my blood and Amlodipine to lower blood pressure.*

In later tests I was taken off the treadmill very quickly as electrical traces showed the same fault. My blood pressure was deemed high. I had an **ultrasound echocardiogram** *through the chest wall, and an* **electrocardiogram**, *with lots of electrodes attached to my chest and other parts. These showed that my heart was the right size and shape, with all valves working properly.*

I was referred for a **cardio-angiogram** *(coronary angiogram) to check my heart arteries (Photo 16.1). I was warned that the procedure carried a risk of serious injury and a one-in-a-thousand risk of death. It is used to identify the location and severity of blockage of the heart arteries.*

A thin catheter was inserted in my groin into the main artery in my right leg. I was lying on an X-ray machine bed with high technology all around me. The tip of the catheter was pushed up to the **openings of the coronary arteries**. *I was worried that the consultant kept correcting the registrar operating the catheter. Radiographic-contrast iodine solution, which is easily visualized in X-ray images, was injected into each coronary artery in turn, while moving X-ray pictures were taken through my chest.*

I spent about 30 minutes on the X-ray table. After the catheter was removed, I spent several boring hours in bed, drinking five litres of water to flush the iodine out in the urine. Eventually the consultant came to tell me that my heart arteries were fine and gave me a written summary. I loved the part which said, 'Discharge status: alive'.

If there had been blockages or narrowings, treatments could have included medications, balloon angioplasty, coronary stenting, atherectomy ('roto-rooter'), or coronary artery bypass surgery. Atherectomy means using a catheter to gouge out blocking materials, while the balloon method involves inflating a small balloon to widen the relevant heart artery. The widened part could have a stent (a mesh tube) inserted to keep it open.

A close friend had a cardio-angiogram after symptoms of **angina**, *sharp chest pains when walking or carrying heavy loads. She was found to have a minor narrowing of some heart arteries. She was treated by being put on statins to reduce cholesterol, as well as low-dose aspirin to thin the blood.*

I later convinced my doctor that I did not have high blood pressure. I had had 'white coat hypertension', a raised value due to anxiety in a medical environment.

Bernard Lamb

The most drastic treatment for heart disease is a **heart transplant**, with a good rate of success, but there are insufficient donor organs available. Even with good tissue matching, recipients have to take strong immuno-suppressing drugs which make them more susceptible to infections, and organ rejection eventually starts to occur. The longest survival was nearly 31 years by Tony Huesman (1957–2009) in America.

The thumb-sized **Jarvic 2000** is a promising implanted **artificial heart** pump. The first UK patient, Peter Houghton (1938–2007), operated on at 62, lived for seven years before death. By 2001, 15 Americans had had the pump fitted to keep them alive whilst waiting for a transplant. As the heart needs much energy, artificial hearts (unlike pacemakers) cannot be powered by a chest battery and need an external electricity supply via through-the-skin cabling.

Treatments for heart artery diseases include **coronary artery bypass grafting**, bypassing a narrowed artery to restore blood flow to heart muscles. The saphenous vein from the leg or the chest's internal mammary artery or other blood vessels are used in an operation lasting three to six hours. A heart-lung machine is used to oxygenate and add nutrients to the patient's blood outside the body, returning it through the aorta. The average hospital stay is five days. About 420,000 coronary artery bypass graft operations are performed a year in America, with only 1 to 2% giving complications. The usual operation involves cutting through the breastbone. Minimally Invasive Direct Coronary Artery Bypass Grafting has been developed, with no splitting of the breastbone, no heart-lung machine and with surgery directly on the beating heart. There are more than 30,000 heart bypass operations a year in Britain.

Personal account. *I had a heart attack and a quadruple bypass*

I went to the GP with a persistent cold. She didn't prescribe antibiotics, saying that the symptoms would go away. She took my blood pressure which was moderately high (165/90). Heart medication was prescribed. I was 69, white European, moderately fit, not overweight or diabetic, an average drinker, and could easily walk 10 miles in a day. I had smoked until 11 years earlier.

While setting up scenery for a musical some 5 years after that consultation, I had to take breaks much more frequently than previously. I made an appointment to see my GP the following Tuesday. On Monday evening I had a heart attack at the theatre. Luckily a colleague who had suffered a similar event recognised my symptoms. After a long wait in a freezing ambulance while I was connected to various machines and their data radioed to a doctor somewhere, I was carted off at high speed to St. George's Hospital, Tooting, feeling increasingly very ill, vomiting, and shivering continuously. The heavy pain in my chest and back was intense. I didn't realise how serious my condition was.

At 11.30 p.m. I arrived at the cardiology department feeling really bad, still shivering violently, though by half past midnight a stent had been inserted in a completely blocked heart artery by means of a catheter in my left groin, providing massive relief, and I was installed in the Cardiac Care Unit. Over the next two days I was subjected to many tests. The good news was that there appeared to be little damage to the side of the heart affected by that blocked artery; the bad news was that three other arteries had blockages from 50% to 80%, needing bypassing.

About 2 months later I had the operation. The surgeon even bypassed the stent! I was warned that the next few weeks would be unpleasant, and they were. The first few days after the operation — wired up and plumbed in to all sorts of stuff — I felt somewhat detached, not apprehensive, and relieved when progress reports were positive. The staff attitude was a major plus and even the young tea lady cheered me up. The highlight of week one was having a shower on day 4! But realising that even after a week I could barely walk the length of the hospital corridor, and after a couple of weeks at home I could get to the end of the road but not up a slight incline to collect the newspaper, was depressing. Things did improve and now I can walk 10 miles again! I was lucky to be treated so quickly and professionally.

Kip Punch

The **Interheart study** (Rosengren *et al.*, 2004) found that 'psychosocial factors' (mainly **stress**) increased the risk of a heart attack two and a half times, while smoking increased it three times. Depression also increased it.

In the brain, atherosclerosis plaques, hardening of the arteries and blood clots are the main causes of **strokes.** Either the clot blocks an artery, starving an area of oxygen and nutrients, or the obstructed blood bursts though the vessel wall, causing a **cerebral haemorrhage.** Damage is not restricted to that area because blood-deprived areas of brain release **glutamate** which over-stimulates adjacent neurones, damaging them and causing them to release more glutamate, damaging larger areas.

The **harm done by a stroke or cerebral haemorrhage** depends on which brain areas are damaged, and how extensively. If a key area is destroyed, say the control of breathing, the stroke is fatal. Strokes are the

third most frequent cause of death in America. Damage might result in one or several functions being impaired permanently or temporarily, such as sense-perception, muscular control, feeding, mental abilities, language or speech. Loss of speech or mobility is very frustrating. Survivors may gradually regain some lost functions, especially if the brain can form new nerve pathways. New neurones are not produced, but new connections are possible.

Research from the American Heart Association showed that stroke survivors lose a month of healthy life for every **15-minute delay in receiving an anti-clotting drug.** If a clot has blocked a blood vessel, **Alteplase** can rapidly dissolve it. Ambulance staff cannot administer this as a brain scan is needed to determine whether the stroke was caused by a clot or a bleed; the latter requires a different treatment.

TIAs are **transient ischaemic attacks** (TIAs) or **'mini strokes'.** They are caused by a **temporary disruption in blood supply** to part of the brain by a plaque or a clot, resulting in a lack of oxygen there. The symptoms are similar to those of strokes. There may be a numbness or weakness in the arms and legs, and speech and visual disturbances. The difference is that — unlike with a stroke — the effects may last for only a few minutes and are usually over within 24 hours. **TIA symptoms** may include distortion of the face, including the eyes and mouth, with sufferers unable to smile or raise their arms; speech may be slurred. Emergency treatment is required.

In the heart's blood vessels, atherosclerotic plaques cause insufficient blood supply to the muscles, **myocardial ischaemia.** As the plaques get bigger and progressively impede blood flow, the heart becomes unable to get enough oxygen during exercise. This causes pain in the chest, **angina pectoris**, and often in the left shoulder and arm. Angina is most pronounced after exercise, a heavy meal or strong emotion. Typically the pain eases after a few minutes rest. There may be sweating, nausea, vomiting and shortness of breath. According to the British Heart Foundation, more than two million people suffer from angina in Britain, with 330,000 new cases a year, especially among the elderly.

Angina can be treated with drugs (such as **nitroglycerine**) which dilate the blood vessels, and by rest. Symptoms of breathlessness, chest pain, and an inability to cope with physical exertion, are initially temporary but growth of plaques in the heart's blood vessels can give complete blockage, causing a heart attack.

The atherosclerotic plaques may become the site of a **blood clot**. Platelets do not normally stick to the smooth cells lining blood vessels, but if a growing plaque breaks through the vessel lining, platelets stick to exposed collagen, eventually forming a blood clot called a **thrombus**. This may partly or wholly block the blood vessel, or it may break into the blood, when it is called an **embolus**. This may later clog a smaller blood vessel, perhaps in the heart (causing a heart attack) or the lungs (a pulmonary embolism). At first aid courses, one is taught **cardio-pulmonary resuscitation** (CPR), to press down hard near the bottom of a heart-attack sufferer's breastbone, pumping 100 times a minute.

A lack of oxygen or damage to the heart nerves may cause uncoordinated contractions in the ventricles, **ventricular fibrillation**. If detected quickly enough, it may be treated with **electrical defibrillation**, applying a strong current to the external chest wall. If successful, it restores pacemaker activity to the sinoatrial node, restoring a co-ordinated beat of different heart muscle cells. If the restart is long delayed, lack of oxygenated blood and glucose may damage the brain, heart and other organs.

Atrial fibrillation (AF) is the most common **arrhythmia** (problems with the rate or rhythm of the heartbeat). People with AF may not feel symptoms, but even then it can increase the risk of stroke five-fold. It can cause palpitations, fainting, chest pain, shortness of breath or heart failure. AF affects 500,000 people in the UK, becoming more likely with age.

An **electrical signal** spreads from the top of the heart to the bottom, causing it to contract and pump blood. The signals begin in the sinoatrial (SA) node in the right atrium and travel through the right and left atria, causing them to contract and pump blood into the ventricles. The signals then travel down to the atrioventricular node between the atria and the ventricles, then make the ventricles contract, pumping blood out of the heart.

In atrial fibrillation, the electrical signals do not begin in the SA node but come from another part of the atria or in nearby pulmonary veins. Those signals are abnormal and may spread throughout the atria in a **fast, disorganised way**, causing them to fibrillate. They flood the atrioventricular node with electrical impulses so that the ventricles also beat very quickly, but less so than the atria, resulting in an uncoordinated fast, irregular rhythm, reducing heart output.

The episodes may be short, with symptoms that come and go, or be prolonged, needing treatment. Treatment can be with drugs or by defibrillation using an electric current to the heart at a specific moment in the cardiac cycle. A modern, more permanent, treatment is ablation, passing a catheter into the heart and using heat to destroy the aberrant initiating nerves around the pulmonary veins. In 2004, the then **Prime Minister Tony Blair** had 'catheter ablation' to cure heart palpitations.

A **heart attack** may not be immediately fatal, but cause death later from complications, such as heart muscle damage or rupture, or the heart may be so damaged that it cannot pump enough blood to sustain life, with progressive heart failure. Heart attacks may be followed by an almost full recovery if undamaged muscles can grow and compensate. Some areas of heart muscle may receive blood from more than one artery, so if one is blocked, vessels from another artery may expand. Blockage of the left coronary artery is most damaging as this supplies about 85% of the heart muscle, including the left ventricle which pumps blood round the body. Treatment of damaged hearts by injecting **stem cells** to help regrow muscle appears promising.

Heart murmurs can be heard with a stethoscope through the chest or back wall. Their location indicates which of the four valves is malfunctioning, and the timing and kind of sound (whistling or swishing) indicate the type of fault.

Here are some students' statements about a diversity of heart problems.

- *My maternal grandmother has her heart on the right side of her chest, as did her father; this causes her no problems. My mother has her heart in the middle, with no complications.*
- *I have a friend who is* situs invertus *(body organs on the opposite side, e.g., heart on right side).*
- *My brother passed away at the age of 16 from a congenital heart defect.*
- *I have a mitral valve prolapse. This is not serious and affects 3% of people. No one else in the family has a heart defect. My doctor says I can climb the highest mountain.*

- *I have a heart defect — strangely shaped tricuspid valve, with only two flaps. It has been studied by echo location. I cannot do sports and will need an operation later.*
- *We have varicose veins in family (I'm going to get them young! i.e., now!).*
- *I have heart murmurs.*
- *I have high blood pressure, osteoporosis and anorexia.*
- *I have heart problems.*
- *I have low blood pressure.*

If people's blood is **too thin**, they risk excessive bleeding; if it is **too thick**, there is the risk of a stroke. The **blood-thinning drug Warfarin** is taken by 1.2 million people in Britain to reduce the risk of blood clots, strokes and heart attacks. Rob Cleaton, 27, had a leaky aortic valve replaced with a mechanical one. His immune system recognised it as alien and promoted blood clotting. It is essential to get the Warfarin dose exactly right, so every two weeks he tests his blood in a portable device. A specialist nurse advises him on adjusting the dose. Doses vary widely, from 0.5 to 20 mg, once a day.

Long-distance flights raise the risk of strokes and **deep-vein thrombosis** (DVT). Restricted movement causes blood to thicken and to pool in the legs, sometimes forming clots which can break off during or after the flight. The clot, called an **embolus**, can travel to an artery in the lungs and block blood flow, causing a fatal pulmonary embolism. Preventative measures include frequently moving around, in-seat exercises, drinking plenty of water, not drinking much alcohol, taking aspirin before flying, and wearing compression socks or stockings. In Britain, 1 in 1,000 people is affected each year, with about 33,000 deaths a year from DVT or pulmonary embolism.

Aged 24, Camilla Bull had a **leg pain** and thought that it was an aching muscle. She had been sitting extensively, writing her PhD thesis. The pain became worse and tests revealed a **deep-vein thrombosis** in her left leg. She was given an injection of heparin to dissolve the clot and then Warfarin. Tests showed that she carried the gene for Factor V Leiden (as do 1 in 20 people) which increased her risk of clots about five-fold.

16.4. High and low blood pressure; varicose veins; fainting

Pressure is measured in millimetres of mercury (mm Hg). Blood pressure is highest at the **aorta** at the exit of the heart and lowest in the two vena cavae which return blood to the heart. Blood enters the wide aorta at an average pressure of 93 mm Hg and flows into several hundred large **arteries** with very little loss of pressure in these wide tubes, e.g., the carotid arteries to the head, or the brachial artery in the arm, where blood pressure is usually measured above the elbow.

The main **pressure drop** occurs as these main arteries split into about half a million narrower **arterioles.** By the end of the arterioles, at the beginning of the ten to forty billion capillaries, blood pressure has fallen to about 37 mm Hg. The capillaries are so narrow, with an average diameter of 7 µm (one micrometre is one millionth of 39.4 inches; a human hair is about 100 µm in diameter) that even red blood cells (diameter 8 µm) have to squeeze through in single file, creating wear as they rub against the walls. Blood pressure drops in the capillaries to about 17 mm Hg.

The **control of the blood supply according to immediate needs**, such as exercise, is partly done by different degrees of **contraction** (vasoconstriction), reducing blood flow, or **expansion** (vasodilation), increasing flow, of different arterioles. For cycling, the arterioles to the legs dilate, with little increase to the arm muscles.

There are several hundred **large veins**, offering little resistance to blood flow. At rest, the easily stretched veins expand to act as **reservoirs**, holding over 60% of the blood volume, but when needed the extra blood is returned to the heart. Many large veins in the limbs lie amongst skeletal muscles; when those muscles contract, they squeeze the veins, helping to impel blood towards the heart. The big veins have non-return valves about every inch (2 to 4 cm). If a person stands still for long time, like a guardsman on parade, gravity causes blood to pool in the legs, with no leg muscle contractions to pump it back up. That reduces the blood supply to the brain and may cause **fainting**.

The legs are a special problem because in a standing adult, their veins have to support the weight of the column of blood from the foot to the heart. If valves lose their non-return function from the weight of blood and effects of prolonged standing, those veins can become distended, giving **varicose veins**. This particularly happens in people predisposed genetically to weak-walled veins, or with weak valves. Women between the ages of 30 and 60 are the most affected. Varicose veins may develop in pregnant women, where the baby puts pressure on the upper parts of leg veins.

Although varicose veins are unsightly, their main **health risk** is if the congested blood forms clots which break loose and block small blood vessels elsewhere. Varicose veins can be tied off and stripped out, or be injected with a coagulating fluid. In 2013, the National Institute for Health and Care Excellence (NICE) gave guidance on varicose veins, affecting 15% of men and 25% of women. The recommendations were injections (first choice), then laser treatment or heat therapy, before considering surgery or compression stockings.

Although **capillaries** are where the main work of blood is done (exchange of oxygen, carbon dioxide, nutrients and wastes), they only hold about 5% (9 fluid ounces, 250 ml) of the blood. They are so enormously branched that few cells are more than 0.1 mm from one, and their surface area for exchange of gases and nutrients between blood and tissues is about 6,460 square feet (0.15 acres, 600 square metres).

Blood pressure is the force applied to the walls of the arteries by contractions of the heart. It is determined by the amount and force of blood pumped, and by the size and flexibility of the arteries, which have elastic walls, acting as a **pressure reservoir**. The left ventricle pumps more blood into the arteries than can immediately flow into the narrower, high-resistance arterioles, so the artery walls expand, then their elastic recoil helps to drive blood into the arterioles even when the heart has stopped contracting.

The maximum pressure in the arteries comes at systole, reaching about 120 mm Hg, falling to about 80 mm during diastole while the heart refills. The resting blood pressure comes as figures such as 120/80, meaning 120 mm Hg at systole, 80 at diastole. Blood pressure depends on many factors, such as physical activity, posture, temperature, diet, health, emotional state (e.g., stress, anger, fear) and drugs. **Normal blood pressure is 100/60 to 140/90.** Mine was 127/79 recently, with a pulse of 64 beats a minute, within the normal ranges.

High blood pressure (hypertension) gives readings above 140/90, with mild high blood pressure up to 200/100, and significant hypertension above 200/100. Values slightly above 140/90 are not a cause for alarm, especially in people over 50. **Low blood pressure** (**hypotension**) is shown by readings below 100/60. **Chronic high blood pressure** affects about 1 in 4 adults in America, more frequently in men than in women, and is twice as frequent in African-Americans as in white Americans.

The sufferer is often **unaware of having high blood pressure**, but it may cause tiredness, confusion, anxiety, excessive sweating, nosebleeds, sickness and chest pains. If untreated, it can lead to heart attacks, heart failure, coronary artery disease, blood vessel damage, kidney damage, strokes and loss of vision. It can cause **aneurysms** of the arteries, with dilations and bulges which can lead to their rupture and serious internal bleeding. That can be lethal in a large artery, especially the aorta. A moderately high blood pressure is associated with a shortened life expectancy while mild elevation is not.

The global consequences of one man's high blood pressure were enormous. In February 1945, President Franklin D. Roosevelt, Prime Minister Winston Churchill and Premier Joseph Stalin attended the Yalta Conference to discuss Europe's post-war reorganisation. **Roosevelt** was so seriously ill with hypertension and hypertensive heart failure that his political judgement was impaired, resulting (according to Dr Le Fanu, *The Daily Telegraph*, 13/7/2004) in 'the betrayal of the Poles, the imposition of communist regimes throughout eastern Europe and the loss of China to communism'.

Causes of high blood pressure include genetic predisposition, appetite suppressants, excessive alcohol or salt intake, obesity, stress, blood vessel problems, kidney disease, diabetes, gout, and drugs, including some painkillers and tranquillisers.

The body has two **main receptors which monitor blood pressure**. One is on the aortic arch, monitoring the main blood flow from the heart; the other is in the neck at the carotid sinus, monitoring the pressure of blood going to the head. They regulate blood pressure at rest and under exercise. Once high blood pressure develops, the main receptors seem to reset themselves, regulating blood pressure at higher than normal levels. High blood pressure stresses the heart, valves and blood vessels. The heart has to pump against an increased resistance. In the eye, fine retinal blood vessels may be damaged, with increasing loss of vision.

Prevention of high blood pressure can be by weight control. Vigorous aerobic exercise to work the heart muscles and blood vessels is helpful. Reducing salt intake helps people with hypertension but has little effect in healthy individuals.

Treatment of high blood pressure often involves drugs. Diuretics may be used to reduce blood volume. Drugs include beta-blockers, calcium channel blockers and ACE inhibitors. Modern drugs are often very effective for high blood pressure. Potassium replacements for dietary sodium may be recommended. The problems often come from the lack of obvious symptoms, with complications arising before high blood pressure is discovered.

Low blood pressure results in an inadequate blood supply. Dizziness and fainting may occur. It may be caused by a weak heart, by insufficient blood volume, or by excessive blood vessel dilation. Causes include shock, stress, trauma, severe bleeding, a sudden change to an upright position, getting out of very hot baths, allergic reactions, dehydration, excess alcohol consumption and various drugs, including diuretics.

The treatments for low blood pressure depend on its causes. In **serious allergic reactions**, the entire body goes into an anaphylactic shock, when there is a huge loss of blood plasma into the tissues and widespread dilation of blood vessels, causing a major loss of blood pressure. Such shocks are often fatal unless the sufferer can be quickly injected with drugs such as adrenaline to constrict blood vessels and open the lungs' airways.

A rare condition is **pulmonary hypertension**, with increased blood pressure in the lung blood vessels with symptoms such as shortness of breath, dizziness, fainting, leg swelling and insufficient blood oxygen. Severe cases suffer from low exercise tolerance and may have heart failure. It affects about 1 in 500,000. In America about 750 people are diagnosed each year with it, especially women between 20 and 40. Normal pulmonary-artery pressure is about 14 mm Hg at rest. A pulmonary artery pressure greater than 25 mm Hg at rest and 30 during exercise is abnormally high.

Jack Waller was diagnosed with pulmonary hypertension just before his first birthday. He had an enlarged heart and was expected to live for only a few months. At the age of four, he went to school, surviving by wearing a **life-saving backpack** all the time. It pumped medication at three-minute intervals through a line beneath his arm into a blood vessel to his heart.

In 2001, Liz Redgate developed almost fatal pulmonary hypertension, becoming breathless, fat and swollen. She was put on Warfarin and a diuretic, and sometimes needed a wheelchair. Dr Jackson, a cardiologist at St. Thomas' Hospital in London, tried as the last resort a new treatment with Viagra, used for male erectile dysfunction. Within three months she was back to normal. She took Viagra four times a day, with no effect on her sex life, but it controlled her pulmonary hypertension.

Reference

Rosengren, A. *et al.*, Association of psychosocial risk factors with risk of acute myocardial infarction in 11119 cases and 13648 controls from 52 countries (the INTERHEART study): case-control study. *Lancet* (2004) **364**: 953–962.

Chapter 17

Blood, Blood Groups; Anaemia, Haemophilia, Leukaemia and other Blood Disorders

17.1. Introduction

Blood constitutes about 8% of our weight, with an average volume of 8.8 pints (5 litres) in women and 9.7 pints (5.5 litres) in men, about 0.9 pints (511 ml) per stone (14 lb, 6.4 kg). It contains the liquid straw-coloured serum (55% of blood volume in men, 58% in women), which transports digested food materials to all parts of the body. The blood from cuts is **bright red** and oxygenated, but deoxygenated blood from a vein is an unattractive **dark purple-brown**. A Canadian man had **green-black blood** from taking a sulphur-containing medicine which produced green sulphaemoglobin. See Chapter 10.14 for **blood blue** with methemoglobin.

The **serum** is 90% water, 7% proteins and 1% inorganic salts, especially ions of sodium and chlorine, with lesser amounts of potassium, calcium and bicarbonate. It carries wastes such as urea to the kidneys for excretion and helps regulate body temperature by its heat-carrying capacity. Its other components are dissolved gases, hormones and nutrients such as glucose, amino acids, fats and vitamins.

Carried round in the serum and occupying 42% (in women) to 45% (in men) of the blood volume, are **red blood cells (rbcs)** which transport oxygen from the lungs to the rest of the body. Rbcs contain haemoglobin (a combination of iron and protein) which also carries carbon dioxide to the lungs. Rbcs are flat and disk-shaped, with a central indentation, giving a large surface for gas exchange. Haemoglobin also transports the gas nitric oxide which helps to regulate blood pressure by relaxing blood vessel walls.

Adults have about 25 to 30 trillion (25×10^{12}; a trillion is a thousand thousand million) red blood cells, 1/3,600 inch (7 to 8 µm) diameter and 1/12,500 inch (2 µm) thick. When mature they have no nucleus, no DNA, and are unable to repair themselves. Their average life span is 120 days, travelling about 700 miles (1,127 km) through the blood vessels.

Old ruptured rbcs are filtered out in the spleen or destroyed in the liver. New ones form in the bone marrow, inside cavities in bones, at the rate of two to three million per second, about 200 billion a day. Bone marrow cells divide about once every 10 hours. There are typically 1,000 times more rbcs than white ones, with white ones having a nucleus, DNA and RNA, averaging 13 days before death or division.

Making up less than 1% of blood volume are the **platelets**, which assist with blood clotting in wounds, and a range of **white blood cells** which fight infectious organisms and combat non-self proteins. Blood is part of the immune system, transporting white blood cells to sites of injury or infection. The same cells in bone marrow which divide to give rbcs also give rise to white blood cells and platelets.

Each crowded millilitre (about one twenty-ninth of a fluid ounce) of blood normally contains about five thousand million rbcs, seven million white blood cells and two hundred and fifty million platelets. Differences in the concentration of these blood components cause human differences, especially anaemia from insufficient rbcs or haemoglobin, and haemophilia from an inability to clot blood.

Each blood component has a series of genes controlling it. For **12 different blood group loci** controlling rbcs, the commonest allele combination is shared by only 2 in 1,000; all 290,303 other combinations are rarer. Even those who are alike for rbc characters differ in their white blood cells (leukocytes) for which there are over 25 million possible combinations for the four main HLA (**human leukocyte antigen**) systems which were discovered from agglutination of white blood cells. Blood groups such as A or Rhesus positive have identifiable subgroups, and there are many other known rbc and HLA groups.

Various **inherited blood disorders** give anaemia which is often mild at birth because **foetal haemoglobin** is still being produced, but this diminishes with time. Foetal haemoglobin is specified by a different gene locus from loci specifying adult haemoglobin. The most serious of these diseases require **blood transfusions** about every two or three weeks, about 30 litres of blood a year. Although this relieves many symptoms, it causes a build-up of iron, with the excess deposited in the heart, liver and other organs, often causing their failure and the sufferer's death in the teens or twenties.

The **serum ferritin test** assesses body iron levels, with more than 1,000 µg (micrograms) per litre suggesting **iron overload**. Having more than 10 blood transfusions can cause overload. Treatment is arduous and long, with iron-chelating agents such as Desferal. This requires the wearing of a pump for subcutaneous injections for 50 to 60 hours a week. An alternative is a once-a-day tablet of Deferasirox (marketed as Exjade) but there are reports of side-effects of kidney failure and reduced rbc number.

By the age of five years in 2015, Amie Rose had had 60 blood transfusions, and has one a month. They are necessitated by the blood disease diamond blackfan anaemia, which affects about 110 people in the UK, making them severely deficient in red blood cells. She had an emergency operation to insert a port from her jugular vein to her heart. For six days a week, a tube from a pump is attached to the port, injecting fluid to reduce her iron overload.

Unless you are too young, too old or in poor health, look up your local blood transfusion service and offer to give blood to save lives, enabling you to find out your blood group. As all blood is tested for diseases such as syphilis, HIV and hepatitis, being asked to give again shows that you are free of these scourges. In Britain one donates about 0.83 pints (470 ml) at a time, up to four times a year.

Blood transfusions are needed for many rbc disorders and for white blood cell diseases such as leukaemias. Someone with **acute leukaemia**

may need four units of red cells and ten platelet transfusions per course for up to six courses of intensive chemotherapy. Many cancers cause platelet counts to drop, giving a risk of spontaneous and life-threatening internal bleeding.

17.2. The ABO and Rhesus blood groups

About one third of the blood transfusions intended to save lives were fatal in the 19th century, from massive coagulation of blood in the receiver. About 1901, Karl Landsteiner discovered why. There are four major types of ABO blood group, depending on the antigen on the red blood cells and on antibodies in the serum. An **antigen** is a substance, usually a protein, which stimulates the production of an antibody by the immune system, with the **antibody** neutralising the antigen, especially in combating bacterial infection.

People of **group O** have two recessive genes, I^oI^o, where I stands for the ABO immunity-type gene and superscript ^{o} stands for O group. There is no antigen on the rbcs but the serum carries anti-A and anti-B antibodies (Plate 1D). Transfusing a person who is blood group O with blood of any other group is a disaster. If group A blood goes into an O recipient, the recipient's anti-A antibody attacks the A antigen on the incoming rbcs, coagulating the blood, usually causing death. I^o is recessive to I^A and I^B, which are co-dominant, with both being expressed in $I^A I^B$ heterozygotes, having neither A or B antibodies but having A and B antigens.

Only transfusions between people of the same ABO group are safe, although group O is a **universal donor** as it lacks the antigens, and group AB is the **universal recipient** as it lacks the antibodies. In emergencies, group O blood could be given to people of other groups because the incoming blood will not be coagulated by the recipient's antibodies, and the donor's anti-A and anti-B antibodies are largely diluted out in the recipient. If transfusions are needed before the recipient's type can be established, group O, Rhesus negative, blood is given. Exact matching is best but **fatal mistakes** occur. In 1999, Hilary Pearce died at Addenbrook's Hospital in Cambridge, England. She was group O, Rhesus positive, but by mistake was given group A, Rhesus positive blood. Within five minutes she died of cardiac arrest.

The ABO locus is on chromosome 9 and the Rhesus locus (actually three closely linked loci, *C*, *D* and *E*) is on chromosome 1. Both control antigens and antibodies, but in different ways, as described below.

17.3. Problems from Rhesus negative mothers

Mother-baby incompatibility arises when a Rhesus negative mother is pregnant with her second or later Rhesus positive baby. The baby's blood is attacked by the mother's anti-Rhesus antibody, causing **haemolytic** (Greek *haima*, blood, *lysis*, dissolution) **disease of the newborn**, with anaemia and jaundice. This usually caused death but survivors often recovered fully.

In the ABO system, the antibodies are produced almost automatically in the young child, but the anti-Rhesus antibody is only produced by Rhesus negative people if they receive Rhesus positive blood. That antibody is **inducible** by transfusion with Rhesus positive blood or by leakage of a Rhesus positive baby's blood into the Rhesus negative mother. With about 16% of white populations being Rhesus negative (autosomal recessive), 1 in 10 pregnancies will be at risk, yet only 1 in 350 pregnancies with the dangerous combination results in an affected baby.

The **first Rhesus positive baby** born to a Rhesus negative woman usually has no problem as the mother has not developed the antibody. Rhesus positive blood from that baby, leaking into the mother just before or during birth, triggers her anti-Rhesus antibody production. The antibody is produced too late to affect that first baby but can affect **subsequent Rhesus positive babies**. Badly affected babies have many of their rbcs destroyed by the antibody, so they are given transfusions of **Rhesus negative blood** (of the same ABO type) **without Rhesus antibodies**, so that the remaining anti-Rhesus antibodies do not harm the transfused blood.

Pregnant women are tested to see if they are Rhesus negative. If they are, the baby's father is tested to identify Rhesus positive fathers-to-be. Homozygotes for the dominant *D* gene always have Rhesus positive babies, but heterozygotes, *D d*, have about equal numbers of Rhesus positive and Rhesus negative babies by a Rhesus negative woman (*dd*). If the Rhesus negative woman has a Rhesus positive baby, the modern treatment to prevent problems with further babies is to **inject her with anti-Rhesus antibody** within 72 hours after the birth.

It seems odd to inject a woman with the substance you want to prevent her making, but it works. The injected anti-Rhesus antibody combines with the baby's leaked Rhesus positive rbcs, preventing them from triggering the mother's own production of anti-Rhesus antibody. ABO blood group incompatibility between mothers and babies does not usually cause problems. Unlike the Rhesus antibody, the A and B antibodies are too large to cross the placenta into the foetus.

17.4. Other blood groups

There are more than 30 other red blood cell group systems, with names such as MN, Duffy, Kidd and Lewis. While they can be used forensically or for tracing paternity, humans do not usually make antibodies to their antigens, so they are not important in transfusions. The HLA groups are not important for blood transfusions but are vital in **transplants** where very close relatives are often best matched.

17.5. Blood group frequencies

In Britain, **group A** has frequencies of about 50% in East Anglia, going down to 38% in 'Welsh' Wales, to 34% in Scotland, and to 25% in western Ireland. From Britain eastwards, the frequency of the **B group gene** rises from 5 to 10% in western Europe to 25 to 30% in Asia. In most populations worldwide, the **O group gene** has a frequency of over 50% (as it is recessive, that does not mean that over 50% of people are of group O) and the B group gene is less than 30%. The Native American groups all have a very low frequency of the B group gene yet different tribes have the highest (82%) and lowest (1.5%) frequencies of the **A group gene**, suggesting that some tribes are of very different ancestries.

There are huge population differences for the Duffy and Rhesus groups. The respective frequencies in Caucasians and Negroids are 3% and 94% for Duffy allele Fy, 42% and 6% for Duffy allele Fy^a, and 55% and 0% for Duffy allele Fy^b. The R_0 allele of the Rhesus group is fairly rare in Whites but has a frequency of about 60% in West Africans. Such allele-frequency differences enabled geneticists to calculate the amount of White ancestry in modern American Blacks. About 4% of genes in Blacks in

Charleston come from Whites, compared with 26% in Blacks in Detroit, reflecting different degrees of racial mixing. About 15% of genes in American Blacks have been introduced from Europeans in the last 275 years. There have, of course, been genes passed from Blacks to Whites.

The world's **commonest blood** group is **O**, the oldest group. About 26 million Britons are group O, as are or were Prince Charles, Charlie Chaplin, Elvis Presley and Al Capone. Group **A** appeared about 25,000 to 15,000 BC. It is the commonest group in Norway, Denmark, Austria, Armenia and Japan. Group **B** emerged between 15,000 and 10,000 BC as tribes migrated from Africa to Europe, Asia and the Americas. The rarest group, **AB**, appeared between 1,000 and 500 years ago as a result of mixing of existing groups. Its frequency is about 10% in Japan, China and Pakistan, and 5% in America, including Marilyn Monroe and President Kennedy.

17.6. Blood groups and disease susceptibility

People of different blood groups have **different susceptibilities to diseases.** **Group A** have more stomach cancer and **group O** have more duodenal ulcers. On the white blood cells, the **HLA antigens** are involved with **immune responses.** Many conditions such as Hodgkin's disease, arthritis, multiple sclerosis, types of cancer and type 1 diabetes have weak associations with particular HLA types. Some HLA types are strongly associated with protection against **malaria**, and others with susceptibility to **coeliac disease** and **spondylitis.**

17.7. Sickle-cell anaemia and thalassaemias — inherited disorders maintained by selection for malaria-resistance

In warmer regions, malaria is often a major killer of adults and children, with strong selection for malaria-resistance. Having genetic **malaria-resistance** in a population often involves the suffering of many **uninfected individuals** from the mutations which confer resistance. According to the WHO, at least 5% of the world's population carry genes for a haemoglobin defect.

In western and central Africa, the Arabian peninsula and parts of southern India, about 25% of people carry the recessive gene for **sickle-cell anaemia**, *a*, which gives rise to an abnormal haemoglobin, **haemoglobin S**, instead of **haemoglobin A**. In those populations, 1 to 3% of individuals are homozygous recessives, *aa*, and usually **die from anaemia**. Their red blood cells go sickle-shaped and are broken down in the spleen, giving fatal anaemia. When blood oxygen levels are low, the rbcs tend to become rigid and misshapen, obstructing the smaller blood vessels, leading to painful localised crises. In America, about 80,000 people (mainly from malarial or formerly malarial regions) have sickle-cell anaemia.

People who are heterozygous, *Aa*, sometimes suffer slight anaemia but are **more resistant to falciform malaria** than are non-anaemic dominant homozygotes, *AA*. This is **heterozygote advantage**, where in malarial regions the heterozygote is the fittest. Selection in favour of the *Aa* type, with one of each allele, helps to keep both alleles in the population, even though the recessive *a* is lethal when homozygous. Because two heterozygotes on mating produce on average one *AA* baby, two *Aa* babies and one *aa* baby, half their children have non-ideal gene combinations because of genetic segregation. **Sexual reproduction therefore has a cost**, the **segregation load**, because it cannot maintain 100% of heterozygotes even when these are the fittest. That requires vegetative reproduction!

Where there is no longer selection for malaria resistance, where mosquitoes have been eliminated or where people have migrated to non-malarial areas, there is no heterozygote advantage. **Natural selection** against the recessive homozygote, *aa*, from deaths from anaemia, then reduces the frequency of the *a* allele. This is a well-proven example of natural selection in humans. One can detect heterozygotes by blood analysis and give genetic counselling to reduce the risk of producing sufferers. In Britain, about 12,000 people have sickle-cell anaemia and more than 300,000 are heterozygotes, with **sickle-cell trait**.

In a malarial region, the heterozygote *Aa* has a fitness advantage over *AA* of about 10% and *aa* is lethal. The expected frequencies of the three types at birth, before selection, are *AA* 82.6%, *Aa* 16.5% and *aa* 0.8%. Nearly one per cent of babies would die at birth from sickle-cell anaemia, 82.6% would be malaria sensitive and only 16.5% would have malaria-resistance. If a higher proportion had malaria resistance, there would be more deaths from anaemia.

The malaria resistance story is similar for **alpha-thalassaemia** in southeast Asia and in some African populations, and for **beta-thalassaemia** in Cyprus, Greece and Turkey, around the Mediterranean Sea and in Asia. In **alpha-thalassaemia** there is insufficient production of alpha-haemoglobin chains, so excess beta-haemoglobin chains precipitate out from the blood, causing anaemia, whereas alpha chains are in excess in **beta-thalassaemia**.

People with normal haemoglobin are malaria-susceptible. Normal adult haemoglobin is made up of two alpha chains and two beta chains. Chromosome 16 carries two copies, not the usual one copy, of the **alpha-haemoglobin gene cluster**. In the **beta-haemoglobin gene cluster** on chromosome 11, some alleles work strongly, others weakly and others make no beta-haemoglobin. An individual can therefore have **0, 1, 2, 3, or 4 fully working alpha-haemoglobin genes**, and **0, 1 or 2 fully, partly or non-working beta-haemoglobin genes**, giving a huge diversity of seriousness, from none to death. The mutations are often **big deletions**, completely knocking out that gene's functions, but others involve changed DNA bases, which can result in varying degrees of functional impairment.

Having no working alleles for alpha- or beta-thalassaemia causes death from anaemia. People with one to three working alpha-haemoglobin genes, or only some beta-haemoglobin gene activity, suffer varying degrees of anaemia, from mild to severe. Their rbcs are smaller than normal, and they may have skeletal abnormalities. Severe cases often have enlarged livers and spleens and may need frequent blood transfusions, causing problems from excess iron. A friend in Sri Lanka appeared normal for blood until she became pregnant, when her haemoglobin level dropped so far that she needed hospitalisation and a series of blood transfusions for beta-thalassaemia. She now has a serious iron overload.

Defective alpha-thalassaemia genes have frequencies of 5 to 15% in many southeast Asian populations. In Cyprus, about 6 babies in 1,000 die — usually within minutes of birth — from being homozygous for beta-thalassaemia, and about 16% are heterozygous carriers. People in Cyprus can have blood screening to discover their beta-thalassaemia status, with genetic counselling if necessary. These defective genes can only persist at such very high frequencies if being heterozygous confers a significant advantage, which is increased malaria-resistance.

In 2002, Raj and Shahana Hashmi in Leeds had a son, Zain, with life-threatening **beta-thalassaemia major**. He needed blood transfusions every three weeks, with injections of Desferal five days a week to combat iron-overload. In five years he had more than 100 transfusions. A world-wide search for a perfect genetic match for a bone-marrow transplant failed. A second child was not a good match. The Human Fertilisation and Embryology Authority (HFEA) gave permission for the mother to have *in vitro* fertilisation and **pre-implantation genetic screening of embryos** for being free of thalassaemia and for being a tissue-type match for Zain. From such a baby, stem cells would be taken from umbilical cord blood, multiplied up, and used for a transplant. Unfortunately a pro-life campaign group, Comment on Reproductive Ethics, brought a successful court action, delaying matters until it was too late for Mrs Hashmi to produce a 'saviour sibling' for Zain.

In **pre-implantation genetic diagnosis**, the woman's eggs (extracted after hormonal fertility treatments) are fertilised in the lab with the man's sperm. Under a microscope, a number of three-day-old embryos at the eight-cell stage have a cell detached and tested for the defect of interest. Taking one cell then does no harm. Chromosomes can be assessed visually or by DNA tests. For deleterious mutations, the cell's DNA is multiplied many times by the polymerase chain reaction and then tested for the bad sequence. One or more non-defective embryos is implanted in the woman's womb.

Pre-implantation diagnosis was successful in a case of **Diamond Blackfan anaemia**, whose sufferers do not produce enough rbcs. Charlie Whitaker's parents applied to HFEA for permission, which was denied because of worries about whether the testing harmed the embryo. Charlie's parents went to Chicago where two suitable screened embryos were implanted in Mrs Whitaker. She bore Jamie, a perfect match for Charlie. A successful transplant using umbilical cord stem cells from Jamie was made. Within three months Charlie no longer needed transfusions or nightly injections.

In **beta-thalassaemia trait**, one beta-thalassaemia gene is normal but the other is defective. The people are usually healthy but occasionally have mild anaemia. Its frequency is 1 in 7 in Cyprus, 1 in 10 in Italy, 1 in 12 in Greece, 1 in 20 in India and Pakistan, 1 in 50 in Africa and the West Indies, and about 1 in 1,000 in England, nearly all in people of immigrant descent from malaria-affected areas.

In **beta-thalassaemia intermedia**, both genes have mutations but one is mild. Sufferers do not usually need transfusions but have anaemia, an enlarged spleen and some bone deformities. In **beta-thalassaemia major**, both genes have no or a severely reduced function, causing death within two years if untreated.

For the thalassaemias, prenatal (before birth, after normal fertilisation) diagnosis by **amniocentesis** (tests on embryonic cells from fluid around the embryo at about 16 weeks, sampled through a syringe needle through the woman's abdominal wall) or **chorionic villus sampling** (done at 8 to 10 weeks via the vagina) is available, giving parents a choice of continuing or terminating the pregnancy. In Sardinia, many individuals have mutant genes for beta-thalassaemia. The disease incidence fell from 1 in 250 live births to 1 in 4,000 between 1974 (when population screening and prenatal diagnosis were introduced) and 2003.

17.8. Haemophilia, which affected European royal families

Haemophiliacs have extremely poor **blood clotting**, so much blood may be lost even from a small wound, perhaps giving anaemia or even death. The haemophilias are diseases whose severity varies from lethal to negligible, depending on how badly the particular mutations affect the protein. Pre-implantation genetic diagnosis and prenatal diagnosis are available.

Haemophilia affected several European royal families. In England, Queen Victoria (1819–1901) was an unaffected carrier, having one affected son who died of haemorrhage after a fall, and two or three carrier daughters. Her daughter Princess Alice had a sufferer son who died as a child of haemorrhage after a fall. She also had a carrier daughter who became Queen Alexandra, wife of Tsar Nicholas II. Their son **Alexis** (1904–1918) had haemophilia B. Alexis and his parents were shot by the Bolsheviks in 1918. The present British Royal Family is free of the disease as they descend from an unaffected son of Queen Victoria, who became Edward VII (1841–1910).

Haemophilia A and haemophilia B are X-linked recessive conditions resulting in bleeding in soft tissues and joints, with very poor clotting, and can be fatal. Haemophilia A affects blood clotting Factor VIII, with its functional levels varying from 30% of normal (mild symptoms) to less than 1% (severe). It affects 1 to 2 males per 10,000, and accounts for 82% of haemophiliacs. The much higher frequency of male sufferers than of female sufferers for sex-linked recessive conditions was covered in Chapter 8.3.

Carrier females have less Factor VIII than normal but are usually symptomless. Sufferers can have injections of genetically engineered Factor VIII which normally works very well, but when donated contaminated blood was used as its source, many haemophiliacs developed **HIV**.

Haemophiliacs are obviously discouraged from taking part in contact sports such as rugby. Alex Dowsett at the age of 18 months was diagnosed with severe haemophilia A and spent his childhood being advised by doctors not to push himself too hard. However, in 2015, aged 26 and a successful professional cyclist, he broke the all-comers world record for the distance cycled in one hour, covering 52.94 km (32.9 miles). He has to inject himself with clotting Factor VIII every 48 hours to guard against internal bleeding. Dowsett runs the charitable foundation Little Bleeders to raise awareness of haemophilia. In 2015, the British Government apologised because about 7,500 Britons in the 1970s and 1980s were given contaminated blood (especially in donations from prisoners), causing more than 2,000 deaths from HIV and hepatitis C. Some recipients passed such diseases on to partners and children.

Haemophilia B affects blood-clotting Factor IX; treatment is with injections of this. It accounts for 17% of haemophiliacs. **Haemophilia C**, less than 1% of all haemophiliacs, is autosomal and affects females as much as males.

Female haemophiliacs do not die from menstruation, as the end of a period is from contraction of blood vessels lining the womb, not from clotting. They can have children: all sons will be sufferers and all daughters will be carriers if the father is normal. Medical attention during birth is essential for administering blood-clotting agents.

17.9. Other anaemias and polycythaemia

Nutritional anaemia results from a dietary deficiency of ingredients for red blood cells, especially iron needed for haem. It can be rectified by pills or a better diet, including liver.

Pernicious anaemia gives a progressive reduction in rbc numbers. It is largely dominant, with an incidence of 1 in 7,500, resulting in an inability of the digestive tract to take up enough vitamin B12 from food. It is treated by vitamin B12 injections, bypassing the gut.

Aplastic anaemia is caused by the bone marrow not producing enough rbcs. It can come from toxic chemicals, cancer, or radiation. Repeated transfusions can be given but it is sometimes fatal.

The kidneys control the production of new red blood cells by secreting the hormone **erythropoietin** which stimulates the bone marrow. The kidneys secrete this if insufficient oxygen is reaching them, which is the case in anaemia. If kidney disease interferes with this, **renal anaemia** results.

Haemorrhagic anaemia results from extensive blood loss, from a wound or excess menstrual bleeding. Transfusions may be given. **Haemolytic anaemia** results from excessive rupture and breakdown of rbcs. Its causes include sickle-cell anaemia and malaria.

Primary polycythaemia is caused by **excessive production of rbcs**, with their concentration reaching 11 million per mm^3 instead of the normal 5 million, and they may make up 70 to 80% of blood volume, instead of 42 to 45%. The blood is about six times thicker than usual, giving sluggish flow and reduced oxygen-carrying capacity. Raised blood pressure and diminished muscular performance are common.

Secondary polycythaemia is a natural adaptation which improves the body's oxygen-carrying in response to long periods of low oxygen levels. It occurs in **people living at high altitude** and in people with chronic lung disease or heart insufficiency. The greater number of rbcs gives more oxygen-carrying, but at the expense of more viscous blood, which is harder to pump. Many long-distance runners train at high altitude to increase the oxygen-carrying capacity of their blood.

Personal Account. *I suffer from anaemia, giving me severe period problems*

I suffer from anaemia, a deficiency of red blood cells resulting in reduced oxygen-carrying ability. My mother and younger sister are very anaemic. Anaemia can result from deficiency of folic acid, B12 and iron, all needed for the production of red blood cells. I suffer from low iron and folic acid. Both my parents are beta thalassaemia carriers, so my sisters and I had tests. We were diagnosed with anaemia. I am a carrier.

Anaemia has not greatly affected my life but has disadvantages. I get tired very easily and feel dizzy and light headed a lot, when I am very pale and cold. Anaemia is only a major disadvantage when I menstruate and lose a lot of blood, making me seriously weak so that I cannot lift my head or muster the energy to speak during the first few days when my period is very heavy. I have sometimes fainted then. My younger sister is more seriously affected, fainting almost every time she menstruates and had to spend four days in hospital on a drip because she was so weak.

This condition is mostly genetic but habits play a role. I am vegetarian and don't eat red meat, the main dietary source of iron. I have been vegetarian for seven years and notice the effects of anaemia more now than when I ate meat. I take iron pills and sometimes folic acid, with a varied diet, including a lot of raw green vegetables — a good source of iron and folic acid.

My anaemia is not particularly serious and I have no problem discussing it with anyone. The only time it would affect my career is when I get my period I would probably have to take a day off work. The only serious effect is when I have to miss a day or two of university because I feel too weak to come into college. Iron stores fall during pregnancy (my mum was very weak when pregnant) and I may feel the disadvantages then. If I was feeling very weak and iron supplements were not helping, I would consider eating meat until the baby's birth.

Christina Petridou, Greek, normally active and healthy-looking.

17.10. Leukaemias and lymphomas

'**Leukaemia**' covers disorders characterised by an **excessive number of white blood cells** (leukocytes, Greek *leukos*, white, *kytos*, a case or cell), which is associated with a **decrease in the number of red blood cells**, giving anaemia and affecting the spleen. Leukaemias may be **chronic**, long-lasting, or **acute**, coming to a rapid crisis. Chronic leukaemia takes more than a year from onset to death, with a gradual onset of anaemia, often with marked enlargement of the spleen, liver or lymph nodes. With treatment, some sufferers live for **decades in remission**. With **acute leukaemia**, death occurs within months. Bone marrow transplants, drugs and intensive chemotherapy are used as treatments, with varying success.

As white blood cells proliferate, patients suffer bruising, breathlessness, swelling of the spleen, enlarged lymph glands, persistent infections, paleness, fevers, joint pain and abdominal discomfort. In Britain, about 22,000 people a year develop a leukaemia, with about 450 children aged between two and five getting lymphoblastic leukaemia. A Danish study found that pilots who had flown for more than 5,000 hours had a five-fold increase in **acute myeloid leukaemia**, attributed to cosmic radiation at high altitudes.

Leukaemias may cause white blood cell concentrations to rise up to 71-fold, from $7,000/mm^3$ to $500,000/mm^3$. This **decreases the immune defences** because most cells are defective. The extra white cells are produced at the expense of rbcs, giving **anaemia**, and of platelets, giving **internal bleeding**, so leukaemia patients tend to die from infections or haemorrhages. Leukaemias act like cancers of the blood and bone marrow. They are usually caused by **translocations**, where part of one chromosome gets incorporated into a different chromosome, or **deletions**, where part of a chromosome is lost (Plate 1C).

Chronic myeloid leukaemia (Greek *myelos*, marrow, of bone marrow) gives severe long-term anaemia. Patients usually have a **reciprocal translocation** between chromosomes 9 and 22. The shortened 22 is called a **Philadelphia chromosome**. This translocation moves cancer gene *ABL* from chromosome 9 to chromosome 22 where a rearrangement of *ABL* with the *BCR* gene gives a **hybrid gene** making a protein giving **uncontrolled proliferation of white blood cells** in the bone marrow. It activates **latent cancer genes**. Treatments with anticancer drugs and bone marrow transplants have extended survival times, with about one quarter of chronic myeloid leukaemia sufferers surviving for more than 10 years. The drug Glivec (Imatinib) has produced wonderful treatments for chronic myeloid leukaemia, but is expensive.

Some other leukaemias and lymphomas involve other translocations near a proto-oncogene which can cause cancer if activated. These diseases are not inherited because bone marrow translocations do not affect eggs and sperm. These types of disease tend to run in families.

Here are statements by students about blood diseases.

- *I am a thalassaemia carrier. I have beta thalassaemia trait and do not suffer from anaemia, but am advised to carry my thalassaemia trait card at all times.*
- *I have Beta Thalassaemia.*
- *I am heterozygous for thalassaemia.*
- *I suffer from thalassaemias.*
- *I am a heterozygote for sickle cell anaemia.*
- *I have sickle cell anaemia trait* [she is from a malarial part of Africa] — *cold feet; difficulty in breathing after hard exercise. Not noticeably anaemic. My father had it too.*
- *I lack enzyme G6PD.* [see Chapter 12.20]
- *A common genetic disorder amongst us Parsis is G6PD. I don't have it but several friends do.*

17.11. Too much iron — haemochromatosis

In contrast to anaemias, **haemochromatosis** (blood, *haema*; colour, *chroma*; too much, *osis*) involves **excess iron**, with the skin showing a general darkening, bronzing, slate-blue-grey effect. It affects men five times more frequently than women, in spite of being autosomal, because women lose iron in menstrual blood.

Haemochromatosis is the most common metabolic genetic disorder in America, with a frequency of 1 in 250. In a group of 1,000 Americans, one expects 100 unaffected carriers and 4 sufferers. Symptoms often develop in men between the ages of 30 and 50, and in women over 50 (postmenopausal). The effects include fatigue, loss of appetite and weight, joint pain, abdominal pain, heart disease, loss of sex drive and body hair.

It is usually caused by a recessive gene on chromosome 6, but also has non-genetic causes, especially repeated blood transfusions. **Excess iron is deposited throughout the body**, notably in the liver (where it can cause cirrhosis, liver failure and liver cancer) and in the pancreas. An **enlarged liver and spleen**, with skin bronzing, are easily diagnosed. Blood tests show raised levels of iron, with transferrin saturation and raised serum ferritin, where these are iron-containing proteins in blood serum. **Ferritin** is the body's main iron-storage protein, with its level in the serum proportional to the amount of iron stored in the body. Normal levels show much human diversity, varying from 24 to 336 ng (nanograms)/ml in men, and 11 to 307 ng/ml in women (Mayo Clinic figures). Higher levels can indicate haemochromatosis or haemolytic anaemia, while lower levels can indicate heavy menstrual bleeding, intestinal bleeding or iron-deficiency anaemia.

It is one of the extremely few diseases **treated by old-fashioned bleeding**, of blood from a vein each week for two to three years. Iron-overload can also be treated with **chelating agents** such as Desferal and **dietary restrictions**.

Heterozygous carriers have serum ferritin at three to ten times normal concentrations but are usually unaffected. DNA tests can detect mutant genes, to aid diagnosis and allow genetic counselling.

17.12. Blood analysis results and human diversity

The numerical values for various elements of blood analysis are an object lesson in human diversity. For some factors, there may be a narrow range of values in healthy people, with adverse effects from values which are too low or too high, such as named illnesses. For other factors, there may be a wide range of values in healthy people, yet still with adverse effects from values outside that range, too low or too high. For some factors, medical problems only arise from values which are too low, or only from values which are too high, with only one end of the range of values causing problems. Values may also show differences between the sexes in the normal ranges, as in some instances below, or vary with age. Reference values differ somewhat between different authorities.

If you have a **blood assay**, as in a full health screen, the usual things tested are as follows, with the normal ranges given separately for men and women, if different. Values outside those ranges usually signal the need for further tests in case a medical problem is indicated. If the weird assortment of standard units used means nothing to you, do not worry, as they will mean a lot to the medical practitioners who examine your results. IU = International Units; mmol = millimoles.

Haemoglobin, men, 13.5 to 17.5 grams per decilitre (one tenth of a litre, 100 ml); women, 11.3 to 16.8; this tests for various anaemias (too low; see above) or haemochromatosis (too high; see above).

Red cells, men, 4.2 to 6.0 x 10^{12} per litre; women, 3.8 to 5.3.

Haematocrit ratio (HCT), the ratio by volume of blood cells to plasma, men, 0.37 to 0.52; women, 0.35 to 0.48.

Mean red cell volume (MCV), 80 to 99 femtolitres (1 fl = $1/10^{15th}$ of a litre); this is decreased in some thalassaemias (see above).

Mean red cell haemoglobin (MCH), 28 to 33 picograms (1 pg is one million millionth of a gram).

Mean red cell haemoglobin concentration (MCHC), men and women, 32 to 36 grams per decilitre.

Platelets, 150 to 400 x 10^9 per litre.

White blood cells, 3.8 to 11.0 x 10^9 per litre, including neutrophils, lymphocytes, monocytes, eosinophils and basophils.

Blood urea, men, up to 8.5 mmol/litre; women, 2.5 to 7.0.

Blood creatine, up to 125 mmol/litre. These are **kidney function tests** as urea and creatine are the main metabolic wastes excreted by the kidneys into the urine. Higher values in the blood could indicate reduced kidney excretory functions, as in various kidney diseases.

Uric acid, men, up to 450 mmol/litre; women, up to 350; this is a test for **gout** (Chapter 18.6), which tends to run in families and may be associated with taking a number of drugs, and excess alcohol consumption.

Blood glucose, 3.3 to 6.0 mmol/litre; these are the normal levels after fasting for at least six hours. Higher levels often indicate **diabetes**.

Blood calcium, 2.13 to 2.62 mmol/litre; this level is usually normal in people with osteoporosis but can be changed by a variety of medical conditions and metabolic disorders.

Albumin test, 37 to 52 grams per litre; albumin is a normal blood protein and is involved in transport of calcium around the body.

Vitamin B12, 145 to 900 ng (nanograms)/litre.

Serum folate, from folic acid, 2.5 to 16.9 µg/litre.

Thyroid stimulating hormone (TSH), women, 0.32 to 5.0 mUnits/litre; men, 0.27 to 4.2; higher values indicate an underactive thyroid, which is more common in women and in people over age 50.

Blood bilirubin, 3 to 22 mmol/litre; this measures the amount of blood bile from the breakdown of worn-out red blood cells in the liver.

Enzymes produced by the liver: those tested are usually **alkaline phosphatase** (25 to 120 IU/litre); **aspartate transferase** (up to 50 IU/litre) and **gamma GTP** (men, up to 80 IU/litre; women, up to 50). Levels of these enzymes in the blood are raised above normal in response to various infections and medical conditions. Raised levels are particularly used as an indicator of **liver damage** by excess alcohol consumption, and the return of those enzyme levels towards normal is used to follow recovery of the liver from such damage. Excess body fat, lack of exercise and excessive smoking may also raise the levels of these enzymes.

Cholesterol (total), up to 5.2 mmol/litre. People with significantly higher values are advised to take more exercise, change their diet, and may be put on cholesterol-reducing drugs such as statins, to reduce the chance of heart attacks and strokes (Chapter 16.3).

Cholesterol (HDL) — high-density lipoprotein, 1.3 to 2.2 mmol/litre.

Cholesterol (LDL) — low-density lipoprotein, up to 4.0 mmol/litre.

Total/HDL Cholesterol, up to 6.0. The more HDL ('good') cholesterol in relation to LDL ('bad') cholesterol the better.

Trigylcerides, 0.2 to 1.7 mmol/litre.

Sodium, 133 to 146 mmol/litre.

Potassium, 3.5 to 5.3 mmol/litre.

Recommended reading

Jorde, L. B. *et al.*, *Medical Genetics*, 3rd ed. (2006) Mosby, Philadelphia.

Lamb, B. C., *The Applied Genetics of Humans, Animals, Plants and Fungi*, 2nd ed. (2007) Imperial College Press, London.

Chapter 18

Arms, Legs, Giants, Dwarfs, Arthritis, Left-Handedness

18.1. Introduction

Mutations such as **dwarfism** result in very short legs and arms, while too much growth hormone can cause **giants**. Limbs can be affected by diseases such as leprosy and polio, by frostbite, and damaged in accidents, sport or war. In World War 1 about 41,000 British servicemen lost limbs. A modern trivial disorder is 'iPhone finger', a thumb injury from excessive text messaging. **Repetitive strain injury** can occur through work or leisure; sufferers include typists, butchers, builders and tennis players.

In America the number of **knee replacements** to date exceeds 730,000, and **hip replacements** exceed 350,000. A knee operation was described in *The Sunday Telegraph*, 8/12/2002, by Dr Le Fanu:

> ... *a gentleman in his late sixties told me he had just had both knees replaced — under spinal (epidural) anaesthetic. So there he was, lying fully conscious on the operating table as the surgeon opened up the knee, and cut through the femur and tibia with a hydraulically-driven power tool. You can almost imagine the noise — and the smell of vaporising bone. He then replaced the cracked arthritic surfaces that had been grinding painfully on each other with a couple of glistening semi-circular metallic pieces, which glided effortlessly over a couple of white plastic bits underneath. A few days later and my acquaintance was up and walking about. ... He is now blissfully pain free and has a full range of movement.*

Some determined people cope well with **limb defects**. Dame Alicia Markova was born with **crooked legs**. When she was a child, the surgeon advised her that the choice was between ballet classes or leg irons for life. She took to ballet brilliantly. In China, Qian Hongyan, aged three, had **both legs amputated** after an accident. Unable to afford artificial legs, her parents put her in half a basketball in which she managed to move around with the help of paddles. Her bottom became blistered and sore but she remained cheerful. After wearing out six basketballs, she was given a wheelchair and later had artificial legs.

The Partnership of the **Mouth and Foot Painting Artists** has members who cannot use hands or arms, painting superbly with brushes in their mouth or toes. Trevor Wells is paralysed from the neck down after a rugby injury but uses his mouth to paint excellent landscapes.

The 2012 Paralympic Games in London showcased inspiring competitors who overcame terrible handicaps. For swimming, those with physical disabilities were allocated a category between 1 and 10, with 1 the most severe. Disabilities included single or multiple limb loss, cerebral palsy, spinal cord injuries, dwarfism and joint impairments. Brave 17-year-old British swimmer Eleanor Simmonds set new world records to win two gold medals. Ellie is 4 ft 0 in (1.23 m) tall and an achondroplastic dwarf (see below).

The best-known blade runner is Oscar Pistorius. Although both his legs were amputated below the knee when he was 11 months old, he competed in events for amputees and in those for able-bodied athletes. Georgy Evans lost a leg through an accident at the age of three and is one of 60,000 amputees in Britain. With the help of The London Prosthetic Centre in Kingston, she can now run, ride, scuba dive, ski and wear high heels. The centre makes high-heel legs, low-heel legs, legs for skiing, diving, rowing, etc.

Painful **cramps** and **restless legs** deprive sufferers of sleep. Restless legs syndrome causes burning sensations in the calves and an uncontrollable urge to move the legs. Cramps can affect the calves, feet and thighs, especially at night. A muscle suddenly goes into spasm, causing much pain. Taking quinine sulphate works.

A common problem in older people is **Dupuytren's contracture**, causing one or more fingers or the thumb to bend into the palm. About

10% of men over 50 and of women over 60 are affected. Lady Thatcher and Ronald Reagan had surgery for it.

18.2. Giants, dwarfs and midgets (see also Chapter 3.3)

Gigantism, with a height exceeding seven feet (213 cm) is usually due to over-secretion of pituitary growth hormones. Only rarely is it inherited. A living giant is Lydon Sutcliffe, 7 ft 1 inch (216 cm) tall. He is happy with his height and works as a groundsman at the Wimbledon tennis courts. The 'Alton Giant', Robert Wadlow, born in 1919 in Alton, Illinois, reached 9 feet (274 cm) and 35 st 1 lb (491 lb, 223 kg), which was too much for his joints, making his knees and hips deformed and arthritic. His shoes were 18½ inches (47 cm) long; his arm span was 9 ft 5¾ inches (288 cm). He died aged 22.

    Dwarfism occurs in many forms. The commonest is **achondroplasia,** where the head and body trunk are normal but the arms and legs are very short (Photo 18.1). The incidence is 1 in 15,000, from an autosomal dominant (Chapter 8). The normal allele *d* is recessive and the mutant allele

Photo 18.1.   An Egyptian dwarf with two normal children.

D is dominant. Affected people are heterozygous, *D d*, with the dominant homozygous condition *D D* being lethal before birth.

For dwarfs, **average adult heights** are 4 ft 4 in (132 cm) for males, 4 ft 0.4 in (123 cm) for females, with normal IQs and life span. Dwarfs are fertile but females have small pelvises and usually need a Caesarean section when giving birth. Genetically-engineered human growth hormone allows dwarfs to grow to about 5 ft 4 in (162 cm) for men and 5 ft 0 in (152 cm) for women. They are often in demand for theatres, especially for *Snow White and the Seven Dwarfs*.

The children of two dwarfs (*D d* × *D d*) are expected to be in the ratio of one normal, *d d*, two live dwarfs, *D d*, and one dwarf dead before birth, *D D*. Unusually for a dominant condition, 85% of dwarfs have two normal height parents; such dwarfs result from a **new spontaneous mutation** from *d* to *D*, usually in codon 380 in a gene on chromosome 4 involved in cartilage growth.

Midgets are very short but have fairly normal body/limb proportions. **Pituitary dwarfism** can result from genetic causes or pituitary gland tumours or damage. Treatment is with growth hormones and sometimes with thyroid and sex hormones.

The Dutch midget, Pauline Musters, died aged 19, 23 inches (59 cm) tall, with bust-waist-hip measurements of 18½-19-17 inches (47-48-43 cm). **Calvin Phillips**, born in 1791 weighing 2 lb (0.91 kg), was an American dwarf who stopped growing at 5. At 19 he weighed 12 lb (5.4 kg) and was 26½ inches (67 cm) tall. He died at 21 from progeria (Chapter 21.3). The shortest man measured was a 72-year-old in Nepal, Chandra Bahadur Dangi, who in 2014 was 1 ft 9.5 in (54.6 cm). The shortest woman was measured in India in 2011: Jyoti Amge, 2 ft 0.7 in (62.8 cm) at age 18.

Restricted growth, including dwarfs and midgets, affects about 1 in 10,000 births, with 75% having normal parents. Dr Shakespeare, a dwarf, led a Newcastle University study on the **psychological effects of being small**, published in 2007: 97% of restricted-growth individuals suffered **unwanted public attention**, including name-calling, mockery and even physical violence. Such abuse can damage self-esteem, cause clinical depression and deter people from going out. They were less likely to marry, were twice as likely to live alone as people of normal stature, and more likely to be in lower occupational roles if employed. Worries about passing on their condition and feeling unattractive can be severe and socially crippling.

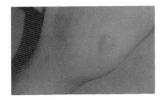

Photo 18.2.   Mild polydactyly; a small pink protrusion only.

18.3. Extra fingers or toes — polydactyly

Polydactyly shows to very **different extents in individuals** (variable expressivity, Chapter 8.2) and may show or not show at all in individuals carrying the gene (incomplete penetrance, Chapter 8.2). About 27% of individuals with a gene for polydactyly fail to show symptoms. Some people with this dominant autosomal gene only have small wart-like protuberances (Photo 18.2) while others have one or more extra fingers, or extra toes, or both.

The mutations may be at one of several gene loci. Polydactyly occurs with a frequency of 1 in 2,000 white Europeans. English film star Gemma Arterton was born with polydactyly. At birth a doctor tied off the boneless sixth digits to remove them. Samantha Evans was born with twelve tiny fingers and twelve tiny toes. They were so perfect that no operation was carried out.

18.4. Lobster-claw feet and hands

Although generally rare, this affects about 25% of Vanyai tribesmen in remote areas of Zimbabwe. Each foot has two large bony toes, one pointing forward and one sideways. They can run and climb trees faster than normal. Their hands are usually normal. It is probably due to a dominant autosomal gene. In other countries, a dominant condition called **split-foot** occurs and is often associated with **split-** or **cleft-hand**. It has very variable degrees and forms of expression, with a frequency of about 1 in 90,000.

American Jason Black, known as 'The black scorpion', is the director of *999 Eyes Freakshow*. He has **lobster-claw hands**, with three distorted fingers on each hand, each forwardly bent, with a deep cleft down the hand's middle.

18.5. 'Flipper arms' (phocomelia), natural or from thalidomide

During the 1950s and early 1960s, many pregnant women were prescribed **thalidomide** as a tranquilliser. Some who took this 27 to 40 days after conception had children with short flipper-like arms or legs. All four limbs could be affected, and more than 10,000 children born between 1956 and 1962 suffered. Sufferers have many practical difficulties, often performing feats of dexterity in bringing up children, even changing babies' nappies with a combination of feet, arms and teeth. It occasionally occurs spontaneously without thalidomide. It is not inherited.

Plate 18A shows a case of asymmetric phocomelia (Greek, *phoke*, a seal; *melos*, limb. The limbs are often flipper-like). The Cambodian beggar lady has one arm missing and one withered arm. Being blind, she felt with one foot when she heard me place a donation in her bowl. She neatly picked up the note between her big toe and the next toe, passing it to her distorted fingers for placing in her clothing.

Janette Cooke was the first thalidomide victim to have a baby. She was born without arms and legs and spent her first four years in hospital in Liverpool. She later learnt to write, type and use a TV remote control with her mouth. She said, "I couldn't ask for a better life or a better husband. I would like to have legs so that I could wear a miniskirt to show them off." She regretted that she could not cuddle her daughter or hold her husband's hand. Janette was awarded £33,000 compensation from the Distillers Company which manufactured thalidomide in Britain and had successfully performed all government-specified safety tests on animals, so it was bad luck that thalidomide had such awful effects in humans.

18.6. Other disorders affecting the limbs

In 1997, it was reported that the Murshidabad area of Bengal had an unusually high proportion of children with deformed legs due to **polio** and to a high concentration of **arsenic** in the water. The scourge of polio is very much less frequent than previously in countries with regular vaccination programmes. **Leprosy** (Chapter 11.6) can lead to loss of fingers, toes and deformed facial features. Between 2001 and 2010, 129 cases of leprosy were diagnosed in Britain; globally there are about 232,000 cases, mainly in southeast Asia.

The red swellings of **chilblains** on hands or feet are painful and itchy. They are caused by exposure to cold and usually heal spontaneously. **Frostbite** is more serious and comes from freezing of body tissues in extreme cold, especially the nose, cheeks, ears, fingers and toes. Prolonged exposure damages muscles, blood vessels, bone and nerves. Blistering is common, and gangrene and loss of fingers, toes or other affected parts can occur.

Gout is an extremely painful condition giving inflammation of the joints, usually starting with the big toes which become red and swollen. The pain is often so great that even the touch of bedclothes can be unbearable. Henry VIII's bad temper was partly due to gout. It is caused by excess uric acid in the blood, causing crystals to form in joints and internal organs. In Britain there are more than 150,000 gout sufferers, mainly men over 40, but even young non-drinkers can get it. Attacks usually last a few agonising days. Drinking beer and heavy red wines such as port, and eating foods high in purines such as oily fish, shellfish, liver, lentils and spinach, are contributory factors. Gout runs in families. Treatments include uric-acid reducing drugs, anti-inflammatories, and drinking six pints (3.5 litres) of water a day. Chimpanzees, turkeys and Dalmatian dogs can get gout.

Common foot problems include blisters from friction with shoes, hard growths such as corns and bunions, verrucas (infectious warty outgrowths), fungal infections such as 'athlete's foot', and painful in-growing toenails. **Twisted and sprained ankles** are usually short-term problems, but fallen arches and **flat feet** are long-term. A **bunion** is a bony deformity at the base of the big toe. Bunions rub on shoes and can blister and become infected. There is a strong genetic influence. Wearing badly fitting, too tight, shoes can cause bunions. 90% of sufferers are women and 50% of American women have bunions. Treatments include surgery, painkillers, orthotics, toe separators and bunion pads.

Even a **small cut on a finger** from a piece of paper can be painful and inconvenient, hurting when we deal with buttons. Imagine the effects of losing several fingers or a limb, or losing their use through arthritis. Velcro, with its many very small hooks and loops, is a great help to the injured or elderly who have to fasten clothing or shoes **using only one hand**. Cyril Smith OBE (1909–1974) was an English concert pianist who suffered a stroke that **paralysed his left arm**, but he and his wife continued to perform as a piano duo, using music written or arranged for three hands.

18.7. High heels

I questioned female undergraduates on **why they wore high heels**. The main reason was to make the girls look taller. Several felt that they looked smarter and better dressed. One said that they gave her a feeling of power. Others mentioned better posture, looking older, or having longer, better-shaped legs. Annabelle said memorably: "I wear platform soles and heels in order to be above the men's smelly armpits!" See also Chapter 5.4.

Women do not like to be taller than their male escorts. In 2001, film stars Nicole Kidman and Tom Cruise divorced each other. Nicole, who is 5 inches (13 cm) taller than Tom, remarked with relief: *"Now at least I can wear high heels!"*

18.8. Short fingers and toes

Brachydactyly is a set of autosomal dominant conditions giving **short fingers and toes**, straight or distorted, with variable expressivity. There is a loss of manipulative ability.

18.9. Miscellaneous finger and toe conditions; webbed fingers or toes

One lady has **asymmetric thumbs**, with thumbs of different length and nail type (Photo 18.3). One thumb is like her mother's two identical thumbs, and the other is like her father's two identical thumbs.

Different autosomal dominants give various symptoms including short thumbs, fingers bent towards the palm, partial or complete absence of fingers, three instead of two bones in the thumb, **synphalangism**, with bony or fibrous union of joints of fingers and/or toes, and **syndactyly**, fusion of the fingers. **Webbed fingers or toes**, with a web or flap of skin between

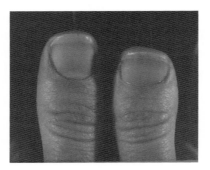

Photo 18.3.   Unequal thumbs.

one or more adjacent pairs of fingers or toes, occur fairly frequently, being mentioned by more than 1% of my students. For example:

- *Two of my toes are partly webbed together.*
- *My younger brother has webbed feet.*
- *The second and third toes on my left foot are slightly webbed. The skin join is higher up than normal. My mum has this too, more on left foot than right.*

The webbing may be a small tag or can join adjacent toes or fingers for all their length, perhaps involving only soft tissues, but in **complex syndactyly** there may be fused bones, blood vessels, nerves and tendons. **Surgical separations** may be made, with skin grafting if necessary.

18.10. Nails

Finger and toe nails are areas for personal choice in length and decoration. Expenditure on nail cosmetic services in America exceeds $10 billion a year. Some women consider it smart and sexy to have very long, talon-like nails, or have extensions bonded on.

Nails strengthen the ends of fingers and toes and grow from the matrix area under the cuticle, averaging 0.1 mm (0.0039 inches) a day for finger nails, and half that for toenails. It takes about 4 to 6 months to replace a fingernail. **Nail-biting** is very common in adults as well as children. Acknowledged biters include ex-Prime Minister Gordon Brown, singer

Britney Spears and hobbit actor Elijah Wood. Treatments include applying bitter substances such as quinine or wearing cotton gloves. **Injury to a nail**, say from shutting a finger in a door, may cause it to fall off eventually. It will usually regrow but may be thicker and less regular than before (Plate 13G).

18.11. Fingerprints

Fingerprints have been used for identification for more than 4,000 years, with finger impressions made in clay for Babylonian legal contracts. Francis Galton developed a system of recording and identifying fingerprints still used today, long after he wrote his book *Finger Prints* (1892). Prints of fingers, palms, feet and shoes are used forensically. Fingerprints are impressions left by the **friction ridges** on our fingers. There are three basic patterns: loop, whorl and arch, which constitute 60–65%, 30–35% and 5% of all fingerprints respectively. Fingerprints can be impressions in soft surfaces but more usually are patterns of sweat. Advances in forensic technique mean that their traces of hormones can reveal the person's sex.

18.12. Joint flexibility, Marfan syndrome, spondylitis, double-jointedness

Joint flexibility differs between individuals and varies with age, disease, accidents and training. **Ballet dancers** are selected for flexible joints, and do stretching exercises to improve flexibility, especially at the hips. **Contortionists** train themselves to have flexible joints, so that for example they can put their feet behind their ears.

Marfan syndrome sufferers have long, thin, spider-like limbs, lax joints, spinal curvature, and heart, blood circulation and eye problems. If their problems are not too bad, and if their co-ordination is good, their height and long limbs often make them good at basketball and volleyball. Their average life span is 45 years. The incidence is 1 in 10,000 in both sexes, with 25% being due to a new dominant autosomal mutation, giving a weakness in connective tissue in the eyes, long bones and around the base of the aorta. Death often results from splitting of the aorta.

Ankylosing spondylitis (Greek *ankylosis*, stiffening of a joint, and *spondylos*, a vertebra) causes chronic inflammation of the joints and spine, giving fusion of the joints between vertebrae and a rigid spine. It is auto-immune, strongly associated with allele *HLA-B27* in the Human Leukocyte Antigen (HLA) system. About 2 in 1,000 men and 2 in 10,000 women develop it.

Double-jointedness means having joints with an unusually large range of motion, or which bend in unusual directions. Double-jointed people do not have extra joints but can bend fingers, toes, elbows, knees or the body trunk at strange angles, e.g., being able to bend their thumbs right back (Photo 18.4).

Shallow sockets allow more movement than deep ones. Someone with a very shallow socket may be able to dislocate the bone from its socket, giving a huge range of movement, then to move it back into its socket. **Extreme flexibility** comes from stretchy ligaments, supple cartilage, smoother than normal bone ends and shallow joint sockets.

Hypermobility (excessive mobility) is more common in children than adults, and in women than in men. Joints may be weaker or easily dislocated. The very flexible joint may have to withstand unusual directions of stress, for which its fibrous capsule, tendons and ligaments may be insufficient.

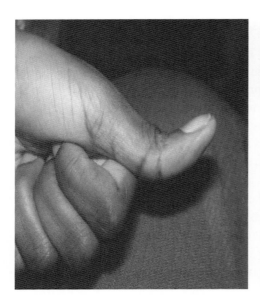

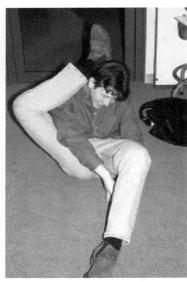

Photo 18.4. A double-jointed thumb.   **Photo 18.5.** Paul, flexible joints.

Joints are well supplied with nerves, so overstretching can cause pain, but joints are poorly supplied with blood, so they heal slowly after injury.

Here are accounts of flexibility from 200 students:

- *My elbows bend forwards as well as backwards — both elbows.*
- *I can bend my knees inwards. I can bend my thumb so that the base is flat; I am the only double-jointed one in family.*
- *Hypermobility results in dislocating joints. All my joints are affected, especially hips, which dislocate regularly, say every 4 months. I can control it by diet, by reducing dairy products* [that is surprising].
- *I have the ability to put my legs behind my neck. It is easy to do and doesn't hurt, but two legs is pushing it* (Photo 18.5). *I have generally flexible joints.*
- *I know someone who can bend their thumbs 90 degrees towards their body. I know someone who can, at will, dislocate his arm, allowing him to spin his hand 360 degrees as well as being able to bend his arm, at his elbow joint, the opposite direction to normal movement and in doing so touch his shoulder.*
- *I am double-jointed in both hands; my mother is in one hand, and my non-identical twin sister is in one hand.*
- *I have double jointed little fingers.*

18.13. Arthritis

Joints become less flexible with **age**, with more pain and less mobility. **Arthritis** involves inflammation of joints. Painkillers may be prescribed, or physiotherapy and/or injections of corticosteroids, e.g., for a 'frozen shoulder' or 'tennis elbow', both of which are painful and inconvenient, lasting months. **Tennis elbow** affects 1.3% of the population and is caused by injury or over-use of the arm, damaging a tendon which joins the forearm's extensor muscle to the humerus.

According to the charity **Arthritis Care**:

This painful, incapacitating and extremely frustrating condition affects nine million people across the UK, 27,000 of them under the age of 25. Arthritis accounts for no less than 1 in 5 visits to the doctor. There is no cure for most of

*the 200 forms of arthritis. The condition can cause a whole mixture of emo-
tions — frustration, anxiety, fear and anger. The effects on an individual can
range from slight inconvenience to a significant loss of mobility, and for every
person with arthritis there will be some good days and some bad days.*

Osteoarthritis can occur at any age but is commonest in older people,
especially the over-60s. It is usually a result of excessive wear and tear, and
can arise from overuse of a joint and be crippling. In the UK each year it
is the reason for **170,000 hip or knee replacements**.

Rheumatoid arthritis is an auto-immune condition where joints
become inflamed. Small joints such as fingers and wrists are most often
affected, but also larger ones such as knees and ankles. **Joint replacement**
may become necessary. Rheumatoid arthritis affects about 1 in 100 people,
being severe in 5 to 10% of cases. Sufferers may have reduced life expectancy
and life quality, often becoming house-bound. About 400,000 in Britain
have rheumatoid arthritis, with three times more females than males; most
sufferers are over 40, although children can be affected. About 16% of
women over 65 have this disorder, which has a genetic predisposition.

Reactive arthritis is a temporary inflammation of joints caused by
illness, especially viral infections. **Juvenile arthritis** was diagnosed in
Sandra Watson when she was nine. At eleven, she became the youngest
person to have both hips replaced. By the age of 51, the disease had ravaged
all her joints, and she had had countless operations.

18.14. Club foot

Club foot is a congenital multifactorial deformity of one or both feet, with
an incidence of 1 in 200; it is severe in 1 in 1,000 live births, occurring twice
as often in males as in females. The affected foot points downward and
inwards, with a variable extent of deformity. It impairs walking. The risk to
children of an affected parent is 1 in 33. The risk for a brother or sister of
an affected boy is 1 in 50, and 1 in 20 for siblings of an affected girl.

One used to see men or boys with club foot walking with a stick and
wearing a built-up shoe, but treatment is better now. **Treatment** is started
in affected young babies. The foot is manipulated to nearer the correct

position and put in a cast. Repositioning and recasting are done every few weeks. When the correct alignment is achieved, it is maintained through exercises, splints at night, and special shoes. Severe deformities need surgery.

18.15. Foot-binding of girls

Foot-binding of young girls was a cruel habit prevalent in China for about a thousand years until about 1915, to give them small, dainty feet. A horrifying account of the foot-binding of her grandmother, Yu-fang, born in China in 1909, was given in Jung Chang's biography, *Wild Swans* (Harper Collins, 1991).

When she was two, Yu-fang had her feet bound by her mother. A piece of white cloth was wound round each foot, bending all toes except the big toe inwards and under the sole. A large stone was placed on top, to crush the arch of the foot. The agonising pain caused Yu-fang to faint repeatedly but the treatment continued for several years. Her feet were bound in thick cloth all day and night to prevent growth. The toenails grew into the balls of the feet, causing further pain. It hurt to walk, and sufferers rarely showed their naked feet to men as they were usually smelly with rotting flesh.

Although Yu-fang was in appalling pain, her mother told her that it was for her future happiness. Normal-size feet on a bride were considered to bring shame on the bridegroom's family. According to Jung Chang, *The mother-in-law would lift the hem of the bride's long skirt, and if the feet were more than about four inches* [10 cm] *long, she would throw down the skirt in a demonstrative gesture of contempt and stalk off, leaving the bride to the critical gaze of the wedding guests, who would stare at her feet and insultingly mutter their disdain.*

18.16. Left- and right-handedness; 'laterality'

Left-handed writers hold their pens and paper awkwardly. **Cack-handed** means 'left-handed' and 'clumsy', from *cack*, excrement (Latin *cacare*, defecate). In some cultures, villagers use the right hand for eating and the left hand for wiping their bottom. In several languages, *left* has **unfavourable**

connotations while *right* is favourable. In German, *rechts* means *right* and *correct,* while *links* means *left* and *awkward;* in French, *droit* means *right* and *honest,* while *gauche* means *left* and *awkward.* In English, *dextrous* (Latin, *dexter,* right) means *skilful,* while *sinister* (Latin, *left*) means *evil.*

There are specialist shops selling items designed for use by left-handers, such as potato peelers, nail clippers and scissors, and lecture-room chairs with writing flaps on the left instead of on the right.

Famous left-handers include Presidents Ford and Truman, baseball star Babe Ruth, tennis players Jimmy Connors and John McEnroe, cricketers Brian Lara and Gary Sobers, inventors Benjamin Franklin, Leonardo da Vinci and Bill Gates, scientist Albert Einstein, artists Michelangelo (ambidextrous) and Picasso, entertainers Judy Garland, Jimi Hendrix and Paul McCartney, footballers George Best and Maradona, criminals the Boston Strangler, Jack the Ripper, Billy the Kid, and Al-Qaeda leader Osama Bin Laden. **About 8% of the population are left-handed with respect to writing,** with the figure stable across all countries and cultures, even in isolated African tribes.

People are not always consistent in **handedness for different tasks.** Some are **ambidextrous,** using either hand for some tasks, or one hand for some tasks and the other for others. Several studies have shown a **decrease in the frequency of left-handedness with increasing age,** for example, 13% of left-handers in 20-year-olds, 5% in 50-year-olds and 0% in those aged 80 plus. Some schools tried to force everyone to write right-handed, but no longer do so. A more sinister (Latin for *left,* as well as meaning *underhand, evil, unlucky, ominous* and *threatening*) interpretation is that **left-handers die earlier.** A study by the University of British Colombia and California State University found that left-handed and ambidextrous women died on average six years earlier than right-handers, while the difference for men was ten years. Table 18.1 shows the effects of parental handedness on that of their offspring.

No combination of parental types yields more than 27% of left-handers or less than 8.5%, so there is no simple inheritance pattern. Having two left-handed parents gives the highest incidence. Table 18.2 shows data on twin handedness from 2,900 pairs of identical twins and 2,589 pairs of non-identical twins. With about 13.5% of left-handed individuals, twins show a higher incidence of left-handedness than non-twins (about 8%).

Table 18.1. Handedness of parents and offspring (McManus, 1991).

Parents' handedness		Left-handed offspring, %	
Father	**Mother**	**Sons**	**Daughters**
Right	Right	10.4	8.5
Right	Left	22.1	21.7
Left	Right	18.2	15.3
Left	Left	27.0	21.4

Table 18.2. Handedness in identical and non-identical twins (McManus, 1991).

Twin handedness	Identical twins	Non-identical twins
Both twins right	75%	75%
One twin right, one left	22%	23%
Both twins left	3%	2%

In both identical and non-identical twins, about 78% of pairs of twins have twins of identical handedness and about 22% have different handedness. This **lack of difference between identical and non-identical twins** means that genetic influences are not strong. The 22% of identical twin pairs with different handedness means that twins with identical genes often differ in handedness. About 25% of identical twin pairs show '**mirror imaging**' for directional features such as head-hair whorl direction and finger-print whorls. Perhaps cases of identical twins with different handedness come from mirroring.

I acknowledge help with section 18.16 from the strongly left-handed Hester Griffiths.

Reference

McManus, I. C., The inheritance of left-handedness. *Ciba Found Symp* (1991) **162**: 251–267; discussion 267–281.

Chapter 19

Kidneys, Urine, Bladder, Cystitis, Police Alcohol Tests

19.1. Kidney structure and function

Your two kidneys lie in the middle-to-lower back, about 4.3 inches (11 cm) long, 2.3 inches (6 cm) wide and 1.2 inches (3 cm) thick, weighing about 5 oz (140 g) each, less than 1% of body weight. They control the body's water balance and excrete water and dissolved wastes down the ureters as urine to the bladder.

Blood from the wide renal artery is filtered in each kidney through about a million nephrons, with selective reabsorption of water and certain dissolved substances in the loop of Henle. The blood retains blood cells and large molecules such as antibodies but smaller molecules such as water, salts, urea and glucose can pass out.

Kidneys regulate the levels of many dissolved compounds or ions, including sodium, potassium, calcium, magnesium, chloride, bicarbonate, sulphate and phosphate, often using active transport, taking energy for reabsorption. Kidneys **regulate blood volume**, the acid/base balance (by adjusting the urinary output of hydrogen ions, H^+, and bicarbonate ions, HCO_3^-). They **stimulate red blood cell production** by secreting the hormone erythropoietin, and activate vitamin D. They **excrete natural wastes** such as urea, uric acid, creatine, and 'foreign' compounds such as pesticides and drugs. Non-waste compounds such as glucose are efficiently **reabsorbed**, although diabetics excrete some glucose.

The kidneys receive 20 to 25% of the heart's blood output and have about 50 miles (80 km) of nephron tube length. The huge volume of about 384 gallons (1,728 litres) a day enters the kidneys at a pressure of 75 mm of mercury.

Human diversity includes some people having **only one kidney** (from a failure in development or if one is removed after disease, cancer or accidents) or **extra kidneys**.

Personal account. *I have four kidneys!*

I have bipartite partial duplex kidneys. My two normal kidneys have **two extra partial kidneys** *stuck to their tops. These extra kidneys are the same width as my main kidneys, but about half their height. When I was born, the valves in my two extra ureters (tubes connecting the kidneys to the bladder) were deficient and I suffered from many urinary tract infections due to back-flow from the bladder. I would often get a raised temperature with these infections and my father took me for walks to reduce my temperature.*

At two and a half years old I suffered a **febrile convulsion** *(caused by high temperature) and my mother rushed me for medical treatment. Again the doctors treated my infection with antibiotics, in a foul-tasting green viscous liquid that I took for eighteen months. The convulsion made doctors investigate further; it is common for young females to suffer frequent urinary tract infections as they have a shorter urethra than males but I was having more than average.*

During tests I had a 'lumbar puncture', removing fluid from the spinal column with a needle. It felt like having a nail hammered into my back. Following an operation to re-implant my duplex ureters, my two partial kidneys are now both correctly connected to my bladder. My grandma was born with only one kidney; I think we have duff kidney genes in my family!

I have a scar across my belly from hip to hip from the operations, and scarring on my kidneys due to infection. I have what looks like an extra belly button from where I had one of five drainage catheters (tubes) inserted for ten days. As a young adult, I do not wear low-cut jeans or crop-top clothes as they would expose my peculiarities.

Helen Plowman

19.2. Water balance and urine

In 24 hours, the kidneys filter out 33 to 40 gallons (150–180 litres) of water from the blood (about 90% passes directly back through the renal vein), but 99% is reabsorbed, leaving about 2½ pints (1.5 litres) as urine. Each urination is about ½ to ⅔ pint (300 to 400 ml), but fluctuates.

Alcohol, tea and coffee are **diuretics**, increasing urine output by chemical action, not just from their water content. Alcohol and coffee have synergistic effects, increasing urine output more together than separately. Many people with high blood pressure take diuretics to reduce the blood volume.

Roughly **60% of body weight is water**, so a typical man has about 68 pints (40 litres) of water in him: 5 pints (3 litres) in blood plasma, 3 pints (2 litres) in red blood cells, 20 pints (11 litres) between cells and 40 pints (23 litres) inside cells. Our water intake and loss are about 4¾ pints (2.7 litres) a day. About 52% of **water intake** is in drink, 34% in food, and 14% comes from food oxidation. Of our water output, about 59% is in urine, 8% in faeces, 5% in sweat, 14% by other evaporation from skin, and 14% from lungs. These figures vary widely with climate, amount drunk, metabolic condition and health, with excess urine output in diabetics. An inactive person loses about 1/10th fluid ounce (3 ml) of water an hour by perspiration, but when active may lose 1¾ pints (1 litre) an hour. If you get dehydrated, you reabsorb more water.

An **aquaholic drinks too much water**, which can be lethal. In 2008, Andrew Else, aged 51, drank several litres of water and died of water intoxication, hyponatremia (low sodium). Marathon runners sometimes drink too much during a race, with fluid gathering in the lungs, giving breathlessness and nausea. The actor Anthony Andrews, aged 55, drank water during stage performances to counter heat and voice strain. One evening in 2003 he drank eight litres of water. His life was saved by his driver taking him straight to hospital. With **water overdoses**, the sodium level in blood drops and fluids build up in the brain, leading to dizziness, headaches, nausea, confusion, convulsions, unconsciousness and possibly death.

19.3. The bladder

We are **acutely aware** of our **bladder** when it desperately needs emptying. When it is fairly empty, the smooth-muscle walls are highly folded and can flatten out as it fills. When it contains about 9 to 14 fluid oz (250 to 400 ml) of urine, stretch-receptor nerves in the walls tell the brain that the bladder will need emptying, with the message becoming more urgent with further filling.

Although gravity helps transport urine down to the bladder, it is helped by **muscle contractions** in the ureter walls. As the bladder fills, the projecting ends of the ureters into it are compressed and closed, preventing backflow of urine from the pressurised bladder. The ureter muscles can push more urine into the bladder against that pressure.

In babies, the bladder automatically empties when sufficiently full, but the growing child develops conscious control. An internal and an external sphincter control **bladder emptying**, which starts when their muscles relax, with contraction of bladder-wall muscles. Urination happens uncontrollably if the bladder is over-full. People differ greatly in **how often they need to use a lavatory**, even when they consume similar amounts of liquid.

19.4. Kidney and bladder problems; dialysis, stones and cystitis

Even a **quarter of normal kidney function** is sufficient, so healthy people can donate one kidney without harm. People with severe kidney disease need a **kidney transplant** from a matching donor or **dialysis**. A dialysis machine takes blood and removes wastes from it, e.g., by diffusion across cellophane tubing surrounded by a large volume of liquid made up to mimic blood plasma. Urea and other wastes can diffuse out but sodium and other ions are retained. Dialysis is typically done three times a week, for several hours. A newer method is **continuous ambulatory peritoneal dialysis**, using the patient's peritoneal membrane (lining the abdominal cavity) and a permanently implanted catheter into the abdominal cavity. Three and a half pints (2 litres) of dialysis fluid are inserted, then drained off after a time, repeated several times a day. The patient can do most normal activities during treatment, but there is a risk of infection.

The first **successful kidney transplant** was in 1954 when an American surgeon removed a kidney from Ronald Herrick and put it in his identical twin brother Richard who had terminal renal failure. It functioned immediately. An altruistic 86-year-old lady in England donated a kidney to a stranger, saying, "Why do I need two kidneys to sit at home knitting and watching television?"

Acute kidney failure results in the sudden reduction of urine formation, which is very serious if less than 0.9 pints (500 ml) a day is produced. The danger with **chronic** (long-term) **kidney failure** is that up to 75% of the kidney tissues may be destroyed with few or no warning symptoms, so when symptoms appear, it may be too late for treatment. By **end-stage renal failure**, 90% of function is lost, with death quickly ensuing unless swift treatment is given. **Kidney failure** has many metabolic and physical consequences, including heart, nerve and mental problems, bleeding, ulcers, vomiting, swelling, weakness and infections. Causes include bacterial or viral infections, cancer and physical damage.

Kidney stones occur in about 12% of men and 4% of women, particularly those aged 30 to 60, when minerals or acid salts form hard crystals in the kidneys instead of being flushed out in the urine. The stones can block ureters or the urethra, causing renal colic, with severe abdominal pain, and there may be a urinary tract infection. The stones can result from not drinking enough fluids, from certain medical conditions or some medical drugs. The usual treatment is **lithotripsy**. The stones are located with X-rays or ultrasound, and shock waves are used to break them into fragments which pass out in urine.

Bladder problems. People in stressful situations often empty their bladder more frequently, as do women in advanced pregnancy as the foetus presses down on the bladder. Bacterial bladder infections can be very painful, and urethral inflammation makes urination difficult and agonising.

Cystitis, bladder inflammation, is much more common in women than in men. Often there is a knife-like pain in the urethra during urination, and an urgent need to urinate. With **common (bacterial) cystitis** the urine is often cloudy and smelly, tested with a dipstick for bacteria, or by microscopy. It normally responds well to antibiotics.

Many bacteria are deposited around the anus during defecation. During bottom-wiping or sexual intercourse, bacteria may be moved to the urethra's opening and cause urethral inflammation and/or get into the bladder, causing cystitis. The complaint affects females far more often than males because their urethra is shorter and its opening is much nearer the anus than in males. Self-treatments include drinking large amounts of water (but there is pain at urination), cranberry juice, or bicarbonate of soda in water to make the urine less acid.

With **interstitial (non-bacterial) cystitis**, affecting about 700,000 women in America, there is bladder-wall inflammation without bacteria. Treatment is difficult. If insufficient water is drunk, uric acid crystals may be deposited on the bladder wall, causing irritation. '**Honeymoon cystitis**' can occur after vigorous sexual intercourse.

19.5. Police tests for alcohol

Along with the liver, the kidneys are important for **detoxifying the blood**. People's kidneys have **different efficiencies** in eliminating different chemicals, including **alcohol**. How many alcoholic drinks take someone over the legal limit depends on body size, sex, metabolism, health, the efficiency of their liver and kidneys in processing alcohol, the time over which the drinks are taken, and on any food eaten.

At a National Guild of Wine and Beer Judges' weekend, a police officer member brought along a large **police breath-alcohol meter**, to test those who wished to be assessed after the celebratory Saturday night dinner, before and during which we had all consumed a fair amount of alcohol. As we were not driving home until the following afternoon, the tests were just for our interest, not prosecution. The UK legal limit for breath alcohol for drivers is 35 micrograms in 100 ml of breath. Most judges scored between 30 and 38 units, near to the legal limit. When I only scored 6 on the test and on a retest, the policeman said, "You've not had much of an evening, then." I told him that I had had a similar number of glasses of wine to my fellow judges. Alcohol does seem to pass through me more quickly than through most others, as I am often the first at a wine tasting to go out for a pee. In the **UK**, the **legal limits for alcohol for drivers** are: **breath**, 35 micrograms in 100 ml breath; **blood** (which must be taken by a qualified medical practitioner), 80 milligrams in 100 ml blood; **urine**, 107 milligrams in 100 ml urine in the second sample. The bladder may contain alcohol-free urine when the person starts drinking, so the police demand two specimens within one hour, discarding the first.

In **America, blood alcohol** is calculated as 2,100 times that in breath, or 1.33 times that in urine. All states designate for drivers a blood or breath alcohol level constituting a criminal offence. There is a second offence of driving 'under the influence' or 'while impaired' which is usually charged where the blood alcohol concentration is over 0.08%. The limit for commercial drivers in New York and California is 0.04%. California has a limit of 0.01% for drivers who are under 21 or on probation for drink/driving offences.

Chapter 20

Lungs, Breathing, Asthma

20.1. Lung structure and function

Our lungs sit above the diaphragm, protected by the ribcage. The right lung has three lobes and weighs about 22 oz (625 g) while the left lung has two lobes and weighs about 20 oz (565 g). Their surface area for oxygen and carbon dioxide exchange is about 645 to 755 square feet (60 to 70 square metres), nearly as big as a tennis court, with more than 350 million small sacks per lung, with very thin walls containing blood vessels.

When you breathe in, your ribcage rises at the top and expands forwards, and your diaphragm flattens downwards, expanding the pleural cavity around the lungs, sucking air in. At rest, one takes about 16 breaths a minute, about 9 pints (5 litres) total, rising to about 530 pints (300 litres) under strong exercise.

The air is 21% oxygen and 0.04% carbon dioxide, while air breathed out is 16% oxygen and 4% carbon dioxide. People having **panic attacks** often **hyperventilate**, breathing too rapidly so that they expel too much carbon dioxide, which reduces blood acidity. Treatment is breathing into a bag, reducing the carbon dioxide loss yet allowing oxygen uptake.

20.2. Lung problems

In America, **lung disease** is the number three killer, causing one in seven deaths, killing about 361,000 a year out of about 25 million sufferers. About 11 million Americans have chronic bronchitis, 3 million have emphysema, 6 million children and 12 million adults have asthma, and 342,500 have lung cancer. In year 2000, American deaths from pneumonia with influenza were 65,300, from pneumonia alone, 63,500, and from influenza alone, 1,765 (National Center for Health Statistics Report).

Cystic fibrosis was described in Chapter 12.6. **Lung cancer**, tuberculosis and pneumonia were covered in Chapter 11.6 and 11.11.

Asthma (Greek, panting, *ásthma*) affects a quarter of UK children, some fatally. It is a chronic allergic disease causing inflammation of the airways and breathing difficulties. It kills about 2,000 people a year in Britain, having increased greatly in the last 50 years. In 2011, 235 to 300 million people worldwide had asthma, with 250,000 deaths. A friend was a bright and attractive young lady, but asthma killed her in her early 30s.

Of the 5.2 million asthma sufferers in Britain, 40% were **diagnosed as adults**. Celeste Abrahams, aged 32, had no idea that adults could get asthma, and she had not had it. On the way to work one day she started gasping for breath, then felt agonising pains in her chest and passed out. Two colleagues saved her life by giving mouth-to-mouth resuscitation, then paramedics gave cardiopulmonary resuscitation. Her attack was so severe that her heart had stopped beating. She then had acute attacks every few months, needing resuscitation ten times. Celeste eventually learnt to manage her condition with inhalers. (Information from *The Daily Telegraph*, 12/9/2005.)

An **asthma attack** occurs when a sufferer inhales protein particles to which they are sensitised. The airways contract and secrete excess mucus, causing wheezing, coughing fits, shortness of breath and tight feelings in the chest. Asthma has genetic and environmental components. If mild persistent disease is present (more than two attacks a week), low-dose inhaled corticosteroids or an oral leukotriene antagonist are used. For those with daily attacks, higher doses of inhaled corticosteroids are provided. For severe attacks, corticosteroid pills are also prescribed. Hospital treatment and intravenous drugs may be needed.

Suffering children in Britain are given inhalers, from which medicines go straight to the lungs. **Reliever inhalers** (usually blue) help to relieve symptoms when attacks happen, and **preventer** inhalers (usually brown, red or white) help to protect the airways and reduce the chance of symptoms. Relievers quickly relax muscles surrounding the narrowed airways, allowing the airways to open wider, making it easier to breathe. Preventer inhalers are used every day to protect airway linings and to reduce their swelling and stop them being so sensitive. They are usually prescribed for children and young people using their reliever inhaler three or more times a week.

20.3. Air pollution, emphysema, particulates and deaths

Air pollution, especially traffic fumes, promotes asthma attacks. According to the WHO in 2014, **air pollution kills about seven million people a year**. The **London Great Smog of 1952** killed more than 4,000 people. The Clean Air Act of 1956 greatly reduced air pollution by introducing 'smoke-control areas' in which only smokeless fuels could be burned. By shifting domestic heating towards cleaner coals, electricity and gas, it reduced the amount of smoke pollution and sulphur dioxide.

Big cities and industrial areas have the most pollution. One can see **smog** as a yellow or brownish haze. In California in 1967 there were three stages of smog alert, with the most serious one closing down industry and halting traffic. The ozone/petrochemical interactions made my eyes water. When I went up Mount Wilson I was horrified to see below me the thick smog haze which I breathed in daily. There was a lot of emphysema there.

Emphysema damages the lung's air sacs, leading to a progressive shortness of breath and can be lethal. Smoking and air pollution are its main causes. The inner walls of the air sacs rupture, making one air space instead of many small ones, reducing the surface area. Treatment cannot reverse the damage.

The **aggressive gases** nitrogen dioxide, sulphur dioxide and ozone are all health hazards and so are '**particulates**', very small particles suspended in the air. While air normally contains pollen, fungal spores, bacteria, dust and moisture droplets, the term 'particulates' is used for ones of human origin, such as sooty particles from diesel exhausts. **Particulate pollution** causes 22,000 to 52,000 deaths per year in America and about 370,000 premature deaths in Europe. It causes asthma, lung cancer, cardiovascular disorders, respiratory diseases, birth defects and premature death.

The **particle's size** largely determines how far into the respiratory tract it penetrates. Particles bigger than 10 micrometres may be filtered out in the nose or brought up from the bronchi by ciliary action. Particles smaller than 2.5 micrometrres can enter into the lung's gas-exchange regions and penetrate arteries, causing plaque deposits and eventual blockage of blood vessels, leading to heart attacks. Very small particles (less than 100 nanometres) can pass through cell membranes and migrate into other organs, including the brain. Particles in the exhaust from diesel engines are typically in the size range of 100 nanometres (0.1 micrometre). These sooty particles also carry cancer-causing chemicals such as benzopyrenes on their surface.

The WHO estimates that, worldwide, particulate air pollution causes about 3% of mortality from cardiopulmonary disease, about 5% of mortality from cancer of the trachea, bronchus and lung, and about 1% of mortality from acute respiratory infections in children under five. Research published in *The Lancet* (Nawrot *et al.*, 2011) showed that traffic exhaust is the single most preventable cause of heart attacks, causing 7.4% of all attacks. In Britain, particulates are estimated to kill about 29,000 people a year, more than alcohol and obesity combined, and ten times more than those killed in road accidents. Vehicles contribute half the nitrogen dioxide and 80% of the particulates in London air, with diesel-engined ones emitting 95% of the nitrogen dioxide and 91% of particulates.

Reference

Nawrot, T. S. *et al.*, Public health importance of triggers of myocardial infarction: a comparative risk assessment. *Lancet* (2011) **377**: 732–740.

Chapter 21

Development from the Fertilised Egg; Sexual and Later Development

21.1. Development from the fertilised egg

The fertilised egg takes four days to become a ball of 64 cells. By days five to six this has become a **blastocyst**, a hollow ball of cells which implants in the womb. Its inner cell mass at one end becomes the baby while the outer layer of cells later forms the **chorion** and then the **placenta**, both of which supply nutrients and remove wastes. At days 14 to 15, the embryo is about 1/10th inch (0.25 cm) long.

At 18 to 19 days, the **nervous system** starts to develop as a groove on the back. Growth is rapid, with **differentiation of tissues**. At 21 to 22 days, the umbilical cord develops blood vessels connecting the embryo to the developing placenta. At 26 to 27 days, there are **limb buds**, a **heart** and a **head** with **mouth** and **eyes**. By four weeks, the heart is beating and the embryo is nearly 3/16th inch (0.5 cm) long. Having missed her period, the woman is usually conscious of being pregnant.

At around eight weeks, the embryo looks human, about 1 inch (2.3 cm) long, known as a **foetus**. The weight increases from 1 ounce (28 g) at eight weeks to 7 pounds (3.2 kg) at term. The foetus is 5½ inches (14 cm) long by week 16, when the woman's 'baby bump' is visible. From 20 weeks, the **heartbeat is detectable**. After five months, a foetus has fully formed nails, eyelids, sweat glands and fine head hair.

In Britain between 1995 and 2006, the percentage of babies born between 22 and 25 weeks who survived rose from 40 to 53%. However, the

Royal College of Paediatrics and Child Health said that this increase in survival should not be used as a reason to continue treatment where babies were left with a non-existent quality of life. Many such premature babies have a number of serious defects which cannot be cured later.

With intensive care, a foetus of 24 weeks can sometimes survive if born prematurely or delivered by Caesarean section. By 28 weeks most foetuses have turned head-down. Some babies fail to turn and are born legs first, a **breech birth**, usually without harm.

The **placenta** eventually weighs about 1 pound (450 g), reaching 8 inches (20 cm) in diameter. It supplies glucose for energy, carbohydrates and fats, amino acids for proteins, minerals for bones, and delivers oxygen and removes carbon dioxide. Although the foetal and maternal blood systems do not connect, there is a large area of close contact. This allows passage of some antibodies from mother to child but also allows transmission of diseases such as German measles, HIV and various drugs.

By week 36, the top of the womb has been pushed up as far as the lower ribs, and the woman has to lean backwards for balance. In the last month of pregnancy, there are increasingly stronger **uterine contractions**. The weight of the baby on the mother's internal organs can cause incontinence. She often has backache and feels very tired.

At an average of 266 days (38 weeks) after fertilisation, the **baby is born**, with an average weight of 7 pounds (3.2 kg) in the West, lighter in some societies and if part of a multiple birth. Labour takes two hours to three days. The placenta is expelled 20 to 30 minutes later. Immediately on birth, the baby has to make the **profound change** from living in a liquid environment, with everything supplied, to **breathing in air** and **feeding through the mouth**. The umbilical cord will shrivel naturally but is usually cut. Very **premature babies** have many problems. Their chest, diaphragm muscles and lung spaces are underdeveloped.

For up to three days after giving birth, the mother's breasts produce **colostrum**, rich in proteins, antibodies and hormones, including a hormone to stimulate the baby's intestines to produce milk-digesting enzymes. During **normal milk production**, about **1¾ pints (1 litre) of milk are produced per day**, typically 87% water, 7% lactose, 4.5% fat, 1% protein and amino acids, and 0.2% minerals, including calcium, sodium and potassium. Kent *et al.* (2006) found that infants breastfed on average

11 times in 24 hours (range: 6 to 18), and a feed averaged 76 grams (0 to 240 g) of milk.

Breastfeeding depends on the **mother's willingness and ability**. Some women breastfeed and enjoy the bonding with their baby. Others do it but dislike it, complaining of sore nipples and discomfort. Some cannot breastfeed physically or psychologically but would like to. Others feel it is degrading and will not breastfeed although capable of it. The **duration of breastfeeding** is diverse, with up to six months the norm in most of the West, but sometimes one or even two years is acceptable.

21.2. Sexual development before birth

The sex chromosomes are both Xs in females (Plate 1C), with one X and a shorter Y in males (Plate 1A). The Y carries few functional genes apart from the *SRY* region (sex-determining region of Y) which bears the crucial testis-determining factor, *TDF*. The embryo is initially sexless but around week six the *TDF* gene, if present, triggers 'it' to become male by inducing the genital ridge to produce two testicles, whose sex hormones make the Y-bearing embryo male. A lack of Y and *TDF* causes the genital ridge to produce two ovaries which make female sex hormones, producing a female.

21.3. Later development

From around the age of eight, the genitals start to mature. The skeleton grows, the torso lengthens and broadens. During the **growth spurt**, teenagers' arms and legs grow by as much as 10 cm (4 inches) a year. Physical maturation is completed in the early twenties. In **boys at puberty, the voice** 'breaks', settling down about one octave lower.

During adolescence, the brain is flooded with hormones which affect moods and cause the nerves in various parts of the brain to develop fatty white myelin sheaths which enable them to transmit more complex signals. The visual areas of the brain are not fully myelinised until the mid-teens and develop at the same time as the emotional centres become more active. Sex hormones stimulate sexual physical abilities and an interest in sex.

The developmental disorder **progeria** causes premature ageing of the skin and face, a small face, dwarfism, baldness and other problems.

Sufferers age almost ten times faster than normal. Fransie Geringer was only nine but progeria gave him the wrinkled face and bald head of a man of 90. Hayley Okines, at five, looked older than her 29-year-old-mother. She died in 2015, aged 17, with the body of a 104-year-old. At the age of 14, she wrote an autobiography, *Old Before My Time*. The disorder occurs in 1 per 8 million live births, causing death in the teens to early twenties. It is genetic, from a new mutation in the *LMNA* gene. Sufferers seldom live long enough to reproduce.

In the **elderly**, **falls** become common. The over-75s are 11 times more likely to be admitted to hospitals after falls than those aged 60 to 64. For **ageing**, see Chapter 22.1. Shakespeare, in *As You Like It*, has Jacques, in his '**Seven Ages of Man**' speech, say, "Last scene of all,/ That ends this strange eventful history,/ Is second childishness and mere oblivion,/ Sans teeth, sans eyes, sans taste, sans everything."

Reference

Kent, J. C. *et al.*, Volume and frequency of breastfeedings and fat content of breast milk throughout the day. *Pediatrics* (2006) **117**: e387–e395.

Chapter 22

Longevity, Ageing, Birth and Death Rates, Immigration, Population Structure

22.1. Longevity, ageing and death

In the Bible (Genesis 5: 27) Methuselah is said to have lived to 969 but about 128 years is probably today's **maximum**. The environment and chance are more important for longevity than are genes, which account for 30% of the variation. **Genes** influencing disease-susceptibility affect longevity. **Intelligence** increases longevity.

Individuals age differently. At a school reunion of men aged about 71, some were young-looking, with an eagerness for life, while others looked old, wrinkled, bent, fat, leaning on sticks. Some personalities had changed greatly but others had not.

Professor Harper, director of the Oxford University Institute of Population Ageing, said that working-class men averaged 12 years of retirement life, compared with 22 for professional middle-class men. In 2014, the ONS put the **healthy life expectancy** for women as 64.1 years and 63.4 for men. With life expectancies of 83.4 for women and 79.2 for men, that means 19 years of ill health for women and 16 for men.

Health improvements have made Britain a **centenarian capital**; only Japan, France, Italy and Spain have proportionally more. According to the ONS, Britain had 13,780 centenarians (85% of them women) in 2013, quadruple the number 30 years previously. At a Royal Society of Medicine meeting in 2014, a speaker told us that with longevity increasing about two years every ten years, our longevity had increased by 24 minutes in the first two hours of the meeting! In India, life expectancy has doubled from 32 to 64 years in 50 years. **Volunteering** improves health and happiness, encouraging older people to stay active and have more social contacts, reducing depression and typically increasing longevity by several years.

In British towns until the 1830s, half the babies died before the age of five. **Drinking water contaminated with sewage** caused dysentery, typhoid, diarrhoea and cholera. Piped chlorinated water, proper sewers and separation of drinking water from sewage have reduced infant mortality to less than 1%. In Zambia there is **20% mortality** before the age of five.

The **quality of life** matters. One hears sentiments such as, "I'd rather smoke, and eat and drink what I like, rather than have an extra two years misery in a nursing home." An **epitaph** from ancient Rome states: *balnea, vina, Venus corrumpunt corpora nostra; sed vitam faciunt balnea, vina, Venus* — baths, wine and sex wreck our bodies but make life worth living. Our earlier actions affect us later, as in my limerick:

> *There once was a wily old peasant*
> *Whose life had been wicked but pleasant;*
> *He repented at last*
> *For his sins of the past*
> *Were spoiling his sins of the present!*

In 2014, Oxford University and the Wellcome Trust reported that **mental health problems were as deadly as smoking. Reductions in life expectancy**, in years, included bipolar disorder, 9 to 20; schizophrenia, 10 to 20; recurrent depression, 7 to 11; smoking 20 cigarettes a day, 8 to 10; drug or alcohol abuse, 9 to 24.

Giants tend to die young. Robert Smith, 8 ft 5 inches (256.5 cm), died in 2014 from a heart attack at 24.

According to a World Health Report in 2002, the **top ten causes of death** worldwide were malnourishment, unsafe sex, high blood pressure, smoking, alcohol, bad water and sanitation, iron deficiency, smoke from indoor fires, high cholesterol, obesity. In the developing world, 170 million children were malnourished, causing three million deaths annually. In developed countries, more than a billion were overweight, with 300 million obese, causing at least 500,000 deaths a year. HIV/AIDS was the fourth biggest killer, with 40 million infected (70% of them in Africa) and 2.9 million deaths annually.

The Centers for Disease Control and Prevention in 2004 reported on **infant** (under one year of age) **mortality rates in America**. Black women had the worst rate, 1.4%, compared with 0.6% for non-Hispanic white women. For multiple births, mortality was 3%, compared with 0.6% for single births. Infants born early, at 34 to 36 weeks' gestation, had rates three times higher than for those born at 37 to 41 weeks.

In 2003, the WHO said that **traffic** killed four times as many as wars, and that far more people committed suicide than were murdered. Injuries killed more than five million in 2000, one tenth of all deaths, with 90% being in poorer countries. **Causes of fatalities** were: road accidents, 1.3 million; suicide, 815,000; inter-personal violence, 520,000; poisoning, 310,000.

War is an obvious cause of death. The treatment of **military prisoners** differed between races during World War 2. Prisoners of war (PoWs) in Britain and America were treated well, but more than a third of 35,000 US servicemen in Japanese captivity died, compared with 1% in German PoW camps. The **atom bombs** dropped on Japan killed 150,000 to 250,000 people from blasts and fires, with more deaths later from radiation sickness, burns and disease.

Natural disasters such as floods, famine, fires, earthquakes, tsunamis and epidemics cause many premature deaths. In 2005, the 7.6 Richter Scale earthquake in Pakistan killed more than 73,000 people, injured 128,000, and destroyed the homes of three and a half million.

Centenarians give widely diverse **reasons for their longevity**. Kamato Hongo died in Japan aged 116. She slept for two days and then stayed awake for two days. Jeanne Calment died in France in 1997, aged 122. She was mentally alert, smoked, drank port and feasted on chocolates. Some who live to a great age leave **many descendants**. Lizzie Bolden, born to freed slaves in Tennessee in 1890, died in 2006 aged 116, when she had 40 grandchildren, 75 great-grandchildren, 150 great-great-grandchildren, 220 great-great-great-grandchildren and 75 great-great-great-great-grandchildren.

Suicide caused 31,000 deaths in America in 2004, with 750,000 failures. 75% of attempts were by women but 75% of successes were by men. Lithuania and Russia have the largest proportion of suicides, with Britain 20th and America 50th. Hanging is a common method but in America firearms are used in more than 65% of cases. Women often use sleeping pills or painkillers.

According to the American Association of Suicidology, one in three who attempts suicide unsuccessfully tries again. In America, 9% of university students said they had seriously considered suicide. Anaesthetists and doctors have a high suicide incidence and success rate. **Reasons for suicide** include bereavement, love troubles, deep depression, a feeling of having no future or way out of poverty, or having a nasty terminal illness. Some loving couples make suicide pacts as neither wants to live without the other.

The elderly are increasingly on **multiple medications**, up to 20 pills a day. It is a nightmare for those running **old people's homes** to ensure that each resident gets the right dose of the right pills at the right time. Some residents **just want to die**, having had enough of life. Each day may mean being asleep for hours in front of the TV, feeling that there is nothing to live for, with reducing senses and mental faculties, yet some are happy.

Problems frequently affecting the elderly include:

- Frailty, falls, broken bones;
- Frustration at no longer being able to do many everyday activities, needing help for dressing and undressing, doing up shoes, eating, managing stairs and going to the toilet;
- Feeling useless, a burden to others;
- Fear of the process of dying;
- Multiple medical conditions, with some pills having unpleasant side effects such as constipation or wind;
- Cataracts, forgetfulness, dementia, joint pains, stiffness, difficulty bending down, tough toe nails, diabetes, cancer, heart disease, strokes;
- Loss of independence; leaving one's home and possessions to go into a care home or hospital, becoming dependent on others;
- Loneliness; many friends and family members have died, moved away, or don't have the means or mobility to visit;
- Lack of mobility, restricting going out, making life monotonous;
- The huge stigma attached to the smell of urine or faeces, the bother of incontinence devices and the shame of lacking full bodily control;
- Being cut off from conversation, TV and radio by deafness, or poor eyesight making reading difficult;
- Suspecting that beneficiaries are waiting for one to die.

In Gilbert and Sullivan's operetta *Patience*, Lady Jane sadly sings about ageing:

Silvered is the raven hair, Spreading is the parting straight,
Mottled the complexion fair, Halting is the youthful gait,
Hollow is the laughter free, Spectacled the limpid eye, —
Little will be left of me In the coming bye and bye!

Fading is the taper waist, Shapeless grows the shapely limb,
And although severely laced, Spreading is the figure trim!
Stouter than I used to be, Still more corpulent grow I, —
There will be too much of me In the coming bye and bye!

Painfully arthritic knees and hips are often treated by **replacement surgery**. Hip replacement is frequently performed and adds on average 6.5 quality-adjusted life-years. Queen Elizabeth the Queen Mother had one hip replaced at 94, the other at 97.

22.2. Births, immigration, population structure

Population sizes and structures depend on birth rates, death rates, immigration and emigration. In 2004, Latvia's population was shrinking by 1.5% annually. Its birth rate of 1.1 child per woman in 1998 was only half that needed to maintain population size. Britain's rate then was 1.6 and America's was 2.11.

Niger's population rose from 2.5 million in 1950 to 11.5 million in 2005, with an unsustainable **birth rate of 6.8 children per female. Sub-Saharan Africa's populations** are rising because modern medicine enables more babies to survive, fewer adults die of preventable diseases and conception stays high. African mothers give birth to an average of 5.2 children, far above Europe's average of 1.6.

Mass migrations cause huge problems. Mrs Thatcher said, "We are a British Nation with British characteristics. Every country can take some small minorities and in many ways they add to the richness and variety of this country. The moment the minority threatens to become a big one, people get frightened." In London, '**White British**' now make up 47%, **less than half** the population, and even less amongst children.

Mao Tse-tung (1893–1976) had hopes of Chinese world domination and banned contraception, resulting in a population explosion. China's population is now 1.2 billion. His successors felt that numbers were unsustainable and imposed a '**one child per family**' policy in 1979, with millions of enforced abortions and baby-killings to implement it, giving a big change in population structure, with fewer in younger generations. Mass starvation during Mao's 'Cultural Revolution' (1965–1969) killed at least 38 million. Rich people could have more than one child if they paid a big fine. The 'one child' policy is now partly relaxed. Many families wanted their one child to be a boy, so female infanticide and abortion were practised, leading to a **surplus of males**.

The **world population** is about 7.2 billion, having grown since the Great Famine and the Black Death in 1350 when it was near 370 million. Annual births are expected to remain roughly at their 2011 level of 134 million.

Immigration has had major effects on current European populations. In 2010, Eurostat estimated that **France** had 7.2 million foreign-born immigrants, 11% of the population. About 40% of newborns there between 2006 and 2008 had at least one foreign-born grandparent (11% born in another European country, 16% in Morocco, Algeria, Tunisia and Libya, and 12% elsewhere).

In **Britain**, more than five million immigrants have come since 1997, including many Eastern Europeans (especially Bulgarians, Poles and Romanians) coming under EU freedom of movement laws. In figures released in 2015 by the ONS, Britain in 2013 had 734,000 India-born permanent residents and 679,000 Poland-born Poles. Over a quarter of births in England and Wales are to mothers born overseas. Britain is overcrowded and these increases place great pressure on housing, infrastructure, social services including health and schools, and jobs. On the other hand, people of overseas origin (e.g., Filipina nurses, Indian doctors) keep the National Health Service running, and East Europeans make good builders and plumbers.

Recommended reading

Abdulla, A. and Rai, G. S., *The Biology of Ageing and Its Clinical Implication: A Practical Handbook*. (2013) Radcliffe, London.

Chapter 23

Abnormalities of Sex Chromosomes and Autosomes, Down Syndrome, Barr Bodies

23.1. Introduction; Barr bodies

Plate 1A and C show our 46 chromosomes per cell, with 22 pairs of non-sex chromosomes called **autosomes**, plus two **sex chromosomes**. Males have one **X** and a smaller **Y** which carries few genes. Females have two Xs per cell, of which one is inactivated, with nearly all its genes turned off, forming a small **Barr body**. A Barr body's presence (females) or absence (normal males) is useful for sexing foetuses in prenatal diagnosis and adults in sports sex tests. Barr bodies can be seen in stained non-dividing cells from a mouth smear, while chromosomal analysis (**karyotyping**) is usually done on stained dividing white blood cells in culture. Because there is only one active X per cell, having more than two Xs gives more Barr bodies, so XXX females have **two Barr bodies**. XXY males (Klinefelter syndrome) have a Barr body although they are male.

23.2. Differences in sex chromosomes

The presence or absence of a Y determines human sex, not the number of X chromosomes, because the Y carries the **testis-determining factor, *TDF*** (also called *SRY*), which causes an embryo to become male. Without *TDF*, **the embryo becomes female by default.**

Klinefelter syndrome, XXY (47 chromosomes), occurs in 1 in 1,000 male births, with an increased risk from older mothers, and is the commonest cause of male infertility. Sufferers have male genitals and often some breast development, appearing somewhat intersex, with the maleness-causing Y chromosome and the female XX combination. They have long arms and legs, a normal penis and small testes making low levels of testosterone, with few sperm. Most are of normal intelligence but they are ten times more likely than normal to be mentally retarded.

Turner syndrome, with only one sex chromosome, XO (45 chromosomes, no Barr body), is rare because nearly 99% of Turner's embryos abort spontaneously, leaving 1 in 5,000 female live births, with about 14,000 sufferers in Britain. There is a diversity of **subtypes;** about 55% are plain XO.

Individuals are sexually underdeveloped females and normally sterile, with a broad chest, small breasts, a webbed neck and small height, averaging 4 ft 9 inches (145 cm). When I met one, I wrongly assumed from her height that she was about 12, not 17. Their intelligence and life span are almost normal, but there is an increased risk of heart defects, swollen hands and feet, and kidney problems. Their ovaries start to degenerate before birth, after 15 weeks gestation, giving a failure of sexual development. About 10% menstruate and a few are fertile, but they are usually **mosaics** of Turner (XO) and normal (XX) tissues. Young sufferers can be given growth hormones to increase height a little, and sex hormones to improve sexual development, but that does not give fertility.

Personal account. *I have Turner syndrome, so am short and infertile*

I am a fifty-seven year old woman with Turner Syndrome. One X has a part missing and the other has some Y chromosome added to it, so I am not a classic XO Turner's. My only symptoms are being 4 feet 7 inches (140 cm) tall and infertile. My only treatment was HORMONE REPLACEMENT THERAPY during what should have been my fertile years.

The traumatic realization that I would never be a standard 5 feet 4 inches (163 cm) came when the results of tests at University College Hospital arrived. The letter stated that my condition was extremely rare: one in five thousand female births.

The year was 1967 and the letter might as well have stated that 'Your daughter is a freak'. I remember my mother bursting out crying, "How could you do it to her?" (addressing God). She was devastated. I would have coped more calmly if the diagnosis had been given to me at ten or twenty, but not during my teenage years, which are traumatic for anybody.

My philosophy degree later confirmed my belief that each individual is a human being foremost and not, for example, a Woman with Turner Syndrome.

Would I change my height? When I was young, yes, but not now. Ironically, the genetic condition has formed an essential part of my psychological and emotional framework, not as a woman but as a human being.

Jacqueline Davies, BA (Hons), MA

Jacqueline died of a **particular ovarian cancer** which is common in Turner's women who have part of the Y attached to their X. I met a girl who has that attachment and had had both ovaries removed when she was five to avoid this kind of cancer.

XXX women. About 1 woman in 1,000 has three X chromosomes, XXX, but there is no named syndrome as the characteristics are so diverse. Many go undiagnosed. XXX women vary from being physically and mentally normal to being sexually underdeveloped and somewhat mentally retarded. About 75% are fertile and the condition is not passed on. The incidence of XXX and XXY, but not XYY, births increases sharply with the mother's age.

XYY men. Men with one X and two Ys are above average height (average 6 ft 1 inch, 185 cm) and below average intelligence, IQs 80 to 95. They occur in 1 in 1,000 live male births, often have severe acne in adolescence, and are found at higher frequencies amongst violent criminals. Their incidence is three times higher in mentally handicapped males and 20 times higher in male criminals.

46, XX males. Males with the female combination XX occur in about 1 in 20,000 male births, usually with a **translocation** of the sex-determining region of the Y (*TDR*) to an X. They are sterile with small testes but are mentally normal.

46, XY females are very rare, usually with a Y chromosome which has lost the male-determining *TDR* region or has an inactive copy. See Chapter 10.11 for **testicular feminisation**.

23.3. Differences in autosomal chromosome number

Children with autosomal abnormalities can be of either sex. **Down syndrome** (named after John Down, but often called mongolism; Plate 19E shows a typical face) has a frequency of about 1 in 600 live births, with about 60,000 in Britain. More than 60% abort spontaneously and about 20% are stillborn. Sufferers have **three copies** of the smallest chromosome, number **21** (Plate 1A). The babies are usually floppy, with muscle weakness and slow development. Intelligence is reduced, with IQs often 40 to 50. Sufferers have an extra eye-fold of skin, broad hands with odd palmprints, and nearly half suffer from heart defects needing surgery. They have increased susceptibility to infection and a 15-fold increased risk of leukaemia. There may be a short neck, slanted eyes and visual problems.

Their life span used to average 20 years, with many deaths in the first year, and those that reach 35 usually develop Alzheimer's. Yang *et al.* (2002) found that in America their median age at death increased from 25 years in 1983 to 49 years in 1997. Cancers other than leukaemia and testicular cancer were listed on their death certificates less than one-tenth as often as expected.

Males are sterile but some females are fertile. Because low intelligence makes it difficult for them to use contraception or be good mothers, females are often sterilised. Down's is easily diagnosed from symptoms and chromosomal tests. The chance of having a Down syndrome baby is strongly affected by a **woman's age**. The risk is 1 in 1,920 for mothers aged 20, 1 in 885 for age 30, 1 in 139 for age 40, and 1 in 32 for age 45. There is little effect of father's age, with the extra chromosome usually coming in the egg, not the sperm.

About 4% of Down's children come from a **translocation** of one chromosome 21 onto another chromosome such as 14, so the child has 46 chromosomes, not 47. The mother usually has only 45 chromosomes but is not abnormal as she has the right amount of each chromosome but with two joined. If the mother's chromosomes are normal, 46, the chance of another affected baby is not much greater than normal. If she is a translocation carrier, however, the risk of another Down's baby ranges from 10% upwards, depending on which chromosomes are involved.

Abnormal chromosome numbers usually arise from the **failure of chromosomes to segregate properly** in the meioses giving rise to the eggs or sperm. There can be mis-segregations in the early embryo, too, giving **mosaics**, with some cells normal and some abnormal, giving milder effects than in a completely affected individual. About 4% of Down's are mosaics.

Patau syndrome comes from having three of chromosome 13, with a frequency of 1 in 5,000 live births. There are many serious symptoms such as brain failure and cardiovascular defects; half die within one month. **Edwards syndrome** babies have three of chromosome 18, with a frequency of 1 in 3,000 live births, but many more are conceived, about 95% aborting spontaneously. They have multiple malformations, 90% dying within a year.

The **autosomal chromosomes are numbered** from 1, the largest (10 μm long) to 22 in decreasing size, although 21 is actually the smallest, 2 μm (Plate 1A). One does not see individuals with three copies of the large autosomal chromosomes, although they are formed, because the eggs do not implant or the embryo soon aborts. We see many Down's children because the extra small chromosome does not do enough harm to abort all sufferers.

23.4. Aberrations in chromosome structure

About 1 in 2,000 babies has an aberration of chromosome structure, such as a piece missing or duplicated (Plate 1C), usually giving multiple mental and physical symptoms. Deletion of half the short arm of one copy of chromosome 5 gives **Cri-du-Chat syndrome** where sufferers give a plaintive continual cat-like cry. They are severely abnormal physically and mentally, with a short life span.

The commonest chromosome rearrangements, **Robertsonian translocations**, affect about 8 embryos per 10,000, contributing greatly to spontaneous abortions, birth defects and mental retardation. **Multiple translocations** are involved in many cancers.

Reference

Yang, Q. *et al.*, Mortality associated with Down's syndrome in the USA from 1983 to 1997: a population-based study. *Lancet* (2002) **359**: 1019–1025.

Chapter 24

Opinions on the Characteristics of the Chinese, Japanese and English

It is salutary to find out how others see us. With the help of Ji-Eun Kim, a Korean schoolgirl, I conducted a **survey of opinions on 25 characteristics of the Chinese, Japanese and English**, such as politeness, arrogance, being very interested in sex, being kind, etc. The 216 responses were divided into four categories: 123 British (English, Welsh, Scots, Irish), 36 Chinese, 16 Japanese and 41 from other countries in Europe, Africa, Asia and Australia. The term '**Others**' here means anyone other than the British, Chinese or Japanese. Most respondents were undergraduates but ranged from senior schoolgirls to the elderly. Responses from the two sexes were so similar that they have been pooled.

The wording on the questionnaires included: '**Please put Y (for yes) or N (for no) for every attribute for each of the three racial types, in comparison with world averages.** Put Y even if the tendency [to be above average] is very slight'. The Politeness table explanations show how to understand the results. Values between 0 and 100% indicate a **diversity of opinions**, with values of 50% indicating the greatest diversity, 50% yes, 50% no, for that characteristic. Not all data are shown.

For politeness, 31% of Chinese respondents thought that the Chinese were above average; 94% of Chinese respondents thought that the Japanese were above average; 100% of Japanese respondents thought that the Japanese were above average, etc. As seen from the column headed 'Chinese',

Table 24.1. Politeness.

Politeness	Percentages of Yes answers		
	Chinese	Japanese	English
Chinese views	31	94	61
Japanese views	6	100	88
Others' views	66	80	56
British views	80	88	50

there were **hugely different views** (6% to 80% Yes) between the four responding classes as to how polite the Chinese are. The Japanese rated themselves as 100% polite, the English as 88% polite and the Chinese as low as 6%. The British rated the English as least polite and all four categories put the Japanese top for politeness.

Table 24.2. Arrogance.

Arrogance	Percentages of Yes answers		
	Chinese	Japanese	English
Chinese views	36	25	64
Japanese views	81	38	75
Others' views	32	32	76
British views	19	33	68

The Chinese, Others and British rated the English as easily the most arrogant, but the Japanese thought that the Chinese were even more arrogant than the English.

Table 24.3. Being deferential.

Being deferential	Percentages of Yes answers		
	Chinese	Japanese	English
Chinese views	42	61	33
Japanese views	38	63	69
Others' views	64	64	51
British views	59	60	28

Surprisingly, the Japanese put the English as most deferential, but the Chinese, Others and especially the British, put the English as least deferential.

Table 24.4.   Racially proud.

Racially proud	Percentages of Yes answers		
	Chinese	Japanese	English
Chinese views	44	61	89
Japanese views	75	50	94
Others' views	76	85	78
British views	76	80	66

The Chinese and Japanese put the English as very racially proud. The Others and British put the Japanese as most proud.

Table 24.5.   Open.

Open	Percentages of Yes answers		
	Chinese	Japanese	English
Chinese views	19	31	72
Japanese views	19	31	44
Others' views	20	17	71
British views	20	26	41

All put the Chinese as low for openness and the English highest.

Table 24.6.   Inscrutable.

Inscrutable	Percentages of Yes answers		
	Chinese	Japanese	English
Chinese views	67	61	44
Japanese views	50	56	44
Others' views	65	70	35
British views	67	69	37

While all four respondent categories put the Chinese and Japanese as fairly inscrutable (50 to 70%), the English had some inscrutability (35 to 44%), even to the British. The Chinese even found the Chinese more inscrutable than the Japanese! The Japanese also put themselves as the most inscrutable.

Kind. The British put all three races as equally kind, whereas the Japanese thought that the Chinese were low on kindness (19%), that the Japanese were extremely kind (94%), and that the English were fairly kind (75%). The Chinese thought the Japanese were kinder than themselves.

Table 24.7. Likeable.

Likeable	Percentages of Yes answers		
	Chinese	Japanese	English
Chinese views	56	75	78
Japanese views	13	81	63
Others' views	60	70	83
British views	72	73	85

The Japanese have a poor opinion of the Chinese, who often rated the Japanese better than themselves. Here the Japanese considered themselves very likeable and the Chinese unlikeable. The English generally came out as the most likeable.

Being devious/cunning. The British put all three races as moderately devious, while the Japanese marked themselves low and the Chinese high.

Table 24.8. Trustworthy.

Trustworthy	Percentages of Yes answers		
	Chinese	Japanese	English
Chinese views	53	64	39
Japanese views	31	88	81
Others' views	53	63	55
British views	64	71	63

The Japanese again marked the Chinese low and themselves and the English high. All assessed the Japanese as the most trustworthy.

Table 24.9. Aggressive.

Aggressive	Percentages of Yes answers		
	Chinese	Japanese	English
Chinese views	33	19	75
Japanese views	75	6	69
Others' views	27	24	63
British views	21	28	80

The British, Chinese and Others put the English as easily the most aggressive. The Chinese rated the Japanese low for aggression, but not as low as the Japanese did (6%). Only the Japanese thought that the Chinese were aggressive (75%), very different from the British estimate (21%).

Table 24.10. Physically attractive.

Physically attractiveness	Percentages of Yes answers		
	Chinese	Japanese	English
Chinese views	36	44	69
Japanese views	25	50	50
Others' views	34	44	68
British views	38	41	78

The English did well for physical attractiveness, with ratings of 68 to 78% except from the Japanese who scored them at 50%, the same as themselves. The Chinese put themselves last.

Religious. Only the Japanese gave the English a high score, while the British put the English lowest. The Others rated all three races about equal. The Chinese put themselves as the most religious.

Table 24.11.   Selfish.

Selfish	Percentages of Yes answers		
	Chinese	Japanese	English
Chinese views	61	33	56
Japanese views	75	19	56
Others' views	29	37	41
British views	42	41	61

The British put the English as the most selfish, and the Chinese put themselves as the most selfish. The Japanese put the Chinese as very selfish and themselves as by far the least selfish.

Money-oriented. The only low scores were by the Japanese for themselves and the English. The British put the Japanese as marginally the most money-oriented. The Chinese put themselves as the most money-oriented, as did most of the others.

Happy. The Others scored the English as most happy, while the Chinese thought that the English were the least happy. The Japanese put the Chinese as least happy, and themselves and the English as very happy.

Table 24.12.   Clever.

Clever	Percentages of Yes answers		
	Chinese	Japanese	English
Chinese views	86	78	28
Japanese views	81	75	69
Others' views	90	88	46
British views	88	90	58

For this admirable character, the English came last by a humiliatingly long way, including in the British estimation. The Chinese put the English as very dim. Even the Japanese put the Chinese first for cleverness, as did the Chinese and Others.

Friendly. These results were better for the English, with three 'firsts'. The Japanese were again disparaging about the Chinese (19%) and thought highly of themselves (75%).

Moral. The British, Chinese and Others put the English last but the Japanese put them top, 81%. The Japanese gave the Chinese an extremely low score, 6%.

Interested in the arts. The Chinese and especially the Japanese rated the Chinese as having the least interest in the arts. The Others and especially the Japanese put the English easily top.

Table 24.13.   Very interested in sex.

Very interested in sex	Percentages of Yes answers		
	Chinese	Japanese	English
Chinese views	17	56	89
Japanese views	57	36	64
Others' views	18	28	85
British views	35	47	85

Here is one category with the English way ahead, with three scores in the 80s. The Chinese had the lowest scores, apart from that given them by the Japanese, who put themselves last.

Introspective. The English had the lowest scores. The Japanese gave themselves the highest score and the Chinese the lowest one.

Fond of animals. The English were easily the most fond of animals and the Chinese the least.

Fond of children. This was more equal, with the English just ahead.

Fond of gambling. The English usually came top, although the Japanese put the Chinese as the most avid gamblers.

Conclusions. There was a **huge diversity of views** within each of the four categories of respondents. There were **huge racial differences** for some characteristics. The **English** were scored highest for interest in sex, arrogance, openness, the least inscrutable, the most likeable, most aggressive, most physically attractive, most selfish, definitely the least clever, the most friendly, least introspective, most fond of animals, children and gambling.

The **Japanese** were rated the most polite, most deferential, marginally the most racially proud, most trustworthy, least aggressive, least selfish, most moral and most introspective. The **Chinese** were put as the least arrogant, least racially proud, least open, least likeable (even in their own estimation), least physically attractive, most religious, cleverest, least friendly, least interested in the arts, and least fond of animals and children. They had high esteem for the Japanese, who had low opinions of the Chinese.

Most respondents were **brutally honest** about their own race's shortcomings, especially the British and Chinese. With the Chinese putting the Japanese as the least aggressive and the kindest, it seems that the Chinese respondents were unaware of or ignored the appalling aggression and cruelty of the Japanese troops towards their people just before and during the World War 2. I find it difficult to reconcile the atrocious Japanese behaviour then with the charm and peacefulness of the Japanese people I meet.

Index